KW-053-656

AQA GCSE

MODULAR SCIENCE

FOUNDATION EDITION

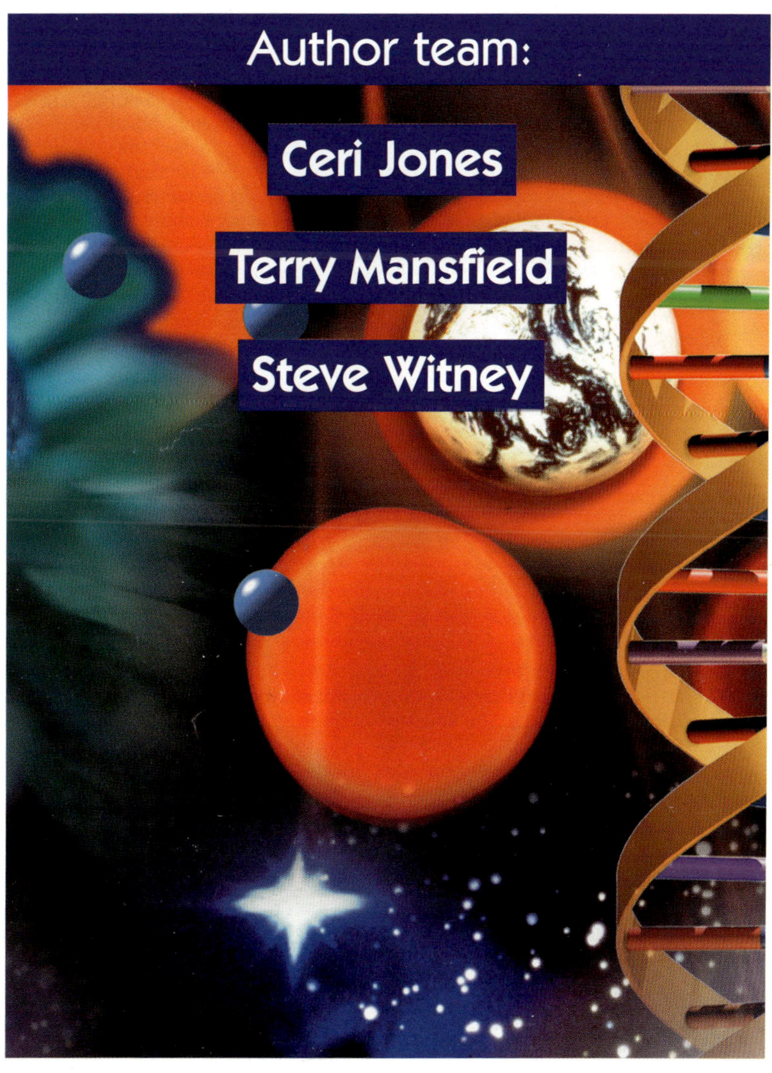

Hodder & Stoughton

Photo acknowledgements

The publishers would like to thank the following individuals, institutions and companies for permission to reproduce photographs in this book. Every effort has been made to trace ownership of copyright. The publishers would be happy to make arrangements with any copyright holder whom it has not been possible to contact:

Actionplus (5, 259 bottom, 269 left, 269 right, 272, 324); Advertising Archives (320); Andrew Lambert (79 left, 83 bottom left, 83 bottom right, 90, 92, 107 top left, top right, bottom left, and bottom right, 128 left, 128 right, 167 left, 167 right, 189 left, 220, 252 left); Associated Press (330 bottom); Bruce Coleman (70 middle, 64 top, 257); Corbis (80 left, 80 right, 98, 132, Adam Woolfitt (97), Amet Jean Pierre (116 right), Bettmann (44, 161, 281, 336 top, 336 bottom), Bob Witkowski (233 bottom), Buddy Mays (202), Chinch Gryniewicz (198 top), George H. H. Huey (107 margin bottom), George Hall (244 bottom, Kim Sayer (88 margin), Lloyd Cluff (205 bottom), Lee Snider (79 right), Lester Lefkowitz (251, 266), Martin Jones/Ecoscene (302), Michael Porsche (116 left), Michael S Yamashita (194), Nik Wheeler (127), Paul Almasy (205 top), Roger Ressmeyer (287), Steve Starr (81 right); Creation (247 top); Fiat (156); Fleetmaster of Luton (258); Hodder and Stoughton (78 bottom, 103 top, 203 left, 333, Murad Sezer/AP (321 top), Nasa (259 top), Natural History Museum (107 margin top), Viktor Pobedinsky, Stringer/AP (330 top); Holt Studios (26 bottom, 28 top, 28 bottom, 49 bottom, 51 bottom); Life File (55 top, 70 right, 217, 244 top, 247 bottom); Martin Sookias Photography (88 top); Museum of London (16); Natural History Museum (55 bottom); PA Photos (321 bottom); R D Battersby (124, 189 right, 203 right, 242 bottom, 245, 314 top); RSPCA Photo Library (51 top); Science Museum (232, 282 bottom left); Science Photo Library (7, 233 top right, 250, 314 left, 329 top, 329 middle, Adam Hart-Davis (242 middle), Alan L. Detrick (103 bottom right), Alfred Pasieka (81 middle, 86, 151), Andrew Lambert Photography (242 top), Andrew Syred (10), Ben Johnson (81 left), Celestial Image Co. (286 bottom), Chris Knapton (227), CNES, 1998 Distribution Spot Image (282 top right), CNRI (14), Conor Caffrey (312), Darwin Dale (224), David Parker (50), Deep Light Productions (310 left), Dept. of Clinical Cytogenetics, Adden-Brookes Hospital, Cambridge (45), Department of Clinical Radiology, Salisbury District Hospital (310 right), ESA (286 top), Eye Of Science (4, 25), Francoise Sauze (283 bottom), George Post (51 middle), Graham Ewens (109 middle), Jerrican/Daudier (235), Jerrican/Fuste Raga (109 right), JC Revy (49 top, 64 margin), John Beatty (57, 63), John Heseltine (65 right), Julian Baum (283 top), Keith Kent (236), Malcolm Fielding, Johnson Matthey Plc (117), Mark De Fraeye (169), Martin Bond (103 bottom middle, 314 bottom), Martin Dorhn (70 left), Martyn F Chillmaid (329 bottom), Mehau Kulyk (226), NASA (26 top, 290 top, 290 bottom, 291 top, 291 bottom), NSRC Ltd (282 bottom right), Peter Scoones (83 top), Royal Observatory, Edinburgh (289), Saturn Stills (15), Sinclair Stammers (54), Simon Fraser (65 left, 67 top, 106, 198 bottom), Simon Fraser/Northumbrian Environmental Management Ltd (103 bottom left), Sue Ford (67 bottom), Tony Hallas (285), Tony Craddock (109 left); Telegraph Library/Getty Images (187); Terry Jennings (195, 200, 252 right); Wellcome Trust (331)

Artworks created by David Graham.

Orders: please contact Bookpoint Ltd, 130 Milton Park, Abingdon, Oxon OX14 4SB. Telephone: (44) 01235 827720. Fax: (44) 01235 400454. Lines are open from 9.00–6.00, Monday to Saturday, with a 24 hour message answering service. You can also order through our website www.hodderheadline.co.uk.

British Library Cataloguing in Publication Data
A catalogue record for this title is available from the British Library

ISBN 0 340 869291

First Published 2003
Impression number 10 9 8 7 6 5 4 3 2 1
Year 2009 2008 2007 2006 2005 2004 2003

Typeset by J&L Composition, Filey, North Yorkshire.
Printed in Italy for Hodder & Stoughton Educational, a divison of Hodder Headline Plc, 338 Euston Road, London NW1 3BH.

Contents

About this book

The contents

The contents of this book are designed to cover all aspects of the knowledge and understanding needed if you are following any of the AQA GCSE specifications (courses) at Foundation Level in:

- Science: Double Award (Modular)
- Science: Single Award (Modular)
- Science: Double Award (Co-ordinated)
- Science: Single Award (Co-ordinated)

The knowledge and understanding required by the Double Award specifications makes up the core content for each of the related separate sciences. This means that you can use this textbook to help you understand the core content if you are following any of the AQA GCSE specifications at Foundation Level in:

- Biology
- Chemistry
- Physics
- Biology (Human)

How the content links with the courses

The contents are divided into 12 **chapters**. Chapters 1–4 deal with the biological aspects, chapters 5–8 deal with the chemical aspects and chapters 9–12 deal with the physical aspects. Each chapter is divided into a number of **sections**. The heading to each section clearly identifies which part of the GCSE course the content addresses. The specification matching grids, in Tables 1.1, 1.2 and 1.3, show how the sections of the book relate to each of the courses.

Specification Matching Grids

Table 1.1: Life and living processes

Content		AQA specification references			
		Co-ordinated		Modular	
Chapter	Section	DA	SA	DA	SA
1 Cells in humans	1.1 Animal cells	10.1	10.1	01	13
	1.2 The digestive system	10.4	10.3	01	13
	1.3 Aerobic respiration – producing energy	10.7	N/a	01	N/a
	1.4 The circulatory system	10.5	10.4	01	13
	1.5 Disease	10.11	10.8	01	13
2 Maintenance of life	2.1 Green plants	10.13	N/a	02	N/a
	2.2 Transporting substances	10.15	N/a	02	N/a
	2.3 How humans respond to changes in the environment	10.8	10.5	02	13
	2.4 Controlling what goes on inside you	10.10	10.7	02	13
	2.5 Drugs and health	10.12	10.9	02	13
3 Variation and inheritance	3.1 What is variation?	10.10	10.16	04	14
	3.2 What about sex?	10.14	10.14	04	N/a
	3.3 What has X to do with sex?	10.17	10.11	04	14
	3.4 Messing about with genes	10.18	10.12	04	14
	3.5 Evolution	10.19	10.13	04	14
	3.6 Hormones and the menstrual cycle	10.9	10.6	04	14
4 The environment	4.1 Adaptation and competition	10.20	10.14	03	14
	4.2 Man's effect on the environment	10.21	10.15	03	14
	4.3 Energy transfers	10.22	N/a	03	N/a
	4.4 How nature recycles living things	10.23	N/a	03	N/a

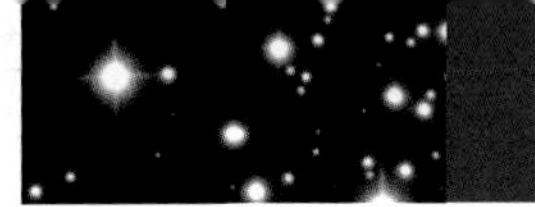

Table 1.2: Materials and their properties

Content		AQA specification references			
		Co-ordinated		Modular	
Chapter	Section	DA	SA	DA	SA
5 Metals	5.1 The Periodic Table	11.11	11.3	05	15
	5.2 The Group I metals	11.11	11.3	05	16
	5.3 Transition elements	11.11	N/a	05	N/a
	5.4 Ores	11.4	11.4	05	15
	5.5 The reactivity series	11.4	11.4	05	15
	5.6 Getting iron by smelting	11.4	11.4	05	15
	5.7 Getting aluminium by electrolysis	11.4	N/a	05	N/a
	5.8 Preventing metals from corroding	11.4	11.4	05	15
	5.9 Neutralisation	11.12	11.4	05	15
	5.10 Making salts from alkalis	11.2	11.4	05	15
	5.11 What makes a substance acidic or alkaline?	11.12	11.4	05	15
	5.12 Making salts of transition metals	11.12	N/a	05	N/a
6 Earth materials	6.1 The use of limestone	11.5	11.4	06	15
	6.2 Crude oil and hydrocarbons	11.3	11.2	06	15
	6.3 Cracking hydrocarbons	11.3	11.2	06	15
	6.4 Fossil fuels	11.3	11.2	06	15
	6.5 Plastics (polymers)	11.3	11.2	06	15
	6.6 The Earth's atmosphere	11.9	N/a	06	N/a
	6.7 The structure of the Earth	12.12	N/a	06	N/a
	6.8 The rock cycle and the rock record	11.10	N/a	06	N/a
	6.9 Earth movements	11.10 12.13	N/a	06	N/a
7 Patterns of change	7.1 Chemical hazard symbols	11	11	07	16
	7.2 How to change the speed of chemical reactions	11.13	11.5	07	16
	7.3 Why the speed of chemical reactions changes	11.13	11.5	07	16
	7.4 Fermentation	11.14	11.6	07	16
	7.5 Reactions using enzymes	11.14	11.6	07	16
	7.6 Energy transfer in chemical reactions	11.16	N/a	07	N/a
	7.7 Reversible reactions	11.15	N/a	07	N/a
	7.8 The Haber process and nitrate fertilizers	11.8	N/a	07	N/a
	7.9 Chemical calculations	11.8	N/a	07	N/a
8 Structure and bonding	8.1 Changes of state	11	11	08	N/a
	8.2 Atoms	11.1	11.1	08	16
	8.3 Structure of the atom – protons, neutrons and electrons	11.1	N/a	08	N/a
	8.4 Structure of the atom – electron energy levels	11.1	11.1	08	16
	8.5 Ionic bonding	11.2	N/a	08	N/a
	8.6 Metallic and covalent bonding	11.2	N/a	08	N/a
	8.7 The Periodic Table	11.11	11.3	08	16
	8.8 Group 1 – The Alkali Metals	11.11	N/a	08	N/a
	8.9 Group 7 – The Halogens	11.11	N/a	08	N/a

Table 1.2: Materials and their properties continued

	Content		AQA specification references			
			Co-ordinated		Modular	
Chapter	Section		DA	SA	DA	SA
8 *cont.*	8.10	Group 0 – The Noble Gases	11.11	N/a	08	N/a
	8.11	Some Halogen Compounds	11.12	N/a	08	N/a
	8.12	Symbols, Formulae and equations	11.7	N/a	08	N/a

Table 1.3: Physical processes

	Content		AQA specification references			
			Co-ordinated		Modular	
Chapter	Section		DA	SA	DA	SA
9 Energy	9.1	Heat energy transfer	12.16	12.10	09	17
	9.2	Keeping your home warm	12.16	12.10	09	17
	9.3	Using electrical energy	12.4	12.4	09	17
	9.4	Paying for electricity	12.4	12.4	09	17
	9.5	Energy efficiency	12.17	12.11	09	17
	9.6	Energy resources and electricity	12.18	12.12	09	17
	9.7	Non-renewable fuel and the environment	12.18	12.12	09	17
	9.8	Using renewable energy resources to generate electricity – one	12.18	12.12	09	17
	9.9	Using renewable energy resources to generate electricity – two	12.18	12.12	09	17
	9.10	Renewable energy resources and the environment	12.18	12.12	09	17
	9.11	Using energy resources – advantages and disadvantages	12.18	12.12	09	17
10 Electricity	10.1	Circuits – starting out	12.1	12.1	10	17
	10.2	Series circuits	12.1	12.1	10	17
	10.3	Parallel circuits	12.1	12.1	10	17
	10.4	Resistance	12.1	12.1	10	17
	10.5	Current–voltage graphs	12.1	12.1	10	17
	10.6	Electricity and magnetism	12.20	N/a	10	N/a
	10.7	Using the magnetic effect of an electric current	12.20	N/a	10	N/a
	10.8	Electrostatics	12.5	N/a	10	N/a
	10.9	Making use of electrostatic charge	12.5	N/a	10	N/a
	10.10	Charges on the move	12.5	N/a	10	N/a
	10.11	Mains electricity	12.3	12.3	10	17
	10.12	Using the mains electricity supply	12.3	12.3	10	17
	10.13	Safety devices and electrical appliances	12.3	12.3	10	17
	10.14	Energy and power in a circuit	12.2	N/a	10	N/a
	10.15	Induced current	12.21	12.13	10	17
	10.16	Generating and transmitting electricity	12.21	12.13	10	17

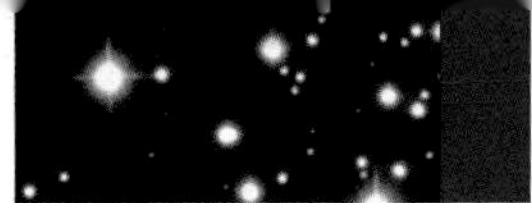

Table 1.3: Physical processes continued

Content			AQA specification references			
			Co-ordinated		Modular	
Chapter	Section		DA	SA	DA	SA
11 Forces and Motion	11.1	Speed, velocity and acceleration	12.6	N/a	11	N/a
	11.2	Motion graphs	12.6	N/a	11	N/a
	11.3	Standing still, moving steadily – balanced forces	12.7	N/a	11	N/a
	11.4	Speeding up, slowing down – unbalanced forces	12.7	N/a	11	N/a
	11.5	Stopping safely	12.8	N/a	11	N/a
	11.6	Falling bodies	12.8 12.19	N/a	11	N/a
	11.7	Work and energy	12.19	N/a	11	N/a
	11.8	The solar system	12.14	12.8	11	18
	11.9	Gravity in space	12.14	12.8	11	18
	11.10	Satellites	12.14	12.8	11	18
	11.11	Outside the solar system	12.15	12.9	11	18
	11.12	The life cycle of a star	12.15	12.9	11	18
	11.13	Life on other planets	12.15	12.9	11	18
12 Waves and Radiation	12.1	Looking at waves	12.9	12.5	12	18
	12.2	Water waves	12.9	12.5	12	18
	12.3	Light waves	12.9	12.5	12	18
	12.4	Total internal reflection	12.9	12.5	12	18
	12.5	Sound waves	12.11	12.7	12	18
	12.6	Ultrasonic waves	12.11	12.7	12	18
	12.7	Electromagnetic waves – one	12.10	12.6	12	18
	12.8	Electromagnetic waves – two	12.10	12.6	12	18
	12.9	Waves and diffraction	12.10	12.6	12	18
	12.10	Sending information	12.10	12.6	12	18
	12.11	Seismic waves	12.12	N/a	12	N/a
	12.12	Parts of the atom	12.23	12.15	12	18
	12.13	Why do some atoms produce radiation?	12.22 12.23	12.14 12.15	12	18
	12.14	Dangers of radioactivity	12.22	12.14	12	18
	12.15	Uses of radioactivity	12.22 12.23	N/a	12	N/a
	12.16	Models of the atom	12.23	12.15	12	18

What is in each chapter?

At the beginning of each chapter is a list of **Key Terms** used in that chapter. When used for the first time in the text these key terms are emboldened. All the key terms together with their meanings are also found in the **Glossary** starting on page 341.

The contents of each chapter are divided into several **sections**. Each section concentrates on one topic. For each section a reference is given linking the topic to each of the specifications (courses), single award, double award, modular and co-ordinated.

Throughout the book are a number of **Did you know**? boxes. The information in these boxes will not have to be learnt, but is provided to give you further background and extra interest in the topic.

At the end of each section is a set of **Topic Questions**. At the end of each chapter is a **Summary**. The summary provides a brief analysis of the important points covered in the chapter.

The topic questions are included to help you understand what you have read in the section. Do not worry if you have to go back to read the section again when you try to answer the questions. Reading the section again to answer the topic questions will help you learn the work. By writing out the questions as well as the answers you will start to build up a set of revision notes.

At the end of each chapter are some **GCSE questions** taken from past AQA (SEG) or AQA (NEAB) examination papers. Answering the GCSE questions will help to give you an idea of what is wanted when you sit your written papers at the end of the course.

Ideas and evidence in Science

You will find that many sections contain information which is marked with a bell and a vertical stripe in the margin. This is material to support the 'Ideas and Evidence in Science' part of your course. It will provide you with information about:

- how scientific ideas were developed and presented,
- how scientific arguments can arise from different ways of interpreting the evidence,
- ways in which scientific ideas may be affected by the contexts in which it takes place (for example, social, historical, moral and spiritual) and how these contexts may affect whether or not ideas are accepted,
- the problems science has in dealing with industrial, social and environmental questions, including the kinds of questions science can and cannot answer, uncertainties in scientific knowledge, and the ethical issues involved.

Each of the 'Ideas and evidence' contexts needed for whatever course you are following is included in this book. A guide to these contexts and where they fit in your Double or Single Award course is given in Tables 1.4, 1.5 and 1.6.

Table 1.4 Contexts for the delivery of 'Ideas and evidence' in Life and Living Processes (Modules 01, 02, 03 and 04).

Section	DA	SA	Context
1.4	✓	✓	How living conditions and life style are related to the spread of disease
2.5	✓	✓	Why the link between smoking tobacco and lung cancer gradually became accepted
3.2	✓	✓	Why Mendel proposed the idea of separately inherited factors and why this discovery was not immediately recognised
3.4	✓	✓	The economic, ethical and social issues raised by the development of cloning and genetic engineering
3.5	✓	✓	How fossil evidence supports the theory of evolution
3.5	✓	✓	How over-use of antibiotics can lead to the evolution of resistant bacteria
3.5	✓	✓	Why Darwin's theory of natural selection was only gradually accepted
3.6	✓	✓	Benefits and problems caused by the use of hormones to control fertility
4.1	✓	✓	Some of the major environmental issues facing society
4.1	✓	✗	Problems involved with the large scale production of food

Table 1.5 Contexts for the delivery of 'Ideas and evidence' in Materials and their Properties (Modules 05, 06, 07 and 08).

Section	DA	SA	Context
6.4	✓	✓	How the burning of hydrocarbon fuels affects the environment
6.5	✓	✓	How the disposal of plastics affect the environment
6.9	✓	✗	Why Wegener's theory of Continental Drift took a long time to be accepted
7.5	✓	✓	How the use of microbes and enzymes to bring about chemical reactions has advantages and disadvantages
7.8	✓	✗	How benefits from the use of nitrogenous fertilisers need to be balanced with the potential contamination of water supplies
8.7	✓	✓	How early attempts to classify elements systematically led to the development of the modern periodic table
8.7	✓	✓	Why the periodic table gradually became accepted as an important summary of the structure of atoms

Table 1.6 Contexts for the delivery of 'Ideas and evidence' in Physical Processes (Modules 09, 10, 11 and 12).

Section	DA	SA	Context
9.11	✓	✓	The advantages and disadvantages of using different energy sources to generate electricty
11.13	✓	✓	How scientists have tried to discover whether there is life elsewhere in the Universe
12.8	✓	✓	The dangers or possible dangers of exposure to different types of electromagnetic radiation and measures that can be taken to reduce such exposure
12.16	✓	✗	How the Rutherford and Marsden scattering experiment led to the replacement of the 'plum-pudding' model of the atom with the present model of the atom

Some hints about doing well in the final written examinations

Some frequently used command words and what they mean

Before you can answer a question, you need to know what is expected. Question-writers use command words or phrases that inform you of the style of answer they expect you to give. A list of the most frequently used command words and phrases is given below. Question-writers assume that you know the meanings of the words or phrases.

Command word	Meaning
Calculate or **work out**	means that a calculation is needed together with a numerical answer.
Compare	means that a description is needed of the similarities and/or differences in the information that has been provided.
Complete	means that spaces in a diagram, a table or a sentence or sentences need to be completed.
Describe	means that the important points about the particular topic must be provided.
Draw a bar chart	– if the axes are already labelled and scales have been given then the values given must be plotted as bars. – if the axes are labelled but no scales have been given then scales need to be added and the values given need to be plotted as bars.
Draw a graph	– if the axes are already labelled and scales have been given then the values given need to be plotted as points and a line (or curve) appropriate to the points plotted must be drawn.
Explain how or **Explain why**	means that scientific theory must be used to show an understanding of how or why something happens.
Give a reason or **How** or **Why**	means that the answer requires a cause for something happening based on scientific theory.
Give or **Name** or **State** or **Write down**	means that a short answer with no supporting scientific theory is needed.
List	means that a number of short answers are needed, each one being written one after the other.
Predict or **Suggest**	means that the answer is based on a *consideration* of various pieces of information and suggesting, without supporting theory, what is likely to happen.
Sketch a graph	means that a line (or curve) is to be drawn to show a trend or pattern without the need to plot a series of points.
Use the information	means that the answer must be based on the information provided in the questions.

Use your understanding of .... to this is the science topic around which the answer needs to be built.

What is meant by .... means that the answer is likely to be a definition.

Some more hints

Obviously if you want to do well you need to have learned and understood as much as you can. However here are some hints about answering questions.

- Do not rush – no marks are awarded for finishing first. A paper worth 90 marks is designed to allow you about 90 minutes to finish it. This means that you have one minute of time to think and write down 1-marks worth of answer.
- Read each question carefully at least twice before you write down your answer. If you need to do rough working to sort out your thoughts use the gaps in the margins – but make sure you put a line through this rough working.
- Look at the number of marks awarded for each part of each question. Each mark is given for a different piece of information:
  - *1 mark* means that one piece of information is needed.
  - *2 marks* mean that two pieces of information are needed etc.
- Lots of questions ask you to give a reason for something or to explain something. Such answers are usually worth 2 or more marks. Your answers to these should include a 'because' part.
- Do not throw away marks. Marks are often given for:
  - units such as joules, °C, ohms etc. Learn all the units and what they measure.
  - the names and symbols of chemical elements – so learn them
  - equations, such as 'potential difference = current × resistance' – so learn the equations and how to use them in calculations. Remember that all equations need an '=' sign in them.
- If you are writing an answer that needs several sentences, make sure that each sentence is saying something new and is not just rewording an earlier sentence. Make sure the sentences are in a logical order.
- Try to avoid using the words 'it', 'they' or 'them' in an answer. The marker may find it difficult to understand what you mean.
- Take care when you are drawing graphs. Make sure all the points are correctly plotted. When you draw in the line for your points use a pencil with a fine point and try to draw the complete line in one go.
- If you have learned the work you should finish the paper in good time. Go through the paper again and check what you have written – it could save you throwing away some marks for silly mistakes.

Chapter 1

Cells in humans

Key terms

cytoplasm • nucleus • cell membrane • soluble • enzymes • saliva • amylase • protease • proteins • amino acids • hydrochloric acid • bacteria • bile • emulsification • starch • fats • sugars • fatty acids • glycerol • alkali • neutralises • absorb • small intestine • blood • fibre • faeces • aerobic respiration • anaerobic respiration • lactic acid • oxygen debt • diffusion • concentration gradient • plasma • oxygenated blood • deoxygenated blood • atria • ventricles • arteries • valves • mucus

1.1	
Co-ordinated	Modular
DA 10.1	DA 1
SA 10.1	SA 13

Animal cells

Humans have got many different kinds of organs. They are made up of many different types of cell, but all these cells have the following parts:

- the **cytoplasm** – this is where most of the chemical reactions that are needed to keep you alive take place.
- a **nucleus** – this is the control unit of the cell. All of the chemical reactions that take place in the cytoplasm are controlled by the nucleus.
- a **cell membrane** – this is a thin, oily layer. It controls the entry and exit of substances to the cell.

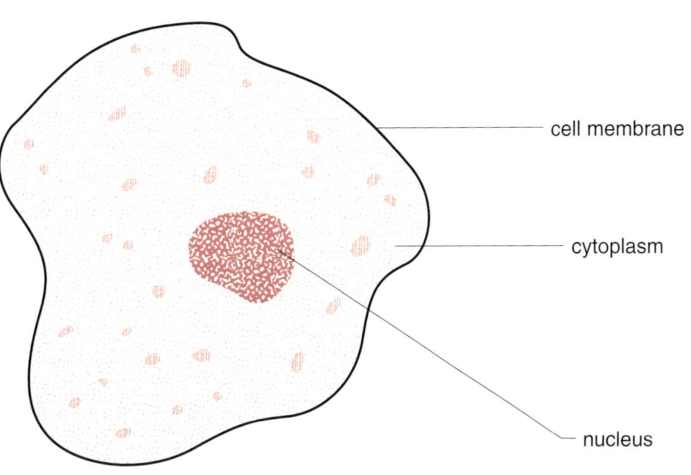

Figure 1.1
A typical cell from a human being

You can think of a cell as a nightclub. The cell membrane is the bouncer, the cytoplasm is the dance floor and the DJ is the nucleus!

In this section we'll be looking at lots of different organ systems and the special types of cell that they have.

1.2	
Co-ordinated	Modular
DA 10.4	DA 1
SA 10.3	SA 13

The digestive system

It's strange to think that some of the food that you eat becomes part of your nails, muscles, hair and bones. Food contains all of the substances needed to make every cell in your body. It also contains the energy needed to keep you alive. To be of use to your body, food must first be made soluble – this is the job of your digestive system.

The digestive system is really a long tube that starts at the mouth and ends at the anus. If you look at Figure 1.2, you'll see how all the important parts are joined together.

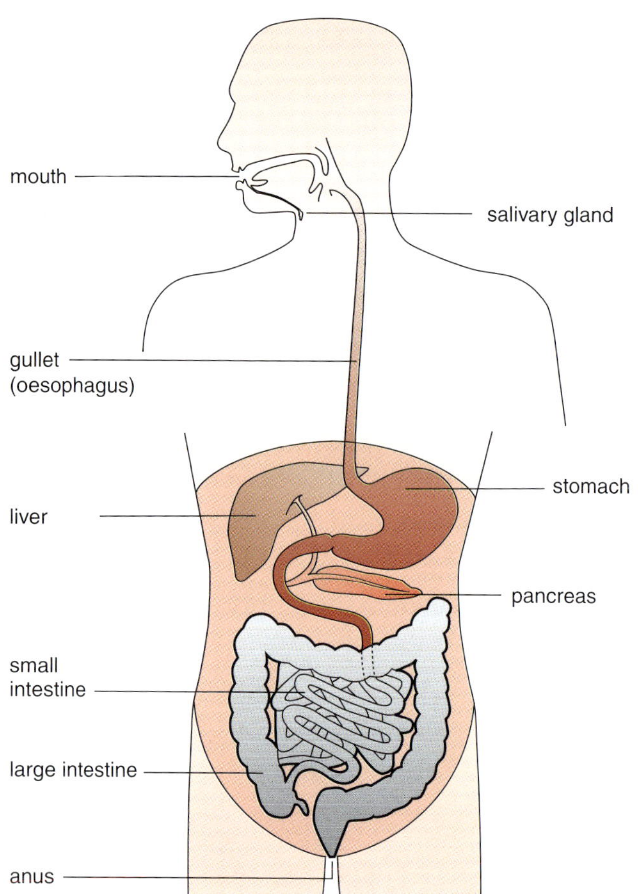

Figure 1.2
The human digestive system

Foods such as protein, starch and fat are insoluble in water. The digestive system makes these **soluble** using chemicals called enzymes. **Enzymes** are like chemical knives that chop up your food into such tiny pieces that it eventually dissolves. The soluble bits are absorbed into the blood; the insoluble bits (like fibre) leave the body through the anus as faeces.

Figure 1.3 shows that starch, proteins and fats are made up of smaller, soluble parts.

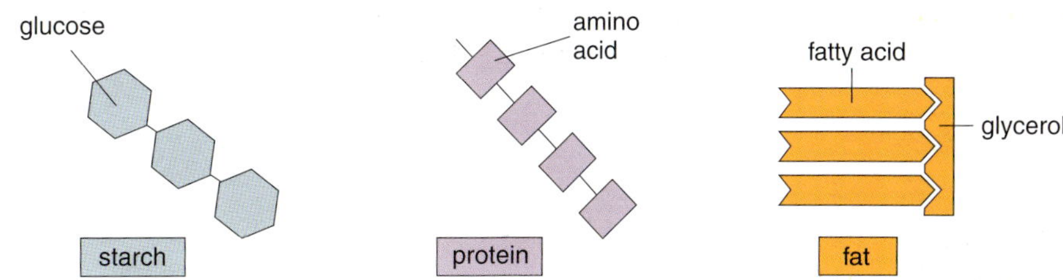

Figure 1.3
Starch, protein and fat are made up of smaller molecules

Your digestive system breaks down starch into sugars, protein into amino acids and fat into fatty acids and glycerol.

The mouth

The salivary glands squirt **saliva** into your mouth whenever you eat something. Saliva contains an enzyme called **amylase**. Amylase breaks down starch (a carbohydrate found in foods like bread, rice and potatoes) into sugars.

The gullet

This is a muscular tube that pushes your food down to your stomach when you swallow.

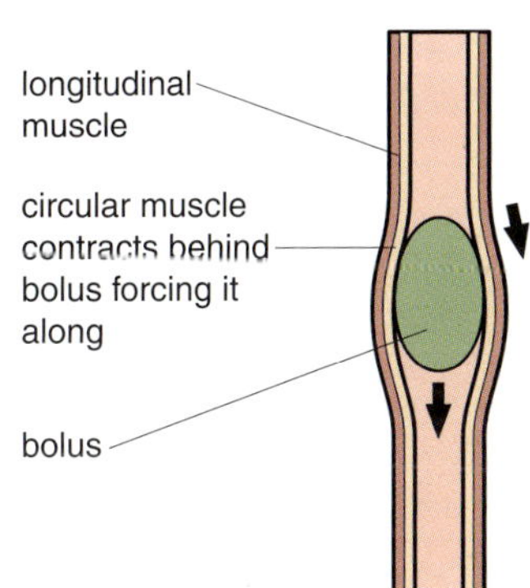

Figure 1.4
Muscles in the gullet push the chewed food along

The stomach

The stomach makes two different substances:

1 An enzyme called **protease**. Protease breaks down **proteins** (found in foods like meat, beans and dairy products) into **amino acids**.

2 **Hydrochloric acid**. The acid kills **bacteria** that are taken in with your food. The protease made by your stomach works best in acidic conditions.

Biological washing powders contain digestive enzymes to get rid of food stains from clothes.

The pancreas

Food doesn't go into this organ, but it makes three types of enzymes which it squirts into your small intestine. The enzymes it makes are:

- **amylase** – which breaks down **starch** into **sugars**.
- **protease** – which breaks down **proteins** into **amino acids**.
- **lipase** – which breaks down **fats** into **fatty acids** and **glycerol**.

The liver

Food doesn't go into this organ; it makes a substance called **bile**. Bile isn't an enzyme but it does help your body to digest fats. It does this by breaking large drops of fats into lots of tiny droplets in a process called **emulsification**. Lots of tiny droplets between them have more surface than a single, large drop. This means that lipase can get to the fat more easily and break it down more quickly.

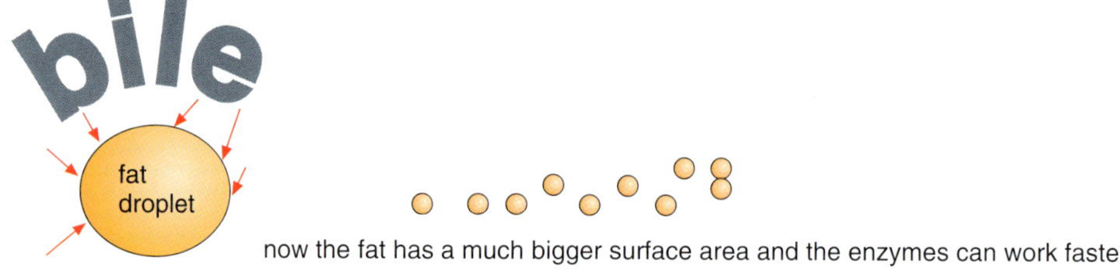
bile
fat droplet
now the fat has a much bigger surface area and the enzymes can work faster

Figure 1.5
Bile breaking a fat molecule down into small droplets

Bile is also an **alkali**. This **neutralises** acid from the stomach because enzymes made by the pancreas work best in **neutral conditions**.

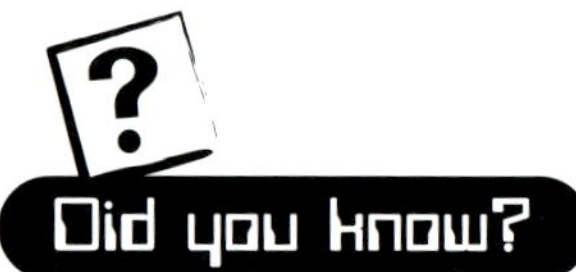
?
Did you know?

Your small intestines are around nine metres in length.

The small intestine

The walls of the **small intestine** itself also make the enzymes amylase, protease and lipase. The walls of the small intestine **absorb** sugars, amino acids, fatty acids and glycerol into the **blood**.

If you look at the walls of the small intestine, you'll see that they're not smooth. They've actually got projections that look a bit like fingers sticking out all over the place. These are called villi and they increase the surface area to help the absorption of the products of digestion.

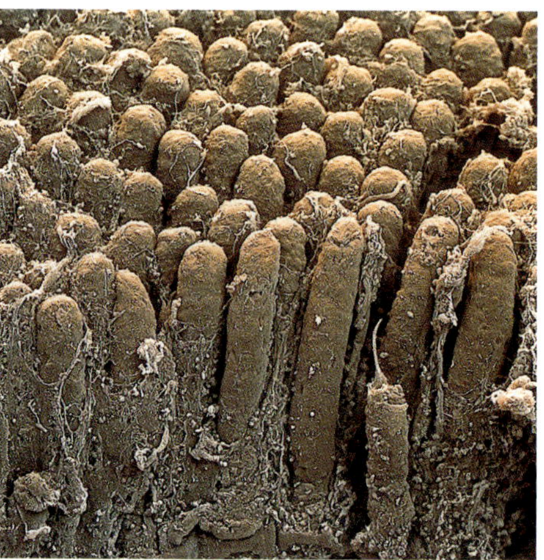

Figure 1.6
Villi increase the surface area of the small intestine

Diffusion in the small intestine

Stuff that's in solution moves from places where it's plentiful (high concentrations) to places where there's less of it (low concentrations). This happens in the small intestine; there's less sugar, for example, in your bloodstream than there is inside your small intestine. The sugar moves from the high concentration in the small intestine to the lower concentration in the blood.

The large intestine

Some of the food that you eat, such as **fibre**, cannot be broken down by your small intestine so it just moves into your large intestine. Your large intestine absorbs water from the stuff that's left and pushes it towards your anus. This leaves your body as **faeces** – commonly known as poo!

1 State the functions of the following parts of a typical animal cell:

   a) nucleus
   b) cytoplasm
   c) cell membrane

2 Explain what the term digestion means.

3 Describe what happens in the following parts of your digestive system:

   a) mouth
   b) stomach
   c) small intestine
   d) large intestine

4 What is the job of the following organs?

   a) liver
   b) pancreas

5 What does the term diffusion mean?

6 What happens to the fibre that you eat?

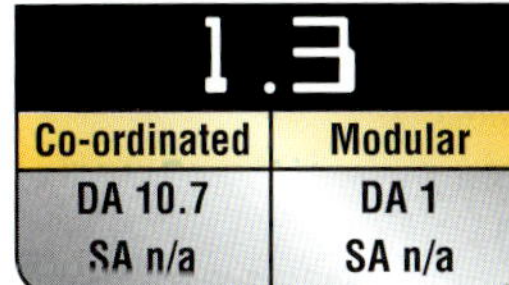

1.3	
Co-ordinated	Modular
DA 10.7 SA n/a	DA 1 SA n/a

Aerobic respiration – producing energy

Your body gets all of its energy from the food that you eat. It can use lots of different kinds of food to make energy but it prefers to use glucose. To release energy from glucose, your body uses enzymes found in the cytoplasm of your cells. The enzymes help oxygen combine with glucose in a process called **aerobic respiration**. Aerobic respiration can be summarised in the following word equation:

glucose + oxygen ⟶ carbon dioxide + water + energy

The energy that's produced during respiration is used to:

- help growth and repair by converting small molecules into larger ones such as amino acids into protein.
- help your muscles contract – so you can move!
- keep your body temperature steady – your body works best at 37°C.

Figure 1.7
This sprinter's leg muscles are using anaerobic respiration. His muscles are burning because of the lactic acid produced – you can see the pain in his face!

Anaerobic respiration

During really hard exercise, your muscles can use oxygen faster than your heart can pump it to them. Your muscles can still produce energy but they use a different process called **anaerobic respiration**. During anaerobic respiration glucose is converted directly into energy **without oxygen**. The problem is that anaerobic respiration releases a chemical called **lactic acid**. As lactic acid builds up in your muscles it makes them burn so you can only make energy in this way for a short period of time.

You may have noticed that you breathe heavily after a sprinting race. This is because the lactic acid that you made during anaerobic respiration must be broken down by your body using oxygen. The amount of oxygen needed to get rid of the lactic acid is called an **oxygen debt**.

The breathing system

If you look at Figure 1.8, you'll see the position of your lungs. They're found in the thorax – the upper part of your body. Your lungs' job is to absorb oxygen into your blood and to remove carbon dioxide from your blood.

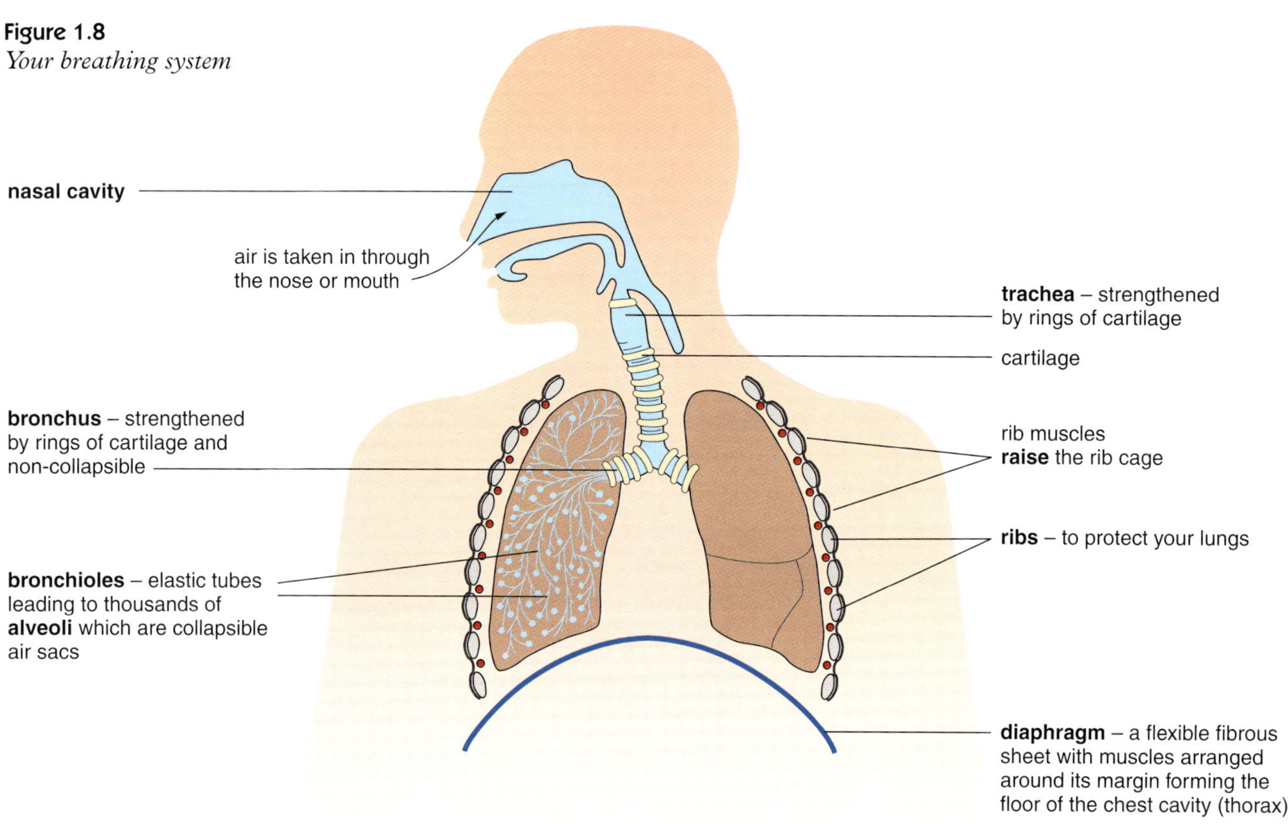

Figure 1.8
Your breathing system

When you breathe in, the air rushes into your body through tubes that keep branching and getting smaller. The trachea (windpipe) branches to form bronchi which themselves divide to form tiny tubes called bronchioles. Bronchioles end up in structures called air sacs or alveoli. The walls of the alveoli are really thin – so thin that oxygen can diffuse (move from a high concentration to a lower one) into the bloodstream.

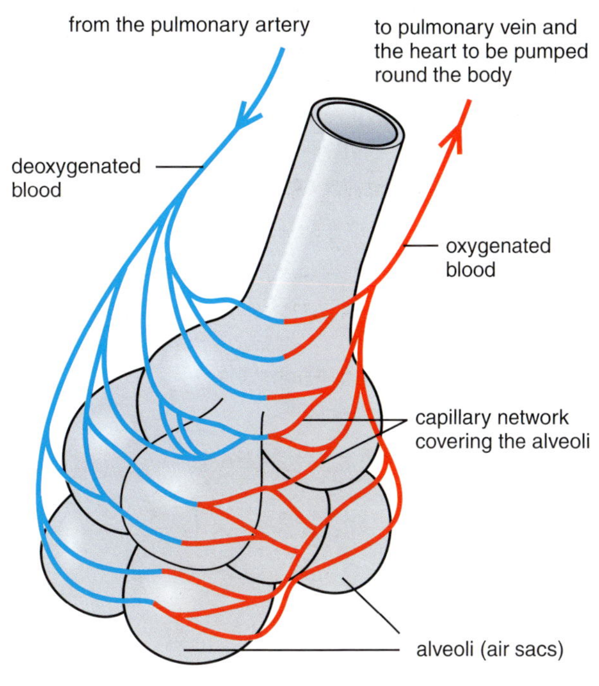

Figure 1.9
Alveoli – where oxygen gets into your blood and carbon dioxide is removed

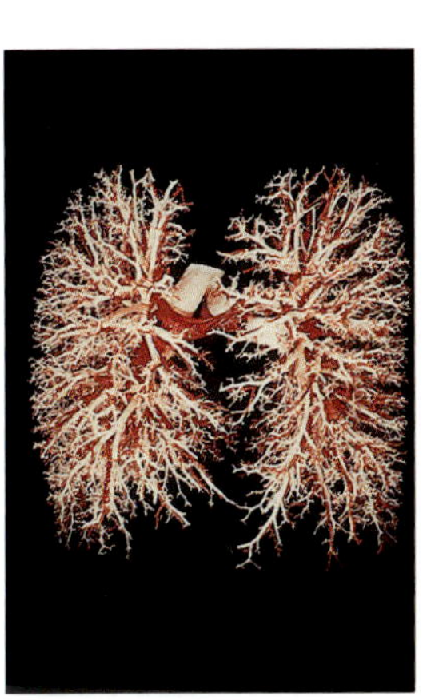

Figure 1.10
This photograph shows the branching of the tubes inside your lungs

Diffusion in the alveoli

Particles of gas always move from places where they're plentiful (high concentration) to areas where there's less of them (low concentration). This movement is called **diffusion**. The bigger the difference in concentration, the faster diffusion takes place. The difference in concentration is known as the **concentration gradient**.

The reason that oxygen diffuses into your bloodstream is that there is a concentration gradient between the air and your red blood cells (it's your red blood cells that carry oxygen around the body). Carbon dioxide diffuses out of your bloodstream because there's a higher concentration inside your blood than in the air.

Figure 1.11
Diffusion of gases in an alveolus

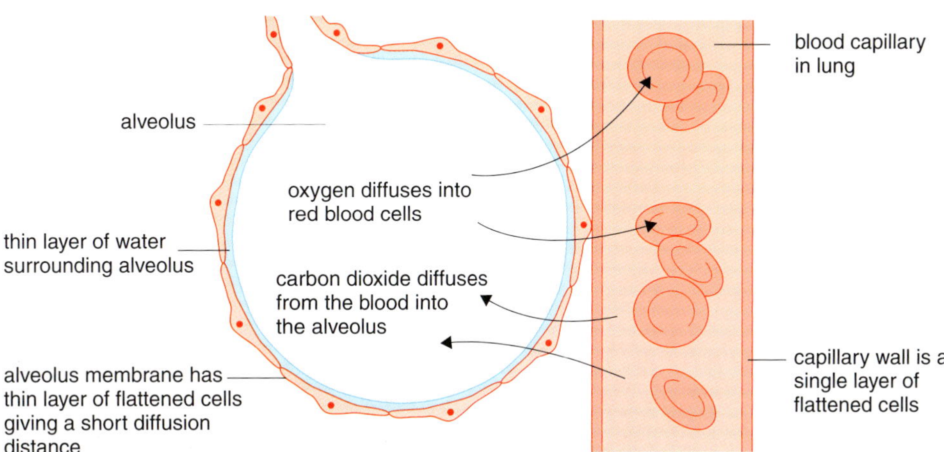

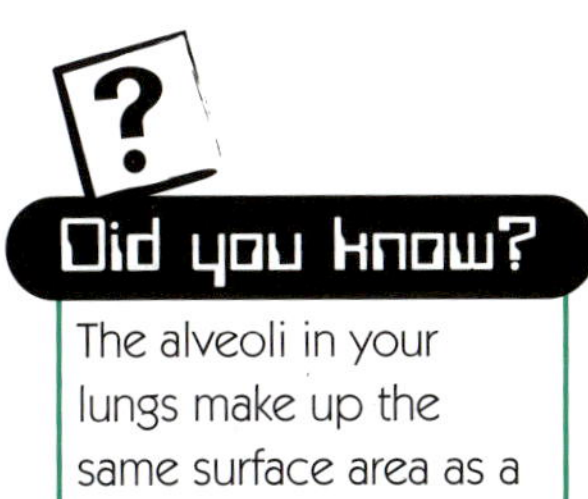

The alveoli in your lungs make up the same surface area as a tennis court.

Ventilation

Ventilation means getting air in and out of your lungs. To do this, you've got to make pressure changes in your thorax. When you breathe in, your ribcage moves up and out and your diaphragm flattens. These two changes reduce the pressure inside the thorax – making air rush in. When you breathe out, your diaphragm pushes upwards and your ribs move downwards. This increases the pressure inside the thorax – forcing air out of your lungs.

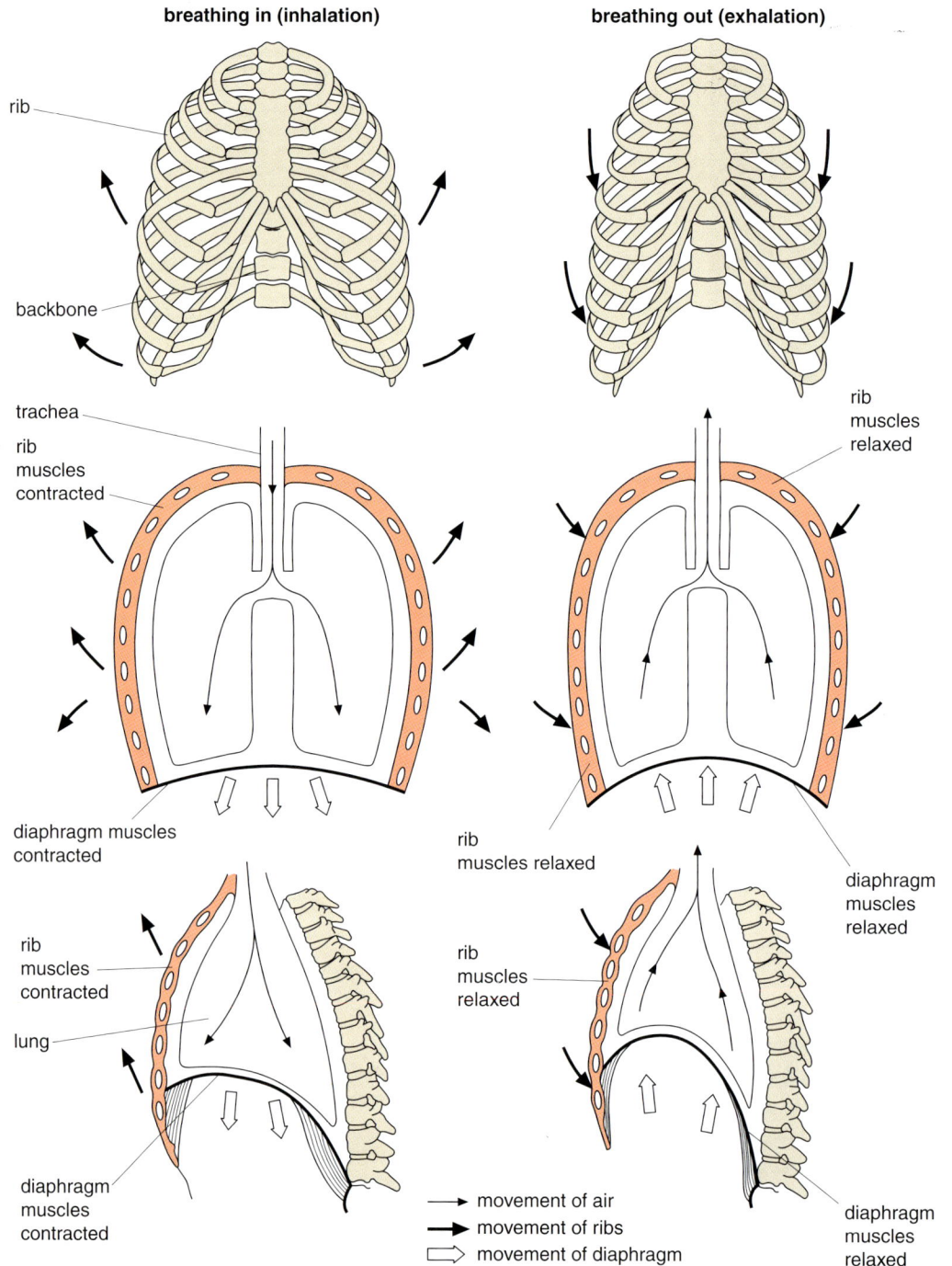

Figure 1.12
Breathing in and out

Topic Questions

1 Write out the word equation for aerobic respiration.
2 What are the differences between aerobic and anaerobic respiration?
3 Explain why you need to respire.
4 Which muscle separates the thorax from the abdomen?
5 Where does oxygen get into your blood?
6 Explain how your lungs are ventilated.
7 Explain why carbon dioxide diffuses out of your bloodstream.

1.4

Co-ordinated	Modular
DA 10.5	DA 1
SA 10.4	SA 13

The circulatory system

All of your body's organs need oxygen and food. They also produce waste products which have to be removed from the body. These substances need to be transported around the body. This is done by the circulatory system which consists of the heart, blood and blood vessels.

The blood

When you cut yourself, the blood that comes out just looks like a red liquid. If you take a closer look, however, you'll see that it's a lot more complex than that. Blood is made up of cells which are suspended in a fluid called **plasma**. These cells, together with plasma, make up your body's transport and self defence system. Figure 1.13 shows the main parts of the blood.

Figure 1.13

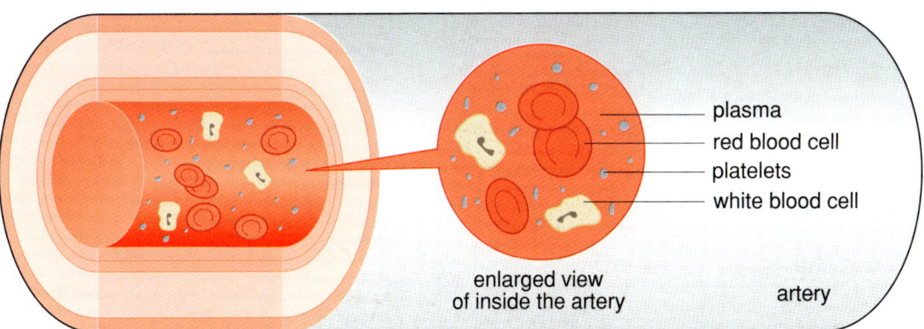

Red blood cells

If you look at Figure 1.14, you'll see a photo of red blood cells which was taken using an electron microscope. Red blood cells are responsible for carrying oxygen from your lungs to your organs.

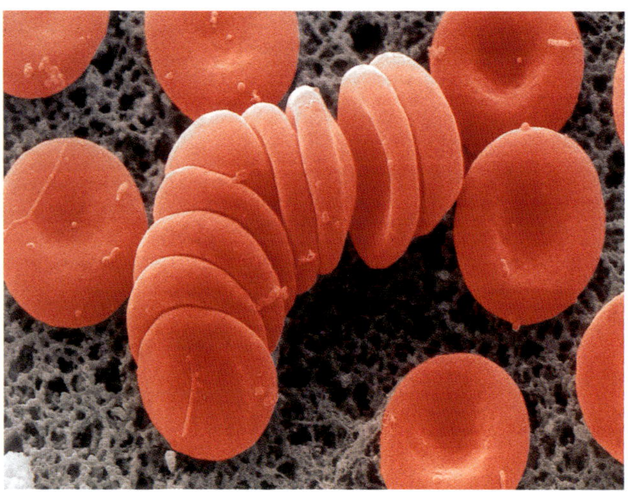

Figure 1.14
This image of red blood cells was taken by an electron microscope

White blood cells

White blood cells are part of your body's defence system against microbes. We'll learn more about these in section 1.4.

Platelets

These are tiny fragments that help your blood form clots when your skin is broken.

The Masai people of Africa drink the blood of their cattle.

Plasma

This is the liquid part of the blood. It transports:

- The products of digestion from the small intestine to the body's organs.
- Carbon dioxide from the organs to the lungs.
- Urea (a toxic waste made by the liver) from the liver to the kidneys.

The heart

The heart is really a muscular pump which lies in between your lungs. Its job is to pump blood around the body.

The heart is made up of two pumps:

- One pumps blood to your **lungs** to pick up **oxygen** and get rid of **carbon dioxide**.
- The other pumps **oxygenated blood** around the whole of the body.

If you look at Figure 1.15, you'll see that the heart's got four chambers. The two at the top are called **atria**; their job is to receive blood. The **right atrium** receives **deoxygenated blood** from the body's organs; the **left atrium** receives **oxygenated blood** from the lungs. The atria pass blood to the **ventricles**. The job of the ventricles is to pump blood out of the heart through vessels called **arteries**. The **left ventricle** pumps **oxygenated blood** to the body's organs. The **right ventricle** pumps **deoxygenated blood** to the lungs.

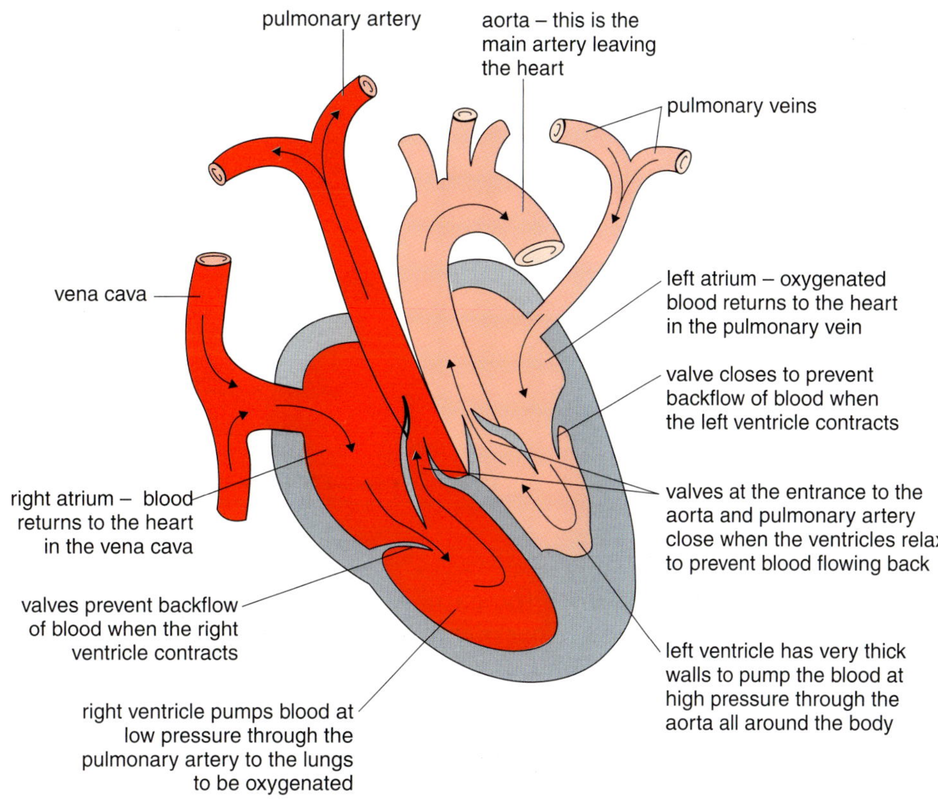
pulmonary artery
aorta – this is the main artery leaving the heart
pulmonary veins
vena cava
left atrium – oxygenated blood returns to the heart in the pulmonary vein
valve closes to prevent backflow of blood when the left ventricle contracts
right atrium – blood returns to the heart in the vena cava
valves at the entrance to the aorta and pulmonary artery close when the ventricles relax to prevent blood flowing back
valves prevent backflow of blood when the right ventricle contracts
left ventricle has very thick walls to pump the blood at high pressure through the aorta all around the body
right ventricle pumps blood at low pressure through the pulmonary artery to the lungs to be oxygenated

Figure 1.15
How blood flows through both sides of the heart

The walls of the ventricles are made of strong muscle tissue. When your ventricles contract, they squeeze blood into arteries at high pressure. The **valves** prevent blood flowing back to your atria.

Arteries

Blood is forced into your arteries at high pressure. To cope with this pressure, elastic fibres allow your arteries to expand with each heart beat. The flow of blood is controlled by muscles in the artery wall.

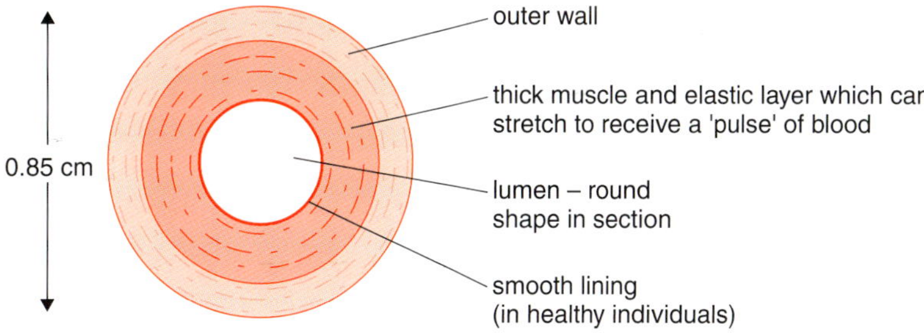
0.85 cm
outer wall
thick muscle and elastic layer which can stretch to receive a 'pulse' of blood
lumen – round shape in section
smooth lining (in healthy individuals)

Figure 1.16
An artery – note the thick muscular walls and elastic fibres

Arteries branch into smaller and smaller tubes, eventually ending up in tiny vessels called **capillaries**.

Capillaries

These are the body's tiniest blood vessels – only one cell thick. Capillaries are a bit leaky, they let dissolved substances pass through their walls. Substances such as glucose and amino acids need to diffuse out of the

blood into tissues in organs. Waste products such as carbon dioxide and urea diffuse into the capillaries from the organs.

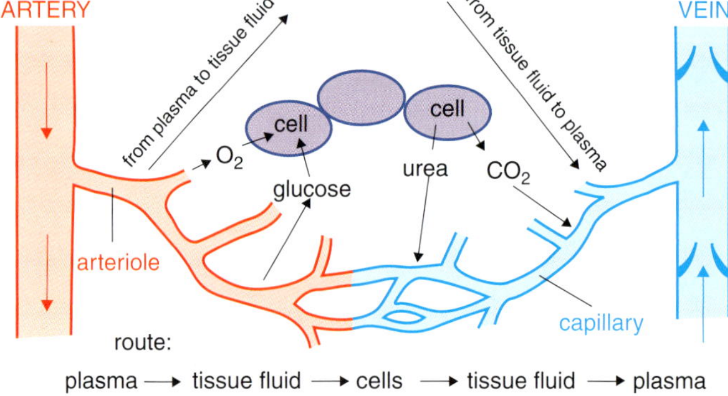

Figure 1.17
Capillaries, where useful stuff leaves the blood and waste enters the blood

Veins

Blood from the capillaries enters the **veins**. By now, the blood is at low pressure. This means that veins need valves to maintain the flow of blood back to the right hand side of the heart.

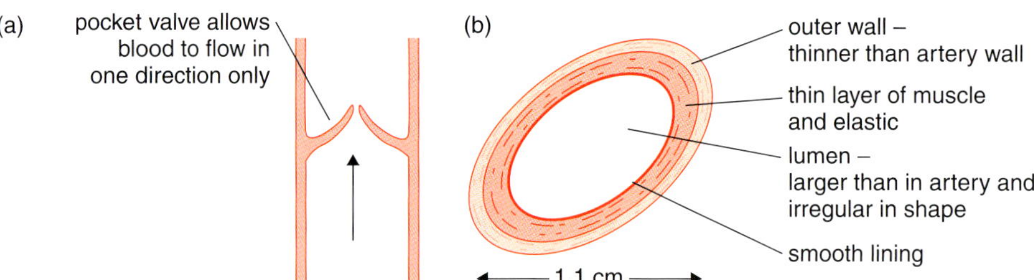

Figure 1.18
Veins don't need thick walls but they do need valves

Topic Questions

1 What are the functions of:

a) red blood cells?
b) white blood cells?
c) platelets?

2 Which substances are transported by plasma?

3 What is the function of:

a) an atrium?
b) a ventricle?

4 Explain why valves are needed in your heart.

5 What are the differences between arteries and veins?

6 Why are capillaries a bit leaky?

1.5	
Co-ordinated	Modular
DA 10.11	DA 1
SA 10.8	SA 13

Disease

Bacteria and viruses are so small that they can only be seen with the aid of a microscope. There are literally millions of them in the air around you, on your skin and inside your body – all of the time! Most are harmless but some of them can cause disease.

Bacteria

These organisms are simply single cells. Like your cells, they've got a cytoplasm and cell membrane. However, bacteria are different to cells in two important ways:

- Bacteria do not have a nucleus – their genetic material just floats in the cytoplasm.
- Bacteria have a cell wall.

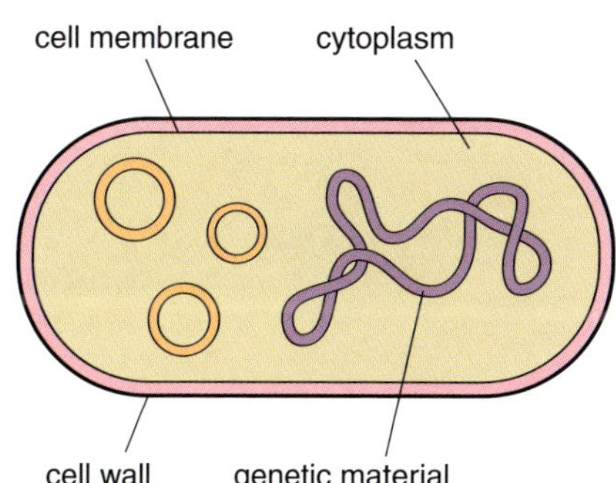

Figure 1.19
One bacterial cell

Some bacteria make poisonous chemicals called toxins. These toxins are responsible for making you ill. Diseases such as tuberculosis, pneumonia, salmonella and cholera are caused by bacteria. Bacteria can live outside of the human body. They survive by breaking down the remains of living things – this includes the dirt and bits of food that humans leave behind. This is why it is so important to keep cooking equipment clean.

Viruses

Viruses are much smaller, you can't see them with a normal microscope. They can only be seen by really powerful **electron microscopes**. Surprisingly, viruses aren't made up of cells – they're simply genetic material surrounded by a protein coat.

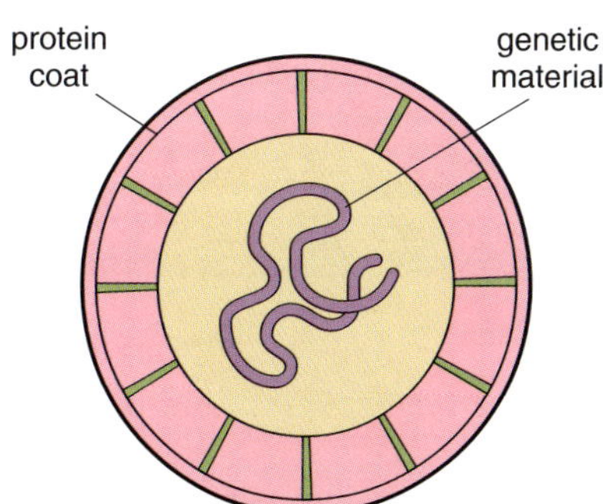

Figure 1.20
A virus

Unlike bacteria, viruses can't live outside the body. They reproduce inside living cells, eventually bursting out of them. When this happens, toxins are released which poison the body making you feel ill. Viruses causes diseases such as the 'flu', hepatitis and AIDS.

Did you know?

Being cold does not give you a cold. Colds are caused by viruses which you breathe in after someone else has sneezed them out!

Body defences

Barriers and traps

Your body defends itself against microbes in a number of ways:

- Your skin acts as a **barrier**, stopping microbes getting in.
- Cells in the lining of your **breathing passageways** make a sticky **mucus**. This traps microorganisms that you may have breathed in. Cells with tiny 'hairs' flick this mucus towards your gullet. You swallow the mucus and the acid in your stomach kills the microorganisms.

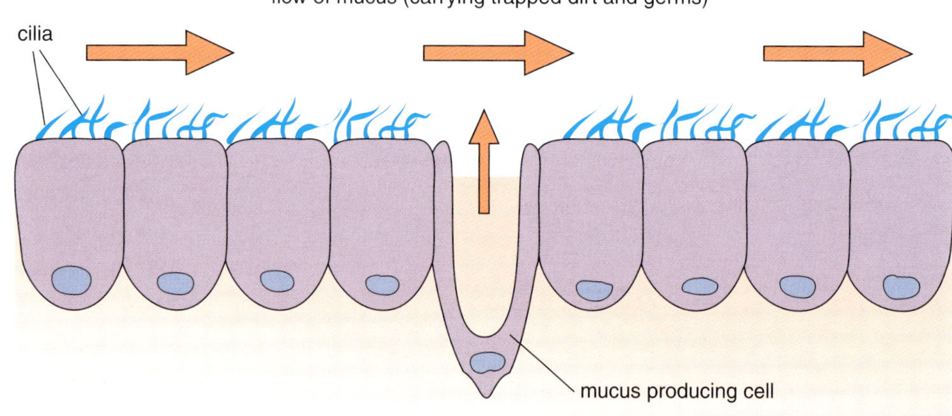

Figure 1.21
Cells in your breathing passageways make mucus which is moved towards your gullet by tiny hair-like structures called cilia

If you cut yourself, your blood quickly produces a clot which you'll know as a scab. The clot stops microbes getting into your body.

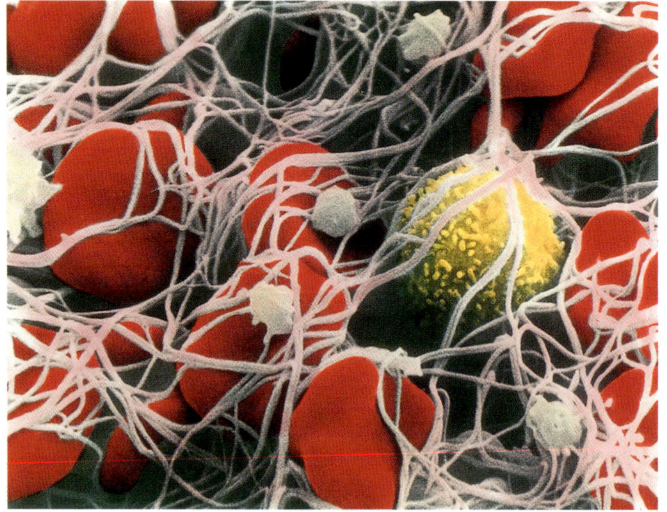

Figure 1.22
This photo shows a clot being formed. Red cells and protein fibres form a mesh which stops bacteria and viruses getting into your body

White blood cells

If microbes get past your body's barriers and traps, white blood cells spring into action. Some of them can wrap themselves around bacteria and viruses, breaking them down.

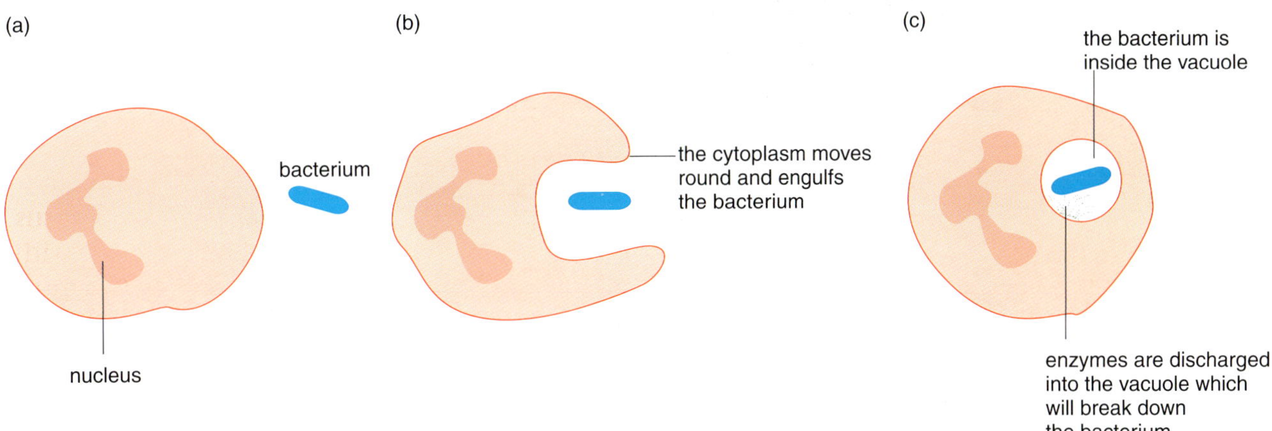
(a)
(b)
(c)
bacterium
the cytoplasm moves round and engulfs the bacterium
the bacterium is inside the vacuole
nucleus
enzymes are discharged into the vacuole which will break down the bacterium

Figure 1.23
A white blood cell ingesting a bacterium

White blood cells also make certain chemicals:

- antibodies – these destroy bacteria or viruses
- antitoxins – these counteract the toxins made by bacteria.

Vaccination

Once your body has been infected by a microbe, it can produce antibodies much more quickly if the same kind of microbe gets into your body for a second time. This means that the microbe is stopped in its tracks before it can make you ill. You are now said to be **immune** to that microbe.

Doctors use your body's natural defences to help protect you against dangerous illnesses such as tuberculosis or meningitis. A **vaccine** can be made by making a solution of a weak or killed microbe. When it's injected into your body, your white blood cells make antibodies to it. **Vaccination** means that you can become immune to a microbe without having to catch the disease first.

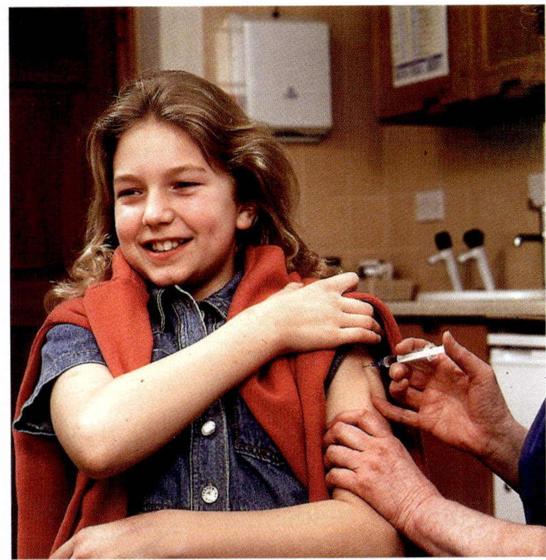

Figure 1.24
This girl is being vaccinated against German measles (Rubella)

? Did you know?

In the eighteenth century it was common for people to die from a terrible disease called **smallpox**. A doctor from Gloucestershire called Edward Jenner noticed that milkmaids caught a similar but much milder disease called cowpox, but rarely caught smallpox. He thought that catching cowpox prevented the milkmaids getting smallpox and set up an experiment to see if his idea was correct.

He injected 8 year old James Phipps with cowpox. Sure enough James went on to develop cowpox but recovered within two weeks. Jenner then decided to test his idea and injected James with the deadly smallpox virus. Luckily for James, he did not go on to develop smallpox. This is because the antibodies that his white blood cells produced to fight cowpox worked against the smallpox virus.

Jenner's methods were adopted by doctors around the globe. The last known case of smallpox was in the 1970s.

Living conditions and health

Diseases are much more likely to spread if lots of people live in crowded and dirty conditions. In Victorian times, there were parts of London called slums. In these slums sewage ran down the street and lots of people shared the same bedrooms. People caught so many diseases that they didn't really expect to live much beyond their thirtieth birthday.

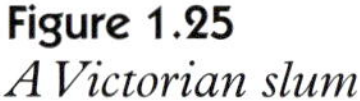

Figure 1.25
A Victorian slum

In Britain nowadays, people's waste is dealt with by an efficient sewage system. People are also much more aware of the need for keeping things clean. Good sanitation (sewage systems) and modern medicine mean that the average British person now lives for 78 years.

Figure 1.26 *Some important dates in the history of public health in Britain*

Date	Milestone
1790s	Edward Jenner developed the process of vaccination.
1860s	Louis Pasteur found out that you can stop milk 'going off' if you heat it up enough to kill the microbes that live in it.
1865	Joseph Lister made the first antiseptic.
1928	Alexander Fleming discovered penicillin.
1948	The National Health Service provides free treatment for all.
1956	Toxic gases are released when coal is burnt. The Clean Air Act banned the use of coal in house fires in London, dramatically reducing the number of deaths from lung diseases.
1960s	Mass immunisation of children against measles, mumps and rubella.

Topic Questions

1. List the differences between bacteria and viruses.
2. What is the type of substance released by microbes which makes you ill?
3. List three ways in which your body tries to prevent the entry of microbes.
4. Explain how white blood cells fight disease.
5. What kind of living conditions help the spread of disease?

Summary

- Most human cells have the following parts:
  - a nucleus which controls the activities of the cell
  - cytoplasm in which most of the chemical reactions take place
  - a cell membrane which controls the passage of substances in and out of the cell.
- Digestion is the breakdown of large food molecules into smaller, soluble molecules so that they can be absorbed into the bloodstream through the wall of the small intestine.
- In the large intestine water is absorbed into the bloodstream.
- Indigestible food makes up most of the faeces.
- Faeces leave the body via the anus.
- Enzymes are catalysts that speed up the breakdown of food.
- Amylase is produced in the salivary glands, the pancreas and the small intestine. It speeds up the breakdown of starch into sugars.
- Protease enzymes (produced by the stomach, the pancreas and the small intestine) speed up the breakdown of protein into amino acids.
- Lipase enzymes (produced by the pancreas and small intestine) speed up the breakdown of fat into fatty acids and glycerol.
- The stomach produces hydrochloric acid which kills bacteria taken in with food. The enzymes in the stomach work best in acidic conditions.

- Bile is made by the liver and stored in the gall bladder. Bile neutralises the acid that was added to food in the stomach. It produces alkaline conditions to help enzymes in the small intestine work most effectively.
- Bile emulsifies fats (breaks large drops of fats into smaller droplets). This increases the surface area of fats for lipase enzymes to act upon.
- The breathing system helps oxygen get into (diffuse into) the bloodstream and removes carbon dioxide from the bloodstream.
- Lungs are found in the thorax (chest). They're protected by the ribcage and separated from the abdomen (belly) by the diaphragm.
- When you breathe in, the ribcage moves up and out and the diaphragm becomes flatter. Breathing out happens when this is reversed.
- The word equation for aerobic respiration is:

  glucose + oxygen $\longrightarrow$ carbon dioxide + water + energy
- Anaerobic respiration happens when there is a shortage of oxygen and makes a waste product called lactic acid.
- Lactic acid must be broken down by oxygen. The oxygen needed is called an oxygen debt.
- The energy that is released during respiration is used:
  - to build up larger molecules using smaller ones
  - to enable muscles to contract
  - to maintain a steady body temperature in colder surroundings.
- Diffusion is the spreading of the particles of a gas, from an area of a high concentration to an area of lower concentration.
- Oxygen diffuses into the blood in the alveoli.
- Blood is pumped around the body by the heart which is made up of muscle fibre.
- Blood enters an atrium of the heart which contracts, squeezing it into a ventricle. The ventricle contracts squeezing blood out of the heart.
- Valves in the heart control the flow of blood.
- The heart pumps blood to the organs through arteries and it returns through veins.
- Humans have two separate circulation systems, one to the lungs and one to all the other organs of the body.
- Arteries have thick walls containing muscle and elastic fibres.
- Veins have thinner walls and valves to prevent the back-flow of blood.
- Capillaries are very narrow, thin-walled blood vessels found inside organs.
- Substances needed by the cells pass out of the blood, and substances produced by the cells pass into the blood through the walls of the capillaries.
- Blood is made from fluid called plasma which carries white blood cells, red blood cells and platelets.
- Plasma carries:
  - carbon dioxide from the organs to the lungs
  - soluble products of digestion from the small intestine to other organs
  - urea (a waste) from the liver to the kidneys.
- Microorganisms such as bacteria and viruses can cause diseases.
- Bacterial cells are made up of a cytoplasm, cell membrane and cell wall. They have no nucleus, genes just float in the cytoplasm.
- Viruses are smaller than bacteria. They're made from a protein coat surrounding a few genes and can only reproduce inside living cells.
- Diseases spread as a result of unhygienic conditions or contact with infected people.
- The skin acts as a barrier to the entry of microorganisms.

- Mucus is made by the breathing organs. It's sticky so that it traps microorganisms.
- The blood makes clots (scabs) that seal cuts.
- White blood cells defend against microorganisms:
  - by ingesting (eating) microorganisms
  - by making antibodies which destroy bacteria or viruses
  - by making antitoxins which neutralise the toxins (poisons) made by microorganisms.

Examination Questions

1 The diagram below shows the human digestive system.

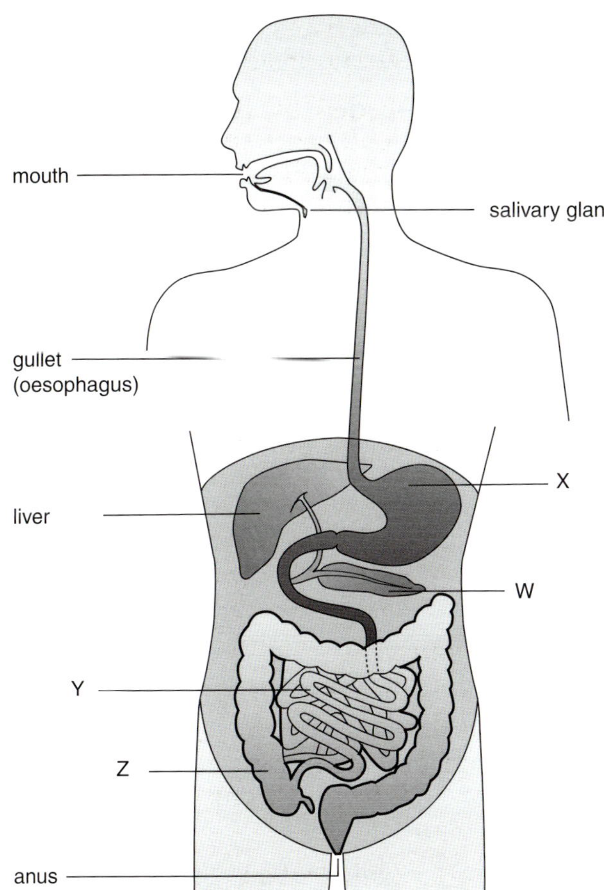

a) Name the parts labelled W – Z. *(4 marks)*

b) In which part of the digestive system are the enzymes that only break down protein produced? *(1 mark)*

c) What is the function of organ X? *(1 mark)*

d) In which part of the digestive system are the products of digestion absorbed? *(1 mark)*

e) What is the function of part Z? *(1 mark)*

f) Name the organs that produce the enzyme 'lipase'. *(2 marks)*

2 The sentences are about respiration. Choose words from the list in the box to complete the sentences that follow.

water glucose carbon dioxide energy mitochondria oxygen

Aerobic respiration is a process in which ___________ is released from ___________. This process occurs in ___________ and uses ___________ absorbed from the air. Aerobic respiration produces ___________ and ___________ as by-products. *(6 marks)*

3 The diagram below shows the human heart.

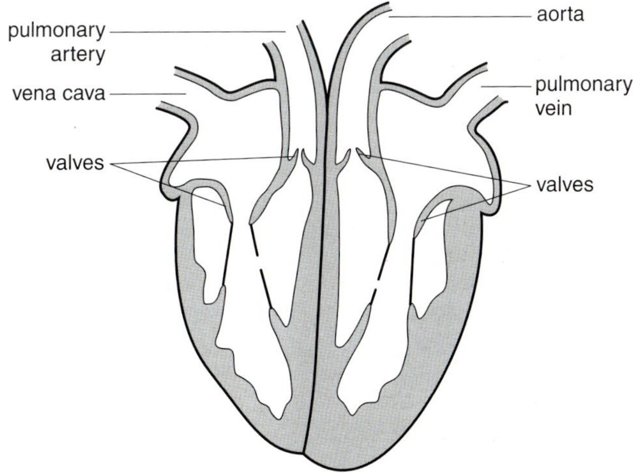

a) Add to the diagram the following labels: Left atrium; Right ventricle *(2 marks)*

b) Explain the function of the valves. *(1 mark)*

c) Name the blood vessel that takes deoxygenated blood to the lungs. *(1 mark)*

d) List two differences that exist between arteries and veins. *(1 mark)*

e) Complete the sentences that follow using words from the list in the box.

left ventricle red blood cells right ventricle left atrium

Deoxygenated blood is pumped out of the heart by the________________. Oxygen is picked up by the ________________. Blood returns to the ________________ and is pumped to the ________________. From here it is pumped around the whole of the body. *(4 marks)*

4 The diagram below shows four parts of the blood.

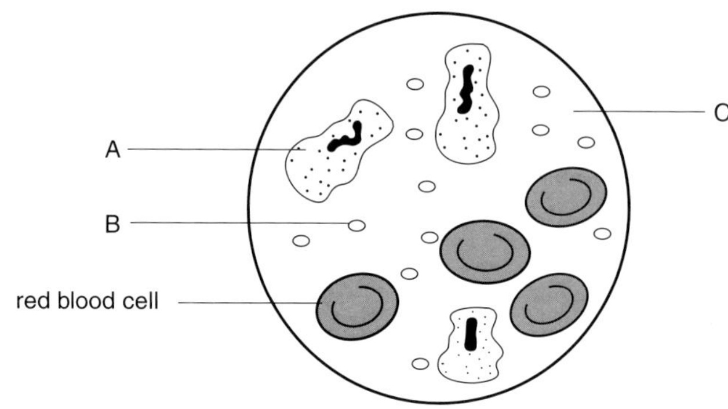

a) Which part fights disease? *(1 mark)*
b) Which part clots blood? *(1 mark)*
c) Which part carries carbon dioxide from the organs to the lungs? *(1 mark)*
d) Which two parts have no nucleus? *(2 marks)*
e) Explain the function of the red blood cells. *(2 marks)*

Chapter 2

Maintenance of life

Key terms

light energy • glucose • carbon dioxide • photosynthesis • chloroplasts • cell wall • vacuole • cell sap • chlorophyll • respiration • palisade layer • stoma • root hairs • xylem vessel • osmosis • phloem • transpiration • neurones • stimulus • receptors • sensory neurones • motor neurones • reflexes • retina • cornea • lens • suspensory ligaments • pancreas • insulin • glucagon • diabetes

2.1 Green plants

Co-ordinated	Modular
DA 10.13	DA 2
SA n/a	SA n/a

Just like animals, green plants are made up of cells. Plants cells have a nucleus, cell membrane and cytoplasm but they also have a few extra bits! Plant cells don't eat, they use **light energy** to make **glucose** from **carbon dioxide** and **water** in a process called **photosynthesis**. This happens inside structures in their cytoplasm called **chloroplasts**. If you look at Figure 2.1, you'll see that they've also got a **cell wall** which helps keep the cell rigid and a large central **vacuole** which contains a fluid called **cell sap**.

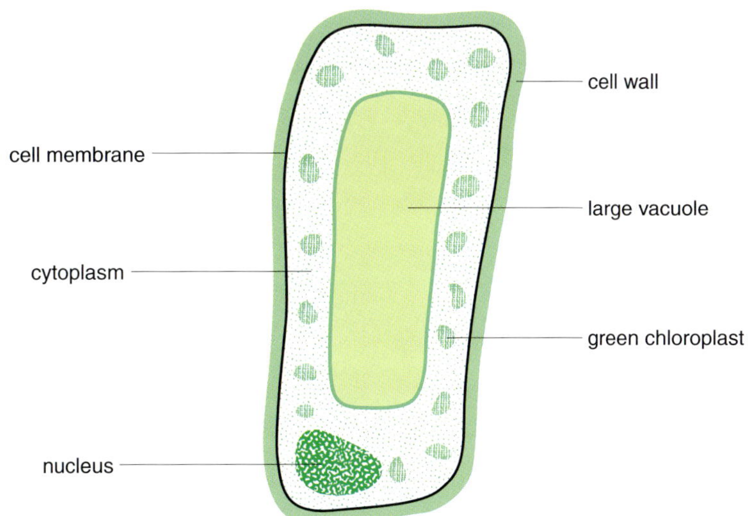

Figure 2.1
A typical plant cell

Chloroplasts contain a green pigment called **chlorophyll** which is where photosynthesis takes place. The word equation for this process is:

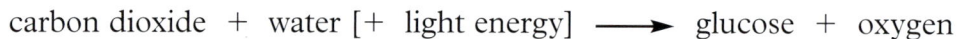

carbon dioxide + water [+ light energy] ⟶ glucose + oxygen

As well as producing glucose which plants then use to make energy in **respiration**, photosynthesis produces oxygen as a by-product.

A plant can store glucose in the form of starch. Starch is sometimes stored in storage organs found beneath the soil. Potatoes and carrots store starch like this.

Leaves

Most photosynthesis takes place in the leaves of plants which have special features that help them absorb light and carbon dioxide efficiently. If you look at Figure 2.2, you'll see that leaves have a **flattened shape** so they've got a large surface area to absorb light.

Although a leaf is very thin, it contains a number of different kinds of cell. If you look at Figure 2.3, you'll see that the very top layer is transparent so that light can get in easily, and below this is a layer called the **palisade layer**. Cells from the palisade layer contain more **chloroplasts** than any other type of cell. This improves efficiency because these cells get most of the light.

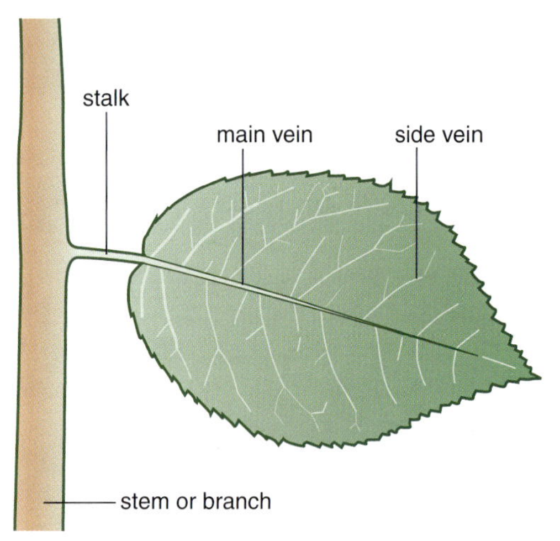

Figure 2.2
The flattened shape increases the surface area for the absorption of light

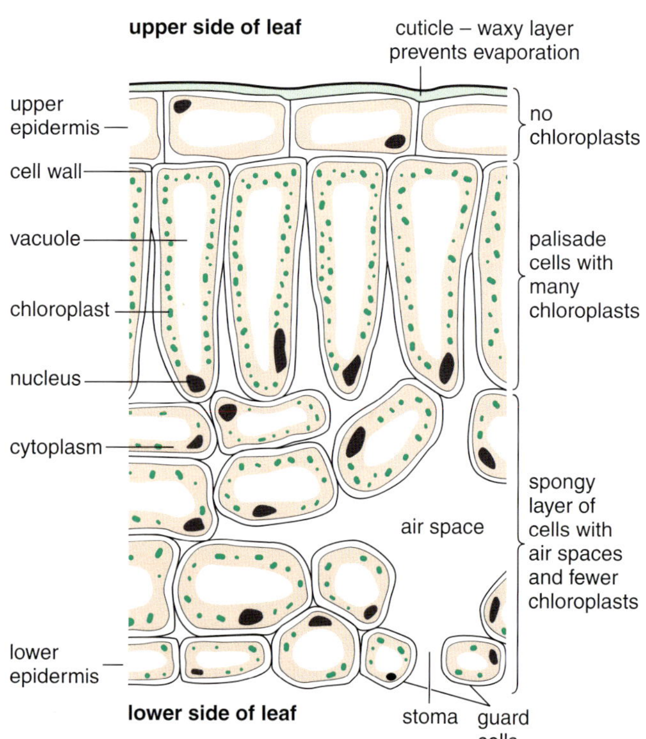

Figure 2.3
The inside of a leaf

If you look at the bottom layer of cells in Figure 2.3, you'll see that there's a gap called a **stoma** – this allows carbon dioxide to diffuse into the leaf (diffusion simply means spreading from a high to a low concentration). The cells above this are loosely packed, which allows carbon dioxide to get to more cells so more photosynthesis can take place.

Things that help photosynthesis

Light

It won't surprise you to find out that the more light a plant gets, the more food it can make for itself during photosynthesis.

Temperature

The colder it is, the slower photosynthesis takes place – you never have to cut the grass in winter!

Carbon dioxide

Since carbon dioxide is one of the raw ingredients for photosynthesis, the more that's available, the faster photosynthesis takes place.

Limiting factors

Any one of these things can limit the rate of photosynthesis. The graph in Figure 2.4 shows what happens to the rate of photosynthesis as the intensity (strength) of the light increases on a cold day and a hot day.

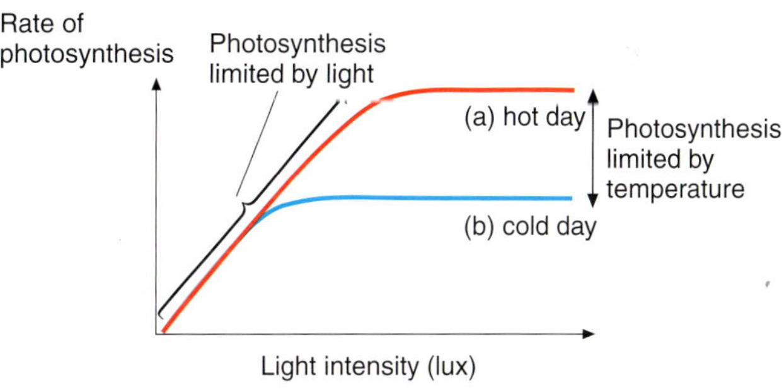

Figure 2.4 *The graph shows that increasing light intensity speeds up photosynthesis, but a low temperature limits how much photosynthesis can take place*

Did you know?

Commercial plant growers keep their greenhouses warm, artificially lit and pump extra carbon dioxide into them so that their plants can photosynthesise at the maximum rate.

Topic Questions

1 List the differences between animal and plant cells.
2 What is meant by the term photosynthesis?
3 Why do plants need to photosynthesise?
4 What is meant by a 'limiting factor'?
5 Where in a leaf do you find palisade cells?
6 Why do the bottom layers of plants have gaps called stomata?

2.2

Co-ordinated	Modular
DA 10.15	DA 2
SA n/a	SA n/a

Transporting substances

Gaining water and minerals

Plants get **water** and **minerals** from the soil by absorbing it through their roots. If you look at Figure 2.5 you'll see the route that water takes to get into the plant. The surface area of the roots is increased by thousands of tiny projections called **root hairs** which directly absorb water. The water then travels into a hollow tube called the **xylem vessel**, which transports water to all parts of the plant.

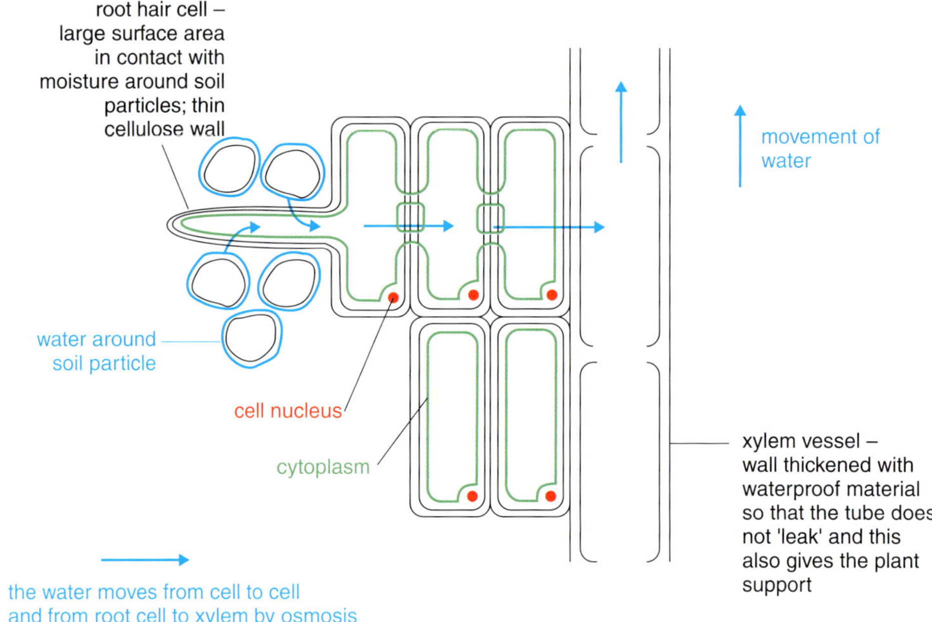

Figure 2.5
Root hairs absorb water, xylem vessels then carry it around the plant

Root hairs absorb water by a special kind of diffusion called **osmosis**. In osmosis, water moves from a dilute area to a more concentrated area through a membrane that lets through water but stops the movement of dissolved molecules. Figure 2.6 shows how osmosis happens.

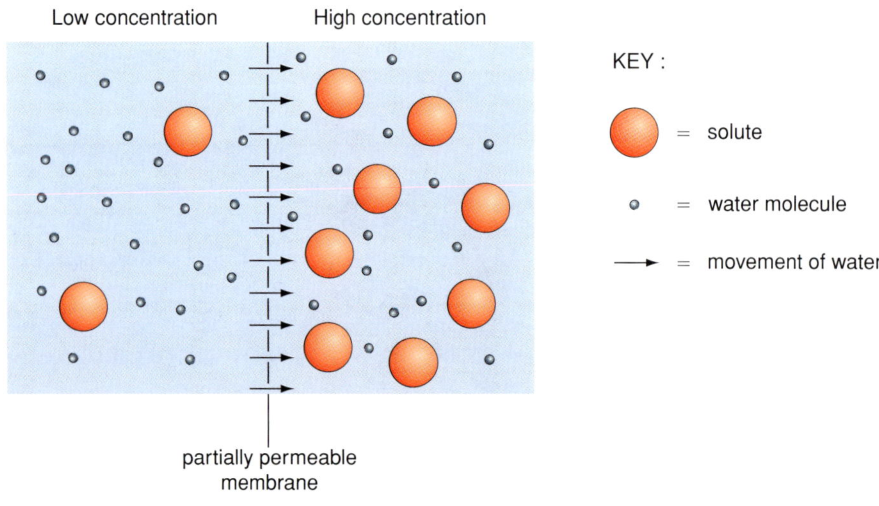

Figure 2.6
The small water molecules can move through a partially permeable membrane, but the solute can't

Did you know?

Fertilisers are often called 'plant food' in garden centres but this is wrong! Plants make their own food (glucose) in their leaves when they photosynthesise.

Minerals such as **nitrates** are needed by plants to help them grow better. Gardeners use fertilisers containing nitrates to help grow really healthy plants.

Moving food

If you look back at Figure 2.5, you'll notice that cells in roots don't have chloroplasts because they can't photosynthesise under the ground! They still, however, need food so they can respire. The food made in the leaves is passed to the roots by special tubes called the **phloem**.

Losing water – transpiration

Plants lose water from tiny gaps in the lower surface of their leaves called stomata (see Figure 2.7). This process is called **transpiration**.

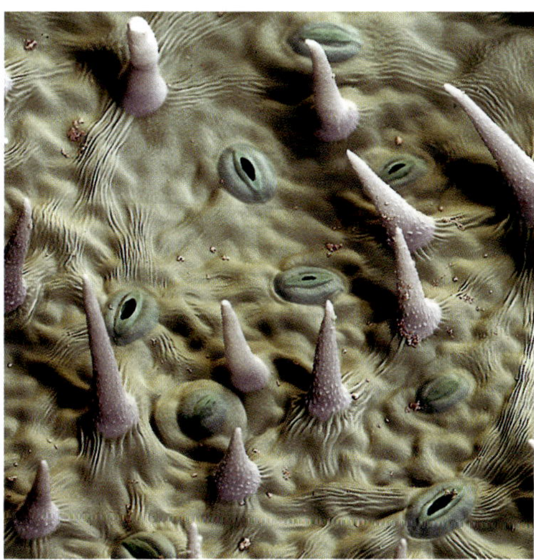

Figure 2.7
The tiny holes in this leaf are called stomata. Water evaporates from these in a process called transpiration

Transpiration takes place most quickly on **hot, dry and windy** days; by contrast it happens really slowly on humid, still and cold days. On really hot and windy days, water can be lost from leaves faster than roots can absorb it. Young plants depend upon water pressure for support, so if too much water is lost they start to **wilt**. Plants can control this water loss by closing their **stomata**. Figure 2.8 shows how a stoma can control transpiration.

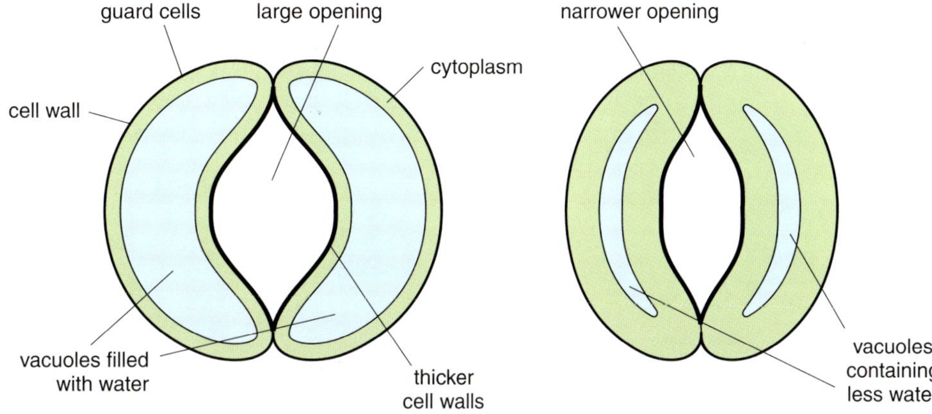

Figure 2.8
When a plant has lots of water, guard cells keep their stomata open; if it's a hot, dry day, the guard cells begin to close the stomata

All leaves have a **waxy layer** (called a cuticle) that prevents too much water loss, but plants that grow in really dry environments have thicker layers. Cacti have extremely thick cuticles to ensure that any water they have stays inside the plant.

Figure 2.9
These plants were grown in space. Their roots and shoots don't know how to grow without gravity

Up, down, wet and light

Gravity

Astronauts wishing to get to Mars would spend a long time in space. They would need to grow their own food but this would be difficult without gravity. Plants are sensitive to gravity, their shoots growing away from it and their roots growing towards it. On Earth this produces a plant which grows the right way up, but the photo in Figure 2.9 shows that in zero gravity, roots and shoots grow in unusual directions.

Light

If you look at Figure 2.10, you'll see that the shoots of plants grow towards the light; this is to make sure that the leaves get as much light as possible.

Figure 2.10
The cress seedlings on the left were lit from above, the ones on the right received light from one side only

Moisture

Roots don't always grow straight downwards. This is because they're sensitive to moisture and grow towards it.

Hormones that control growth

The tips of plants produce **hormones** that control the growth of the plant. If a plant is lit from above, hormones are released **equally** and this produces **even growth**. If the shoot is lit from one side, hormones collect more on the darker side making it grow faster. As the shoot grows, it bends towards the light because of the **uneven growth**. The diagrams in Figure 2.11 show how an unequal supply of hormones creates unequal growth.

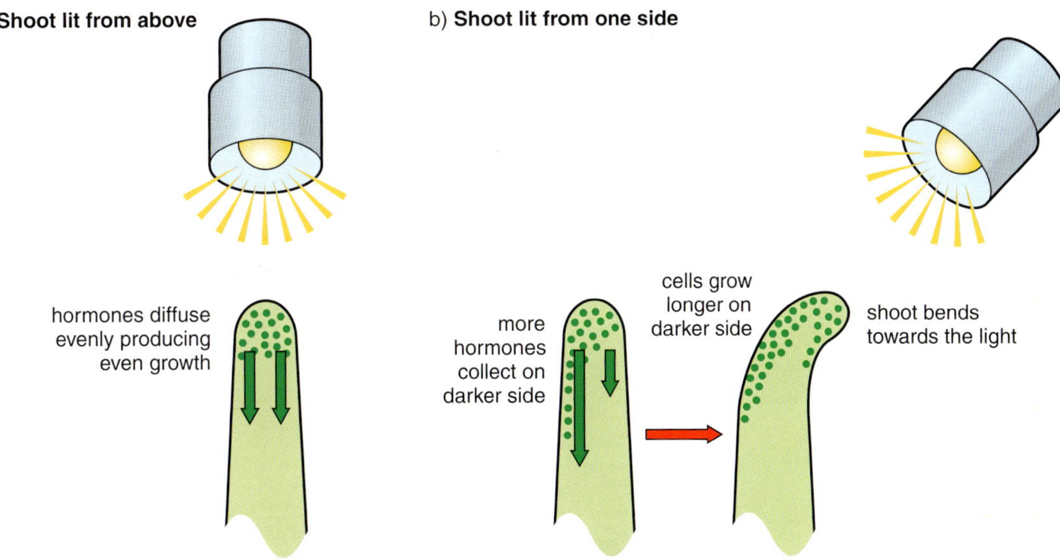
a) Shoot lit from above
b) Shoot lit from one side
hormones diffuse evenly producing even growth
more hormones collect on darker side
cells grow longer on darker side
shoot bends towards the light

Figure 2.11 *How plants respond to light*

Roots of plants are affected by hormones differently than in the stem – here they slow growth down. If a root happens to be growing sideways, hormones collect on the lower side slowing its growth. This makes the root grow downwards, the diagrams in Figure 2.12 show how this happens.

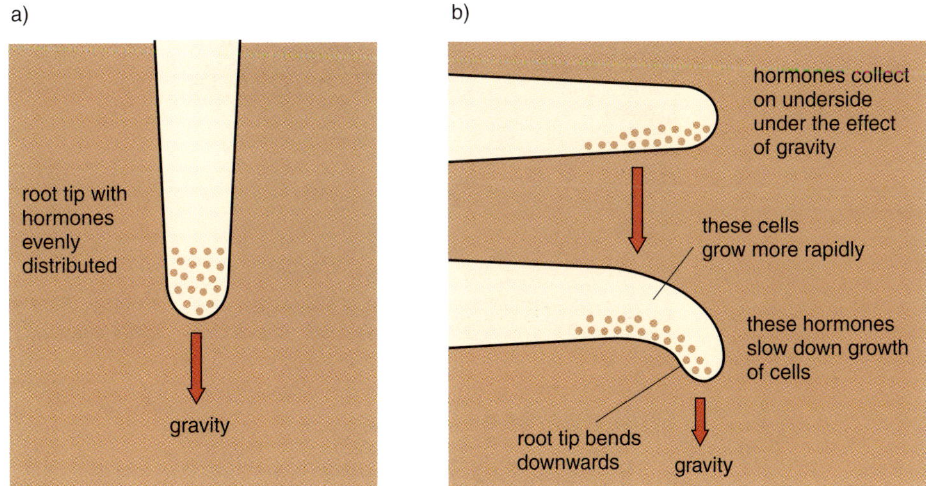
a)
root tip with hormones evenly distributed
gravity
b)
hormones collect on underside under the effect of gravity
these cells grow more rapidly
these hormones slow down growth of cells
root tip bends downwards
gravity

Figure 2.12 *How roots respond to gravity*

Commercial uses of plant hormones

Gardeners can use artificial plant hormones to their advantage.

Weed killer

Unwanted plants such as dandelions burn themselves out and die when hormones are put on them. Grass is not affected as much as broad leaved plants so hormones can help keep lawns in good condition. Figure 2.13 shows how a dock plant responds after hormones have been used.

Figure 2.13
This dock plant dies quickly after the application of hormones

Helping root growth

It's possible to grow some plants simply by cutting off a branch from a mature plant and sticking it into the ground! If a cut stem is dipped into certain hormones, they encourage root growth making the success of the cutting more likely.

Figure 2.14
Rooting hormones make cuttings produce roots very quickly

Fruit ripening

Fruits are really the swollen ovaries of plants. Normally they only ripen (become sweet and large) after fertilisation but the application of hormones can fool plants into ripening without fertilisation. This is how seedless oranges or grapes are produced.

Topic Questions

1 How is the surface area of roots increased to maximise absorption of water?

2 Why do plants need nitrates?

3 How is food transported around plants?

4 By which process does water enter a plant root?

5 What is (i) transpiration and (ii) which conditions speed it up?

6 Explain why plants bend towards the light as they grow.

7 List three commercial applications of plant hormones.

2.3	
Co-ordinated	Modular
DA 10.8 SA 10.5	DA 2 SA 13

How humans respond to changes in the environment

Human beings notice things going on around them by using their senses. Your ears, nose, eyes, tongue and skin are connected to your brain and spinal cord by nerve cells – **neurones**. The brain organises the body's response to changes that take place in your surroundings. A change in the environment that you can detect is known as a **stimulus**.

Your sense organs have special cells called **receptors** which can detect certain stimuli.

- The position of your body and whether it's moving or not are picked up by receptors in your **inner ear**.
- Sound is picked up by receptors in your **ear**.
- Receptors in your eye (in the **retina**) pick up light.
- The presence of certain chemicals is picked up by your **tongue** and your **nose**.
- There are receptors in your **skin** that are sensitive to touch, pressure and temperature.

All of the sense organs mentioned above are attached to your brain by **sensory neurones**. Your brain decides what response to make and sends messages to muscles and glands through nerve cells called **motor neurones**.

Automatic responses – reflexes

If someone flicks their hand towards your face, you blink automatically, without thinking about it. When you walk past a food shop and smell the cooking, your salivary glands automatically squirt saliva into your mouth. When you touch something hot your biceps muscle contracts, automatically pulling your hand away from danger. All of these automatic responses are called **reflexes**.

Reflexes are controlled by your **spinal cord** (or your brain). The diagram in Figure 2.15 shows what happens when you prick your finger.

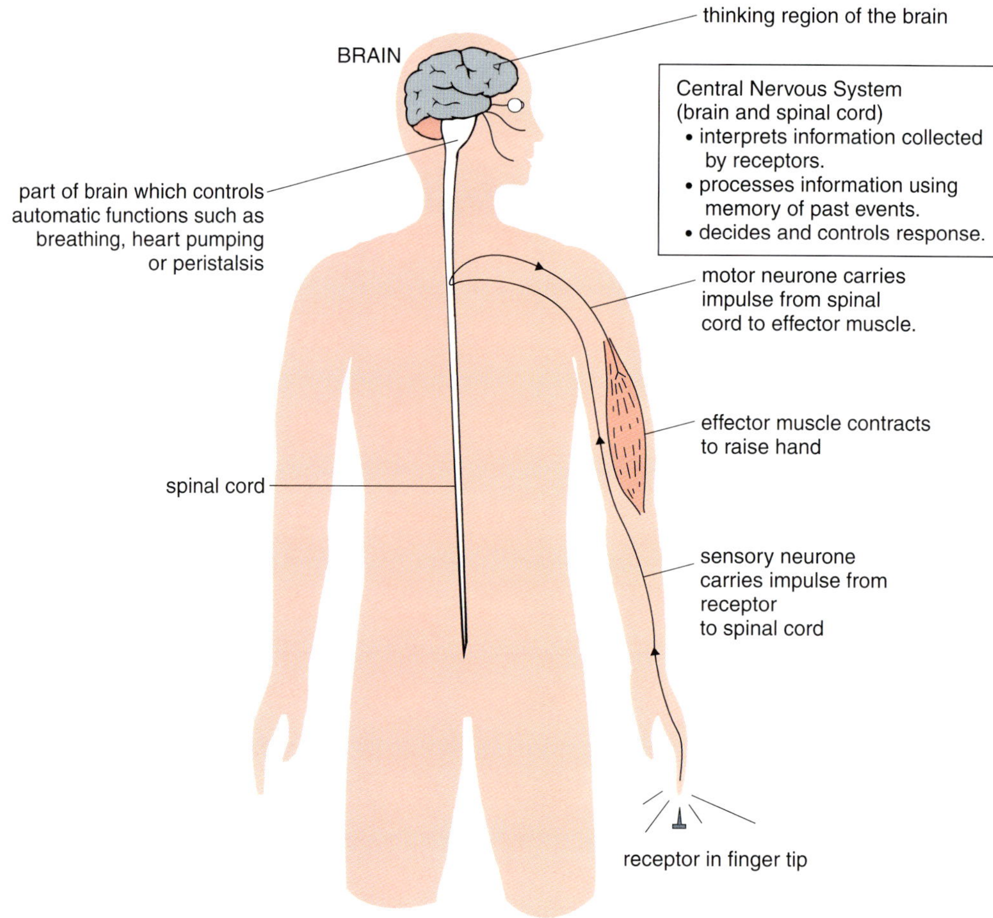

thinking region of the brain
BRAIN
Central Nervous System
(brain and spinal cord)
• interprets information collected by receptors.
• processes information using memory of past events.
• decides and controls response.
part of brain which controls automatic functions such as breathing, heart pumping or peristalsis
motor neurone carries impulse from spinal cord to effector muscle.
effector muscle contracts to raise hand
spinal cord
sensory neurone carries impulse from receptor to spinal cord
receptor in finger tip

Figure 2.15
The reflex arc

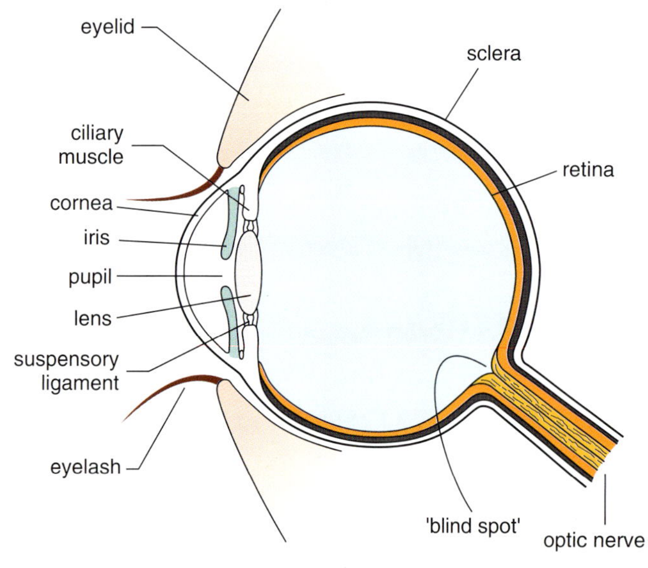

eyelid
sclera
ciliary muscle
retina
cornea
iris
pupil
lens
suspensory ligament
eyelash
'blind spot'
optic nerve

Figure 2.16
What your eye looks like inside

The eye

Light hits objects around you and bounces off their surfaces – this is known as reflection. The job of your eyes is to pick up reflected light. If you look at Figure 2.16, you'll see that light receptors are found in the **retina** at the back of your eye. Light is focused onto the retina by the **cornea** and **lens**. The retina then sends messages (nerve impulses) to the brain along neurones. Your brain makes pictures out of the nerve impulses it gets.

Figure 2.17 shows the jobs of the parts of the eye.

Figure 2.17
The functions of different parts of the eye

Part	Function
Sclera	The tough, white outer layer of the eye.
Cornea	The curved, transparent part of the sclera which bends the light entering the eye.
Iris	The coloured part of the eye. It's actually a muscle that changes the size of the pupil. This controls the amount of light that can enter the eye.
Pupil	The hole in the middle of the iris. Light has to travel through it to get to the retina.
Lens	This focuses light onto the retina. It's like a squashy bag of jelly which can be long and thin or short and fat.
Ciliary muscles	Together with the **suspensory ligaments**, these hold the lens in place. They squash or stretch the lens to help focus the image.
Retina	This is a black layer at the back of the eye which contains receptors for light.
Optic nerve	This contains nerve cells called **sensory neurones**. They carry messages to the brain as nerve impulses.

How does your eye produce an image?

To make an image, a lens has to **focus** light. If you look at Figure 2.18, you'll see that the **cornea** and the **lens** bend the light entering the eye. An image is formed on the retina where the light rays meet. The receptors in the retina then send messages along sensory neurones to the brain.

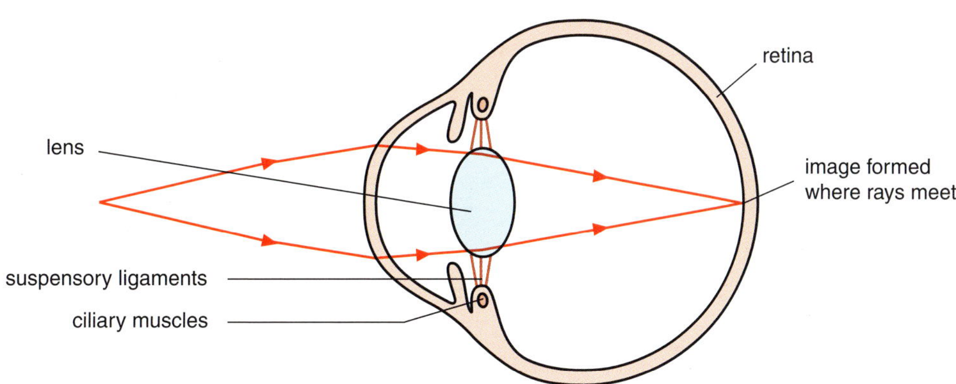

Figure 2.18
The cornea and the lens focus the light so that an image is formed on the retina

On bright days, it's important that the amount of light entering the eye is reduced to avoid damaging the receptors in the retina. In darker conditions, more light must be let in. Your **iris** (the coloured part of the eye) controls the size of the **pupil**. If you look at Figure 2.19, you'll see the difference in the size of the pupil in bright light or dim light.

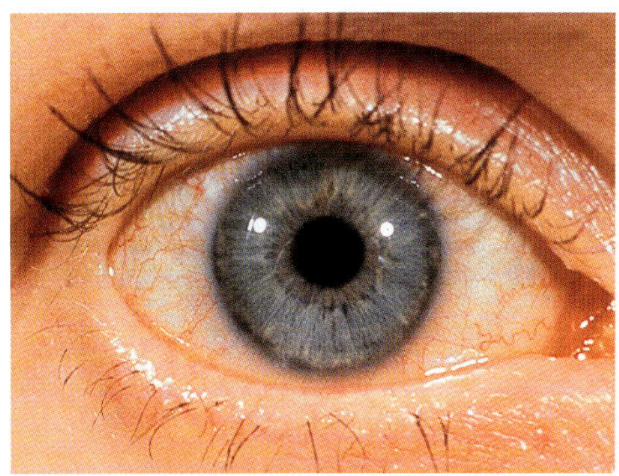

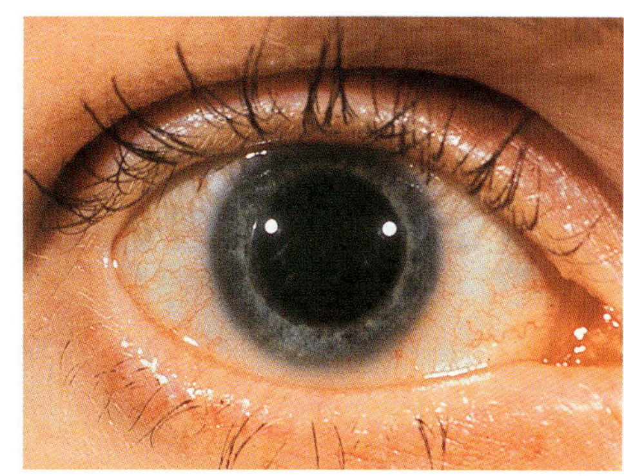

Figure 2.19
In bright light, your pupil gets smaller, in dim light it gets bigger

Did you know?

If you fancy someone, your pupils get bigger so that you can take in more details of their face!

Topic Questions

1 What is meant by the term stimulus?

2 Where will you find receptors to the following stimuli?

a) pressure
b) light
c) sound
d) temperature
e) body position
f) chemicals

3 Explain how messages from receptors get to your brain.

4 Explain what is meant by the term reflex action.

5 Give three examples of reflex actions.

6 Complete the following sentences.

The _______ is a tough white layer that surrounds the eye. Light enters the eye through the _______. The _______controls how much light enters the eye and is held in place by _______ and _______. The _______ focuses light onto the _______. Light receptors are found here which send messages along _______ to the brain.

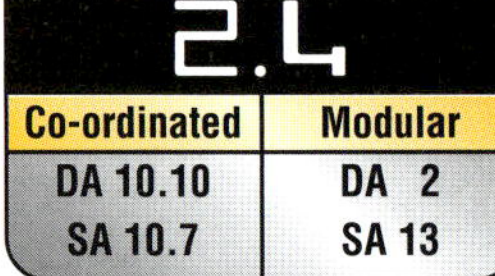

Co-ordinated	Modular
DA 10.10	DA 2
SA 10.7	SA 13

2.4 Controlling what goes on inside you

Your body is a complicated machine. It only works properly if conditions inside are kept pretty much the same at all times. This is tricky because you eat, drink, make waste products and move into places that have different temperatures. These things change conditions inside your body. Your body then has to do something to return the conditions to normal.

Getting rid of unwanted stuff

Carbon dioxide

As you make energy during respiration, you also make a poisonous substance called carbon dioxide. Your body gets rid of this from your lungs when you breathe out.

Urea

During digestion, proteins are broken down into amino acids by your small intestine. Most of the amino acids are then used to help growth or repair. Any amino acids left over have to be dealt with by the liver because excess amino acids can damage your joints. Your liver converts them into urea which has to be removed from your body. Your kidneys remove urea from the bloodstream and it's stored (together with water and salts) in the bladder as urine. When it's full, your bladder is emptied through a tube called the urethra.

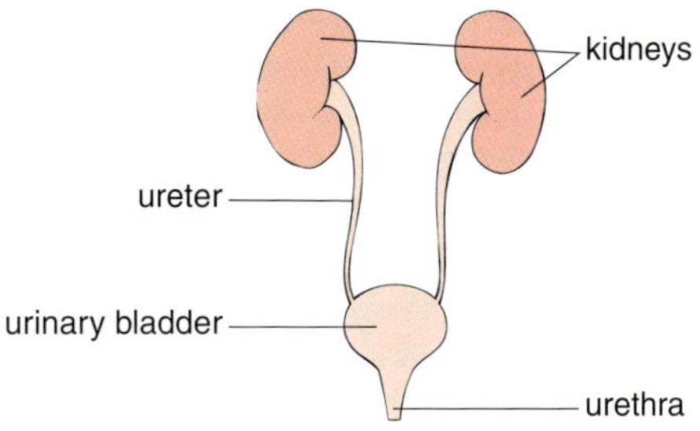

Figure 2.20
Your kidneys remove urea from the bloodstream

Keeping things constant

Temperature

The human body keeps itself at a temperature of 37°C. This is because the enzymes that help chemical reactions in your body work best at this temperature. If you become too hot, these enzymes stop working properly. Sweating puts water onto the surface of your skin; as it evaporates it takes away heat energy so cooling your body.

Water

It's essential that the amount of water in your body stays constant. This is difficult as you lose water from:

- your lungs when you breathe out (if you breathe onto a cold window you'll notice water droplets).
- your skin when you sweat.
- your kidneys when they make urine.

Ions

Ions are produced when salts are dissolved. These are lost from the body from:

- your skin when you sweat.
- your kidneys when they make urine.

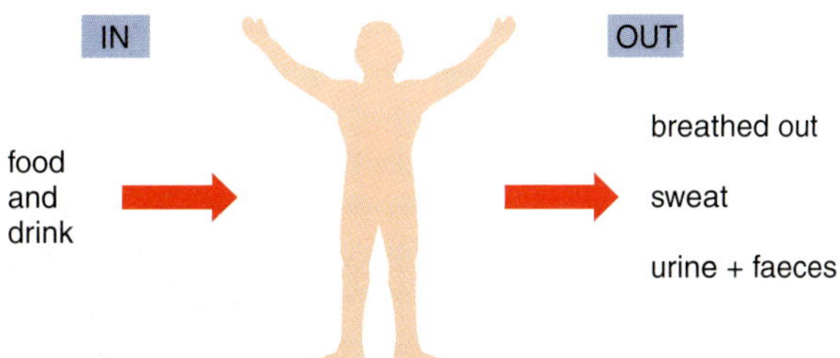

Figure 2.21
Daily water and ion inputs and outputs

Glucose

Your body uses glucose to make energy, in fact your brain can't use any other kind of fuel. If there's too little glucose in your blood, your brain can shut down and you can enter a coma. Glucose is so important that your body converts it into an insoluble substance called glycogen. Glycogen is stored in the liver and can be changed back into glucose when your body needs it.

It's the job of your **pancreas** to check the level of glucose in your blood. If your glucose level is too high, your pancreas releases a **hormone** called **insulin**. Insulin changes glucose into glycogen – this reduces the level of glucose in the blood.

If your glucose level gets too low, your pancreas releases a different hormone – **glucagon**. Glucagon increases the level of glucose in your bloodstream. It does this by changing the glycogen in your liver into glucose.

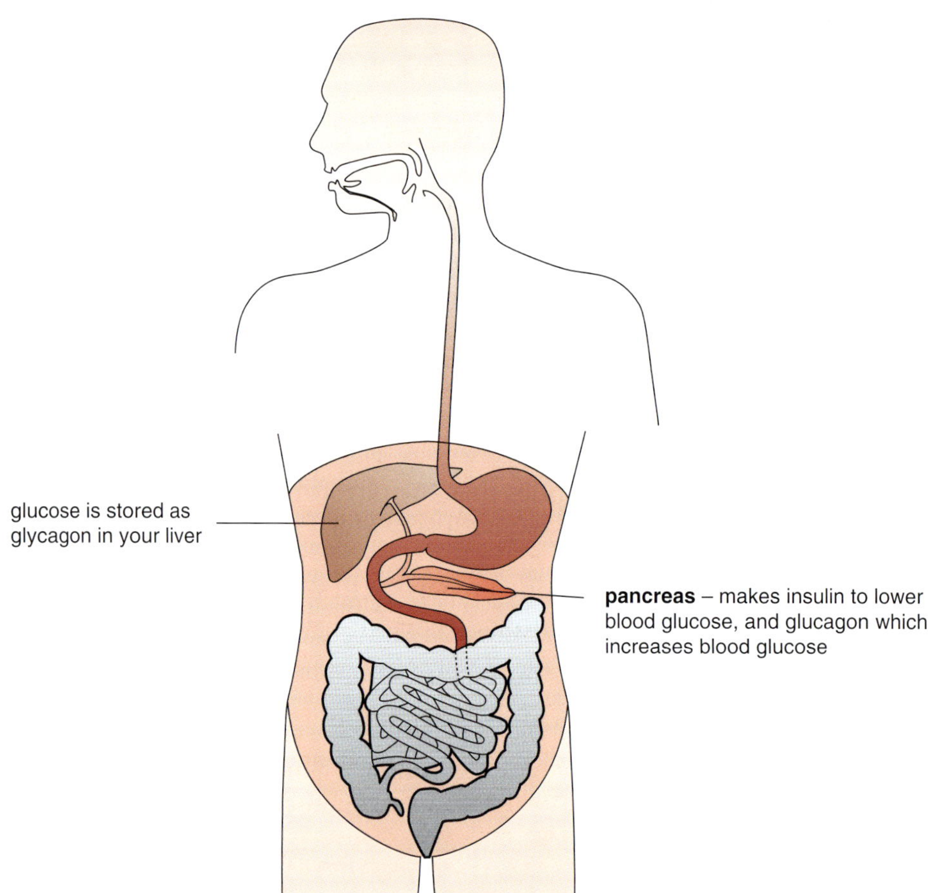
glucose is stored as glycagon in your liver
pancreas – makes insulin to lower blood glucose, and glucagon which increases blood glucose

Figure 2.22 *Controlling and storing glucose in the body*

Did you know?

Some people inherit diabetes from their parents, but people who are overweight sometimes develop diabetes in later life.

Diabetes

Some people suffer from a disease called **diabetes**. Diabetes is a disease in which the pancreas stops making enough insulin. People suffering from diabetes are in danger because their glucose level can increase to a fatally high level.

People who suffer from diabetes can lead perfectly normal lives but they have to be careful to eat the right amount of carbohydrates. They also control their blood sugar by injecting themselves with insulin after eating.

Topic Questions

1 Name the waste product made during respiration.

2 Where is urea (i) made and (ii) removed from the blood?

3 Explain why it is important to keep your blood sugar at a constant level.

4 Which organ monitors the concentration of glucose in your blood?

5 What is the effect of insulin on blood glucose concentration?

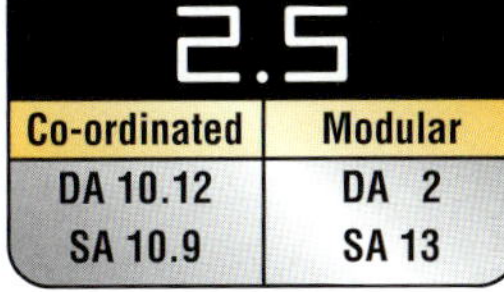

2.5
Co-ordinated
Modular
DA 10.12
SA 10.9
DA 2
SA 13

Drugs and health

A **drug** is a chemical that has an effect upon the body – usually changing the way it works. Some drugs are prescribed by doctors to fight illness but others are taken just for fun. Drugs such as alcohol and tobacco have been used for centuries and are legal for people above certain ages. Some drugs are illegal but many people still take them. Both legal and illegal drugs can have damaging effects upon the body. Some people become dependent upon drugs and find it difficult to live their life without them.

Alcohol

Most adults in Great Britain drink alcohol now and again. Drinking too much has effects upon the body which are commonly known as being 'drunk'. Alcohol is absorbed into the blood by the stomach. It is taken all around the body but has its greatest effects upon the **brain**.

The bad news:
Drinking alcohol and driving is strictly illegal in Great Britain. This is because alcohol, even in small doses, **slows down your reactions**. In larger quantities, alcohol can lead to a **lack of self-control** – people can sometimes do embarrassing or anti-social things. Lots of alcohol can slow people's central nervous system down so much that they become **unconscious**. Some people never recover from this and stay unconscious in a state called a **coma**. Drinking large quantities of alcohol over a long period of time can damage a person's **brain** and **liver**. Some people die from the effects of alcohol.

The good news:
There are some medical reports which suggest that drinking small amounts of alcohol can actually be good for your health. It seems that people who drink small quantities of red wine have healthier hearts than non-drinkers.

Tobacco

The first people to smoke tobacco were the native people of South America. Sailors who went to South America with Christopher Columbus picked up the habit. They brought back tobacco plant seeds, and the habit soon spread throughout the whole of Europe.

The bad news:
Tobacco smoke contains a substance called **nicotine**. Smokers find it difficult to give up because nicotine is addictive. Some people experience **withdrawal symptoms** such as intense craving (for cigarettes and sweet things), irritability and stomach cramps when they try to quit.

Tobacco smoke also contains substances that can damage your lungs and cause the following diseases:

- **Cancer** – more than 90% of lung cancer patients are smokers, most of them have to have surgery to remove their damaged lung.
- **Emphysema** – smokers sometimes cough so violently that this damages the walls of their alveoli. Their lungs end up being less able to absorb oxygen.
- **Bronchitis** – the inner surfaces of the bronchi can sometimes become inflamed. This makes breathing difficult.

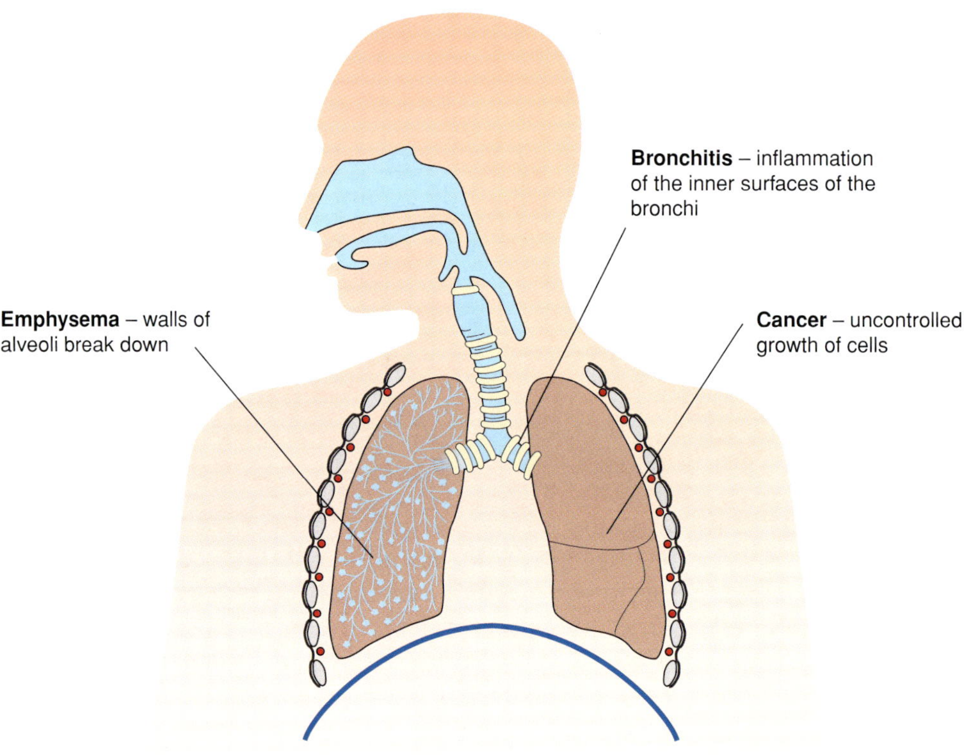
Bronchitis – inflammation of the inner surfaces of the bronchi
Emphysema – walls of alveoli break down
Cancer – uncontrolled growth of cells

Figure 2.23
Smoking can cause lots of damage to your lungs

Did you know?

Smoking makes your skin look older by damaging its collagen. Collagen is the tissue in your skin that keeps it flexible and healthy looking.

The good news:
I'm afraid there isn't any.

Smoking during pregnancy

If you're pregnant, smoking can harm your unborn child. This is because cigarette smoke contains **carbon monoxide**. Carbon monoxide **reduces** the amount of **oxygen** that your blood can carry. This means that a foetus inside a mother who smokes gets less oxygen – this stops normal development. Babies born to mothers who smoke are on the whole less healthy than babies born to non-smokers.

Evidence for a link

The ship's doctor that travelled with Columbus to South America first noticed that smoking damaged health. It wasn't until the late 1940s, however, that a scientific study into the link between smoking and ill health was carried out. Three studies were carried out – two in America and one in Great Britain. All of the studies showed exactly the same relationship.

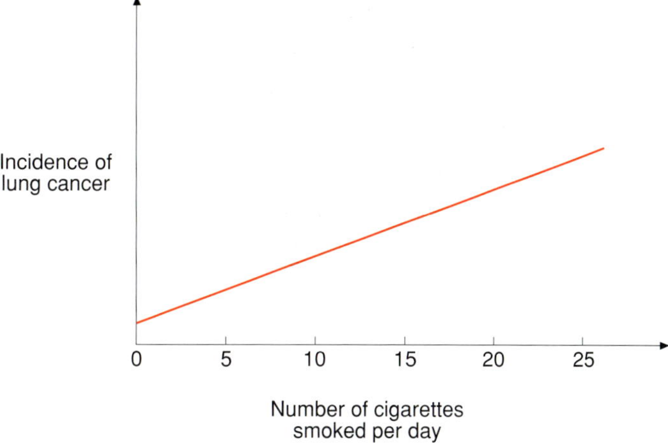

Figure 2.24 *The graph shows the link between smoking and cancer*

All investigations since have shown that the more you smoke, the more likely you are to get lung cancer.

Solvents

A solvent is any liquid that can dissolve something. Water is a useful solvent but it can't dissolve everything. Certain substances in products such as deodorants, air fresheners and glue have to be dissolved by other kinds of solvent. Some of these solvents release a vapour which can be absorbed into the bloodstream by the lungs if they're inhaled.

Inhaling solvents can cause:

- a lack of self-control
- slower reaction times
- unconsciousness – even coma
- sudden death.

Sniffing solvents can damage your **lungs**, **liver** and **brain**.

Topic Questions

1. What is meant by the term drug?
2. Name two drugs that can damage your lungs.
3. Which organs can be damaged by drinking too much alcohol?
4. Name the diseases caused by smoking.
5. Explain why drinking alcohol and then driving is dangerous.

Summary

- Plant cells have the following bits:
  - a nucleus which controls the activities of the cell
  - a cytoplasm where chemical reactions take place
  - a cell membrane which controls the exit and entry of substances to the cell
  - a cell wall which strengthens the cell
  - chloroplasts which absorb light energy to make sugars
  - a permanent vacuole filled with cell sap.
- The word equation for photosynthesis is:

  carbon dioxide + water [+ light energy] ⟶ glucose + oxygen
- The stages of photosynthesis are as follows:
  - chlorophyll (the green pigment in chloroplasts) absorbs light energy
  - this energy is used to change carbon dioxide and water into sugar (glucose)
  - oxygen is released as a by-product.
- Photosynthesis can be limited by:
  - low temperature
  - a lack of carbon dioxide
  - too little light.
- Glucose can be changed into insoluble starch for storage.
- Glucose is used by plants to make energy during respiration.
- Plant roots absorb minerals such as nitrates which are needed for healthy growth.
- Carbon dioxide gets into leaves by diffusion.
- Loss of water vapour from a leaf is called transpiration.
- Transpiration is fastest in hot, dry and windy conditions.
- A waxy layer (cuticle) stops leaves from losing too much water.
- Plants living in dry conditions have a thicker cuticle.
- Stomata are holes in the leaf, they help plants get carbon dioxide from the atmosphere and allow transpiration.
- The size of stomata is controlled by guard cells.
- Young plants wilt if their cells are short of water.
- If plants lose water too quickly, the stomata close to prevent wilting.
- A tissue called the xylem carries water and minerals from the roots to the stem.
- Phloem tissue carries nutrients such as sugars from the leaves to the rest of the plant.
- Osmosis is the diffusion of water from a dilute to a more concentrated solution through a partially permeable membrane that allows the passage of water molecules but not solute molecules.
- The surface area of roots is increased by root hairs and the surface area of leaves by the flattened shape and internal air spaces.
- Shoots grow towards light and against the force of gravity.
- Roots grow towards moisture and in the direction of the force of gravity.
- Hormones coordinate and control the growth of plants.
- Unequal distribution of hormones causes unequal growth rates allowing plants to respond to stimuli.
- Hormones can be used to:
  - reproduce large numbers of plants quickly by stimulating the growth of roots from cuttings
  - regulate the ripening of fruits
  - kill weeds by changing their normal growth patterns.
- A stimulus is any change in the environment that you can detect.
- Different receptors are sensitive to different stimuli:
  - receptors in the eyes are sensitive to light
  - receptors in the ears are sensitive to sound
  - receptors in the inner ear are sensitive to changes in body position
  - receptors on the tongue and in the nose are sensitive to chemicals
  - receptors in the skin are sensitive to touch, pressure and temperature changes.

- Neurones (nerves) take messages to the brain which then coordinates the response.
- Automatic responses are called reflex actions.
- In a reflex action, impulses pass from a receptor along a sensory neurone to the spinal cord or brain, then along a motor neurone to a muscle or gland. The response is brought about by muscles or glands.
- The tough outer layer of the eye is called the sclera.
- The iris controls the size of the pupil and the amount of light getting to the retina.
- Suspensory ligaments and ciliary muscles hold the lens in position.
- The retina contains the receptor cells which are sensitive to light.
- Light from an object enters the eye through the cornea. The curved cornea and the lens produce an image on the retina. The receptor cells in the retina send impulses to the brain along sensory neurones in the optic nerve.
- Carbon dioxide leaves the body through the lungs when we breathe out.
- Urea is made in the liver by the breakdown of amino acids.
- The kidneys remove urea from the blood, storing it in the bladder as urine.
- Water leaves the body from the lungs and the skin (sweating), and the kidneys also get rid of any excess in the urine.
- Ions are lost by the skin when we sweat, and the kidneys remove any excess in the urine.
- Sweating helps to cool the body (enzymes work best at 37°C).
- You drink water to replace lost water.
- The concentration of glucose in the blood is controlled by the hormones insulin and glucagon which are released by the pancreas.
- Diabetes is a disease which results because the pancreas does not produce enough of the hormone insulin.
- Solvents:
  - affect behaviour
  - cause damage to the lungs, liver and brain.
- Alcohol:
  - slows down reactions and can lead to a lack of self-control, unconsciousness or even coma
  - causes damage to the liver and brain.
- People can develop a dependency to drugs and can suffer withdrawal symptoms without them.
- Tobacco smoke contains addictive nicotine and other substances which can cause:
  - lung cancer
  - bronchitis and emphysema
  - disease of the heart and blood vessels.
- Tobacco smoke contains carbon monoxide which reduces the amount of oxygen carried by the blood. In pregnant women, this deprives a foetus of oxygen and leads to a low birth weight.

Examination Questions

1 a) Choose words from the list in the box to complete the sentences that follow.

cell wall cell membrane vacuole chloroplasts photosynthesis

A plant cell has a ________________ which keeps the cell rigid. The ________________ controls the exit and entry of substances to the cell. Inside the cytoplasm are ________________ which make food by the process of ________________. *(4 marks)*

b) Complete the following word equation describing the process of photosynthesis.

water + ______ + light → ______ + oxygen *(2 marks)*

c) The diagram below shows the cells found in a root.

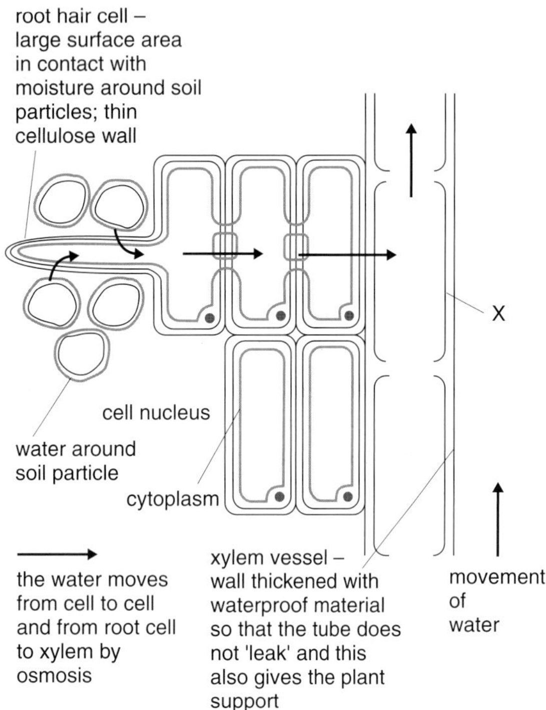

(i) Name the process by which water enters the plant. *(1 mark)*
(ii) Name the structure labelled x. *(1 mark)*
(iii) Explain why root cells don't have chloroplasts. *(2 marks)*

2 The diagram below shows the inside of your eye.

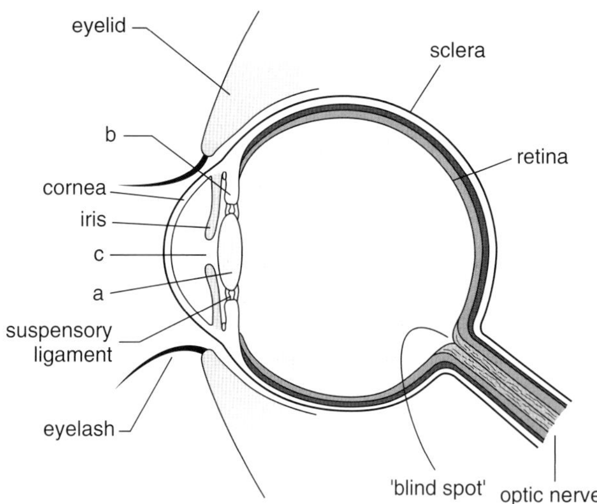

(i) Name the parts labelled a–c. *(3 marks)*
(ii) Describe the function of the part labelled a. *(1 mark)*
(iii) The retina is attached to sensory neurones. Explain the function of the sensory neurones. *(2 marks)*

3 The man in the diagram below has just touched a hot object. He automatically pulled his hand away very quickly.

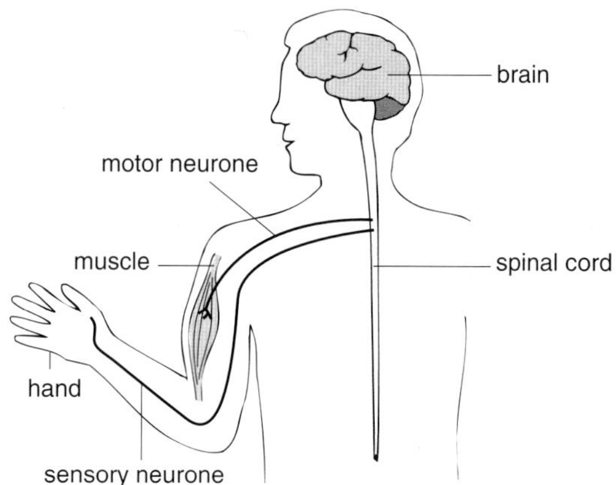

a) What is the term given to this kind of automatic response? *(1 mark)*
b) Where are the receptor cells in this example? *(1 mark)*
c) Name the effector in this example. *(1 mark)*
d) Add two arrows to the diagram, showing the route taken by the nerve impulses. *(2 marks)*

4 The bar chart below shows the effect of smoking upon death rates in men of different ages.

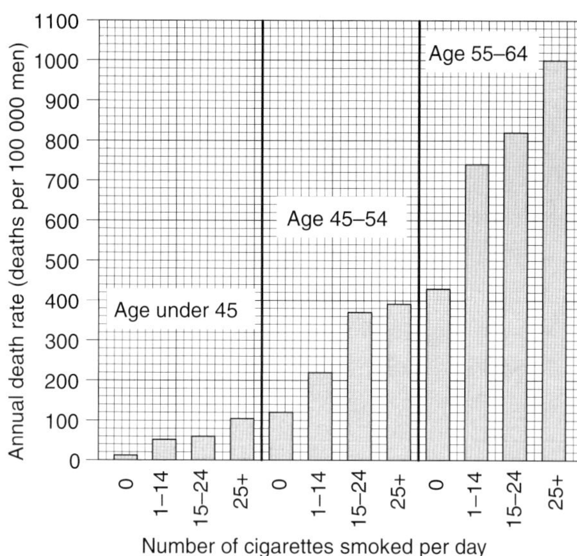

a) Describe the relationship that you can see between smoking and death rates. *(2 marks)*
b) What are the reasons for this relationship? *(3 marks)*

5 Solvents are found in products such as lighter fuel and deodorants.

a) How does inhaling solvents affect behaviour? *(2 marks)*
b) Name two organs that can be damaged by solvent inhalation. *(2 marks)*

Chapter 3

Variation and inheritance

Key terms variation • environment • inheritance • nucleus • chromosomes • DNA • genes • sexual intercourse • fertilisation • asexual reproduction • clones • alleles • dominant • recessive • clone • cross breeding • embryo transplantation • fossils • theory of evolution • natural selection • mutation • beneficial mutations • antibiotics • antibiotic resistant • ovulation • menopause • menstrual cycle • menstruation • sex hormones • oestrogen •

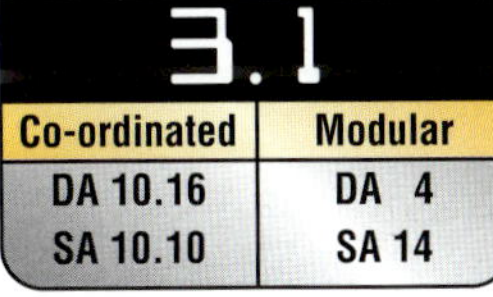

Co-ordinated	Modular
DA 10.16	DA 4
SA 10.10	SA 14

3.1 What is variation?

Everyone is different, even identical twins are different in some ways. The differences between living things of the same type are known as **variations**. Some of this variation is inherited from our parents. You might have noticed that the colour of your eyes or hair is the same as one of your parents, or you might have heard another member of the family comparing you to your mother or father.

Things you're born with, things you get . . .

There are certain features that you **inherit** from your parents, such as your eye colour, nose shape or whether you can smell freesias or not! Other features can change according to what you do and what happens to you in your life – in other words your **environment**. Most of your features are a combined result of your environment and inheritance; skin colour, for example, is inherited but can be changed a little by sunlight.

Figure 3.1

Task

The people in the cartoon above are father and son.

(i) List the features you can see that the son has inherited directly from his father.

(ii) List all the features (in either the father or the son) that have been changed by their environment.

Did you know?

There is a gene that makes people lift their arm and drop it on their noses whilst they're asleep!

Where is this information stored?

Amazingly, all the instructions needed to make up a whole human being are found in the **nucleus** of nearly every cell in your body. Packed really tightly inside the nucleus are 46 **chromosomes** (see Figure 3.2). These are really long molecules made up of a chemical called **DNA**. DNA contains chemical instructions which control how your body develops. These instructions are known as **genes**.

Figure 3.2
The structure of the nucleus

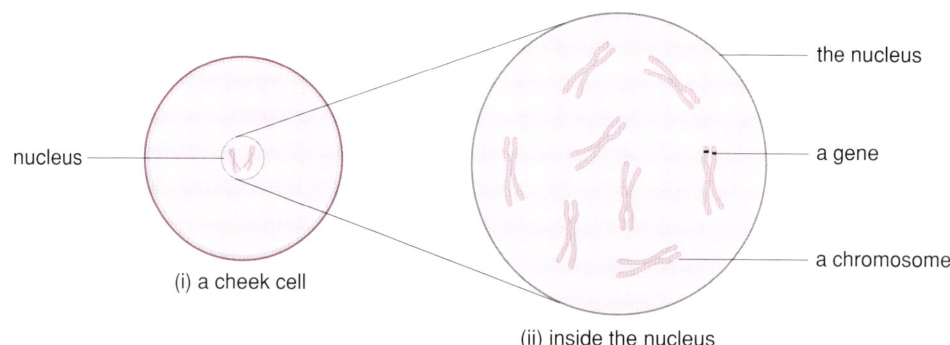

Topic Questions

1 What does the word variation mean?

2 From the list below, decide which features are inherited purely from your parents and which can be changed by the environment.

- body weight
- skin tone
- hair colour
- ability to roll tongue
- sense of humour
- sex
- chin shape
- fitness

3 Describe the meaning of the following words – use your own words if you can!

a) cell
b) nucleus
c) chromosomes
d) DNA
e) genes

4 Arrange the following in order of size, starting from the smallest: organism; gene; cell; chromosome; nucleus.

5 Try to find out which type of cells do not contain a nucleus.

3.2 What about sex?

Co-ordinated	Modular
DA 10.14	DA 04
SA 10.14	SA n/a

Humans reproduce by **sexual intercourse**, which is really nature's way of getting a sperm cell and an egg to join (**fertilisation**). Some living things, however, can reproduce without sex in a process called **asexual reproduction**. In asexual reproduction all the offspring are identical to each other and the parent because they all carry the same genes – they are known as **clones**. Lots of plants reproduce like this (see Figure 3.3), putting out runners (side shoots) that put down roots as they hit the ground.

Figure 3.3 *Asexual reproduction in plants*

Asexual reproduction in plants:
(i) How the strawberry plant reproduces:
parental plant
'runner'
daughter plant grows where runner hits the ground
(ii) Spider plants reproduce in a similar way:
parental plant
'runner'
the daughter plant will be genetically identical to the parental plant

Did you know?

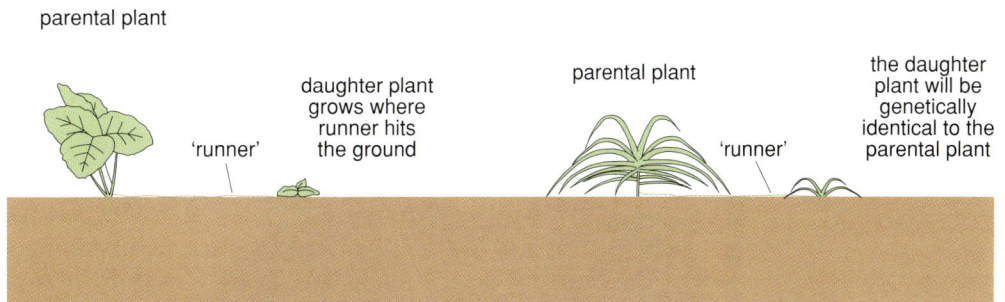

Female aphids (or greenfly as they're better known), do not need male aphids to help them reproduce, they simply give birth to an identical copy of themselves. The amazing thing is that their daughters are born with their granddaughters already inside them, waiting their turn to be born!

Sexual reproduction

All aphids are identical because they contain the same genes, human beings on the other hand are all different. This is because we reproduce sexually. In sexual reproduction a male sex cell (a sperm) and a female sex cell (an egg) join together or fuse. The nuclei of both the sperm and the egg contain genes from each parent and when they fuse (join), the new cell created has genes from both the mother and father. The cell then divides many times (see Figure 3.4.), which is how growth takes place.

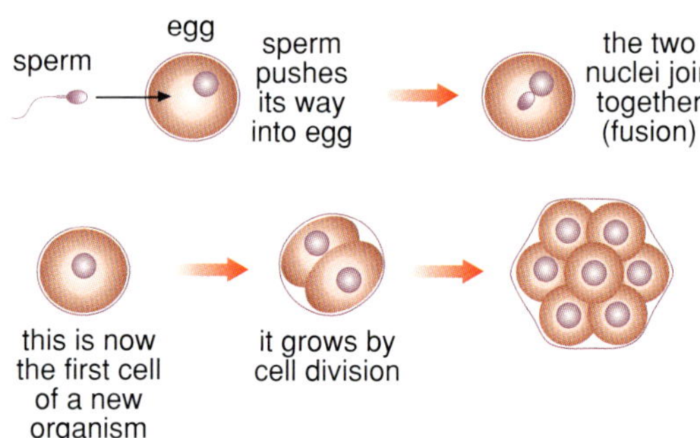

Figure 3.4
Sperms and eggs are known as gametes

Did you know?

Human beings can repair damage to bits of their bodies, making new cells by cell division. A starfish can replace a whole arm in this way!

What are the rules of inheritance?

Gregor Mendel – the man who worked it all out

Figure 3.5
Gregor Mendel

For centuries, people thought that features from parents were passed on in fluids that mixed together during fertilisation. They thought that children inherited a blend of their parents characteristics, for example, if one parent of a child had blond hair and the other had black hair, then the children would have brownish hair. Gregor Mendel was an Austrian monk who noticed that in reality this did not happen. He noticed that brothers and sisters from the same family could have different eye colour or hair colour and that these colours were not necessarily a blend of the parents' features.

He decided to find out how this happened and experimented for a while by breeding different kinds of hamsters together and looking at their offspring. This research was stopped by his bishop who was shocked that one of his monks should be so interested in sex! The bishop did, however, agree to let Gregor experiment with flowers; Gregor wrote in his diary 'apparently the bishop does not know that plants have sex too!' So Gregor turned his attention to pea plants; he had two different kinds – one that produced wrinkly peas and one that produced smooth peas. He decided to breed them together.

According to the 'blending' theory, all of the peas produced should have been half way inbetween wrinkly and smooth.

To Gregor's amazement, however, all of the peas produced in his experiment were smooth!

Gregor then decided to grow a new plant from one of the new smooth peas. When they had grown tall and produced flowers, Gregor took some pollen (male sex cell or gamete) from one of these and put it onto the stigma (female bit) of the same flower and waited for the outcome. When the pea pod had grown, he found a mixture of both wrinkly and smooth peas. When he counted them, he noticed that there were four smooth peas for every wrinkly one that he found.

Gregor thought that this meant characteristics were **inherited separately** and that **blending** of features doesn't take place. We now know that he was right and call the things that carry the information from parent to offspring **genes**.

Unfortunately for Gregor, the importance of this discovery was not realised until long after his death.

More about genes and chromosomes

The photograph below shows the **chromosomes** that are found inside the **nucleus** of a human body cell. Humans have 46 chromosomes which can be arranged into 23 pairs.

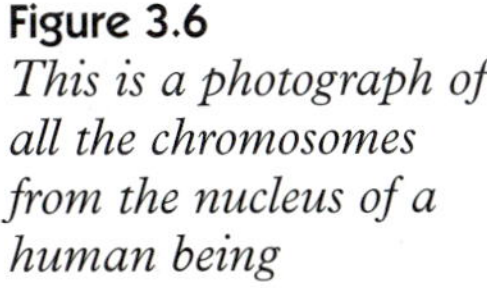

Figure 3.6
This is a photograph of all the chromosomes from the nucleus of a human being

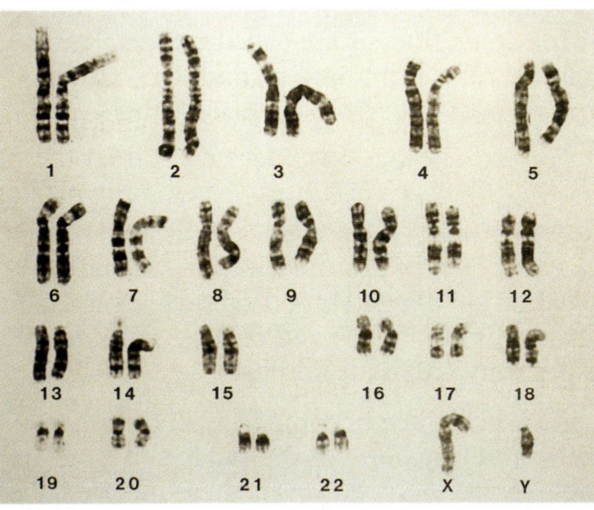

Another funny word . . . allele

Genes are found along the length of each pair of chromosomes. People often inherit different types of the same gene from each parent. Blue eyes and brown eyes, for example are produced by different forms of the same gene. The different forms of a gene are known as **alleles**.

What happens when you've got a mixture of alleles for a particular feature?

Some alleles are stronger than others, these are called **dominant alleles**. You only need one copy of a dominant allele for its effect to be shown – they mask the effect of weaker alleles. Weaker ones are called **recessive alleles** – you need two copies of these in order for their effects to be shown.

This table shows some dominant and recessive features in humans.

Figure 3.7 *Some dominant and recessive features in humans*

Feature	Dominant	Recessive
Hair colour	black	ginger
Eye colour	brown	blue
Tongue rolling	able to	unable to

Genetic diseases

It's not only features such as eye colour and hair colour that can be inherited, some diseases can also be passed down from parents to children by faulty genes.

Huntington's disease

Huntington's disease is an inherited illness that affects the nervous system. It's caused by a dominant allele so people with only one copy of the allele will suffer from the disease. Sufferers will be totally unaware of the fact that they have the faulty allele until they reach middle age. The early symptoms of the disease include twitching, clumsiness and memory loss. The disease progresses quickly and the sufferer begins to lose all their mental abilities eventually dying prematurely.

Cystic fibrosis

Cystic fibrosis is an inherited fault of the cell membranes. It affects the lungs and digestive system of sufferers, causing them to make mucus that's too sticky. This mucus blocks up the outlet of the pancreas so digestive enzymes cannot get into the small intestine. The lungs of cystic fibrosis sufferers get clogged up with mucus which makes them have lots of lung infections. People with this disease die young, often before reaching the age of thirty.

Unlike the allele that causes Huntington's disease, a person can have the allele that causes cystic fibrosis without suffering from the disease. This is because the allele that causes cystic fibrosis is recessive – two copies of the allele are needed before the allele can have its effect. People with one normal (dominant) allele and one allele for cystic fibrosis are called **carriers**. Parents who are carriers for the cystic fibrosis allele can have a child who suffers from the disease even though they do not suffer from the disease themselves.

Sickle cell anaemia

Sickle cell anaemia is an inherited disorder of the red blood cells. Red blood cells are important because they carry oxygen around the body but sufferers of sickle cell anaemia have unusual 'sickle' shaped blood cells which tend to get stuck in blood capillaries. This stops blood flow and reduces the amount of oxygen that tissues get – causing pain and tissue damage.

People with two copies of the sickle cell allele become very ill but people with one copy of the normal gene and one copy of the sickle cell gene are only mildly affected by the illness. Amazingly, however, these people have protection against the disease called malaria and so have a big advantage when they live in countries where malaria is common.

The parasite that causes malaria is carried by the female mosquito.

Topic Questions

1 What's the difference between sexual and asexual reproduction?

2 What happens during fertilisation?

3 What does the word clone mean?

4 Why don't children ever look *exactly* like one of their parents?

5 What does the term inherited disorder mean?

6 In your own words, describe the symptoms of:

a) Huntington's disease
b) cystic fibrosis
c) sickle cell anaemia

7 A friend of yours is scared that she might catch cystic fibrosis. Explain to her why this is not possible.

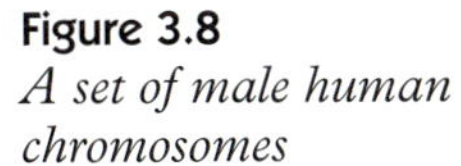

What has 'X' to do with sex?

All sports competitors at the Olympic Games must be tested to ensure that they are the sex that they claim. But how can you *really* know what sex someone is? If you looked carefully at the chromosomes from a woman and a man using a powerful microscope, you would notice a small difference. Women have 23 pairs of chromosomes while men have only 22 pairs, with two others that don't match. The photograph below shows all the chromosomes from a man paired up except for the chromosomes labelled X and Y.

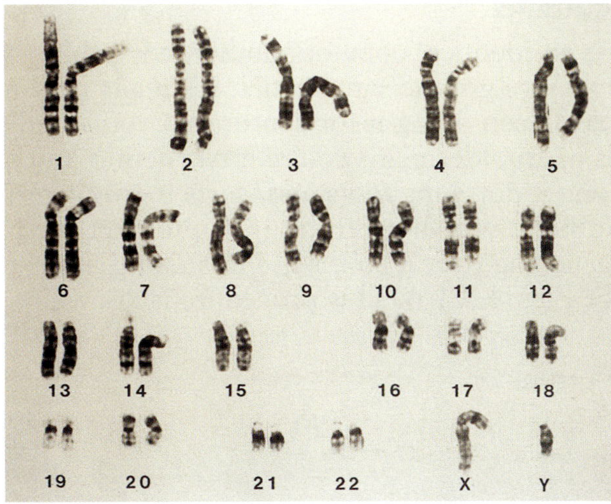

Figure 3.8 *A set of male human chromosomes*

Men have an X and a Y chromosome, women have two X chromosomes – these chromosomes are known as **sex chromosomes**. Only by looking at someone's chromosomes in this detail can you be *sure* which sex they are.

When men make sperm, half of them will carry the X chromosome and half of them will carry the Y chromosome. All of the eggs that women produce carry X chromosomes. During fertilisation there's a 50% chance of a sperm with a Y chromosome or a sperm with an X chromosome meeting an egg (see Figure 3.9).

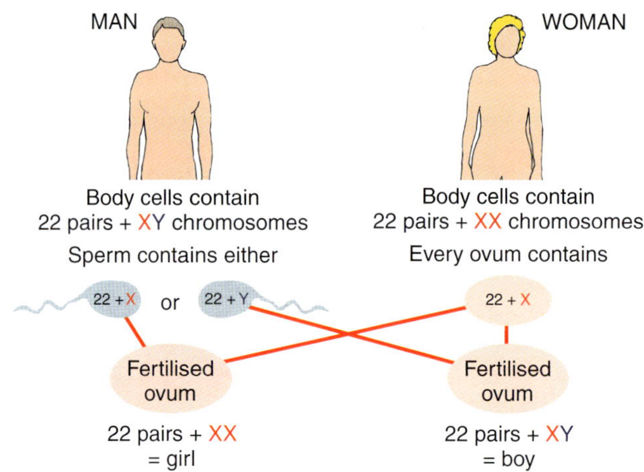
MAN
WOMAN
Body cells contain
22 pairs + XY chromosomes
Sperm contains either
22 + X
or
22 + Y
Body cells contain
22 pairs + XX chromosomes
Every ovum contains
22 + X
Fertilised ovum
22 pairs + XX
= girl
Fertilised ovum
22 pairs + XY
= boy

Figure 3.9
Deciding sex in humans

As you can see, the chances of producing a boy or a girl are equal.

Topic Questions

1 How many pairs of chromosomes do you find in the cells of (i) men, and (ii) women?

2 If a man releases 40 000 000 sperm cells, how many will contain (i) X chromosomes, and (ii) Y chromosomes?

3 What is the probability (chance) that a pregnant woman will give birth to (i) a boy, and (ii) a girl? Explain your answer using diagrams to help you.

3.4 Messing about with genes

Co-ordinated	Modular
DA 10.18	DA 4
SA 10.12	SA 14

1 Cloning plants

Taking cuttings

A **clone** is an identical copy of another living thing. Clones are identical because they've got the same genes. It's really easy to clone plants, and people have been doing it for centuries – you simply cut off one of the branches of an older plant you're interested in and stick it in some soil. Before long it develops roots and starts to grow into an identical copy of the plant that it was taken from. If you're interested in trying this yourself, put a plastic bag over it so it stays in a damp atmosphere until its roots grow (see Figure 3.10). This procedure is known as 'taking cuttings'.

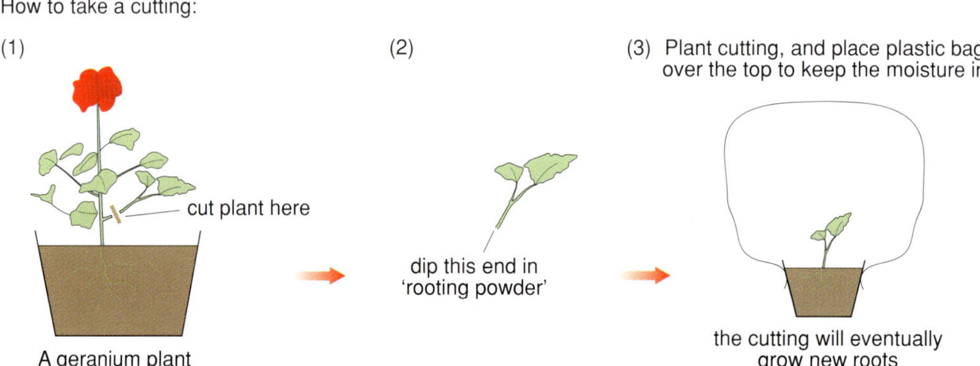
How to take a cutting:
(1)
cut plant here
A geranium plant
(2)
dip this end in 'rooting powder'
(3) Plant cutting, and place plastic bag over the top to keep the moisture in
the cutting will eventually grow new roots

Figure 3.10
Taking a cutting

Tissue culture

Figure 3.11
The photograph shows a Petri dish with lots of young plants growing on it. All of these plants were grown by using tiny amounts of tissue from an older plant

Scientists have found out that you can use tiny amounts of tissue from plants and grow whole new plants from them. They don't grow the young plants in soil, however, instead they use a special jelly which contains all the minerals that the plants need to help them grow (see Figure 3.11).

When the plants have produced roots and shoots, they will be put into a pot with some soil and grown in a greenhouse. In this way thousands of plants can be produced, all of them identical to each other and the original plant that they were taken from.

2 Selective breeding

Plants

Farmers have grown plants for thousands of years, taking seeds from plants that grow naturally in the environment. Wild plants show a huge amount of variation. For example, wild wheat can have any size of stalk from 40 cm to 100 cm tall and their seeds can weigh anything between 0.5 g and 4 g (see Figure 3.12).

Figure 3.12
This picture shows wheat produced from wild seeds   as you can see, they're all different

Farmers have used this to their advantage and selected the best plants to get the best crops. They've also mixed different types of the same plant during pollination to get plants with a certain characteristic they'd like – this is known as **cross breeding**.

This is an example of how a farmer set about using cross breeding to solve a problem.

Farmer Ted normally grows really tall wheat with big seeds that he knows will fetch a good price (see Figure 3.13). The crop isn't perfect, however, as it tends to get blown over in strong winds.

He decides to solve his problem by pollinating his tall plants with a small variety of wheat that has shorter stalks (see Figure 3.14).

Figure 3.13
This type of wheat grows big seeds but has a long stalk that's easily blown over in storms

Figure 3.14
This type of wheat has small seeds but has a short stalk so is less likely to be blown over in a storm

The outcome of this cross breeding is a mixture of different types of plants as shown in Figure 3.15.

Figure 3.15
The circled plant is the one that the farmer was hoping for – a wheat plant with a short stalk and big seeds

Farmer Ted then only breeds from the plant circled in Figure 3.15 which has big seeds but a short stalk that won't blow over in the wind. Eventually, by selecting only the plants with the characteristics that the farmer wants for breeding, a variety of wheat is produced that only produces short plants with big seeds as shown in Figure 3.16.

Figure 3.16
A field of wheat with short stalks and big seeds – as you can see they're all identical

Animals

It's amazing to think that all of the different breeds of dog that we know today have a common ancestor. They have all been selectively bred for different characteristics. The alsatian for example has been bred to be large, strong, agile and easily trained. The British bulldog on the other hand has been bred purely for its good looks!

Farmers have also used selective breeding for many years to get the best milk yields from cattle, chickens that lay large, tasty eggs and pigs that have lots of muscle so they make lean bacon.

Farmers faced different problems in different areas of the world. Australian farmers, for example, needed cows that could tolerate the strong sunlight, tropical heat, frequent drought and voracious ticks found in the northern territories of the country.

The cattle that the original settlers took with them were the Shorthorn breed found in the north east of England. They had a lot of good characteristics: they were fertile, docile and easy to milk. However, high temperatures (above 24°C) cause them to suffer, strong sunlight often causes them to develop eye cancer and they suffer badly from ticks.

Figure 3.17 *The Shorthorn breed of cattle*

Australian farmers cross bred some of these cattle with an Indian breed – the Brahmin breed. These cattle rarely get eye cancer, produce an oily secretion that discourages ticks and have lots more sweat glands than the Shorthorn breed.

Figure 3.18 *The Brahmin breed like it hot!*

As you can see, these cattle have short white fur that reflects the heat away from their bodies. They also have loose folded skin which increases the surface area so heat can be lost better. All of these characteristics mean that the Brahmin cattle can live happily in temperatures as high as 41°C.

Australian farmers cross bred the Brahmin breed with the Shorthorn breed and selected the cattle best suited to the Australian climate for future breeding. In this way they created a new breed with the best characteristics of both breeds and called it the Droughtmaster breed. These cattle are tolerant of high temperatures, repel ticks, are very fertile and are easy to milk.

Figure 3.19 *The Droughtmaster breed – created by selective breeding to suit the climate of Australia*

Is there a downside to selective breeding?

Some people say that even though tomatoes are much bigger than they used to be, they're just not as tasty. This is because the people who bred the tomatoes selected the biggest ones, without taking taste into account. The problem is that the genes that made for really tasty tomatoes may now be lost because the tomato breeders did not keep the seeds of the smaller, tastier varieties.

There is a risk that selective breeding can **reduce the amount of variation** within a population. This means that in the future it may not be possible to do more selective breeding to suit the changes that might take place in our environment. Because of this, plant scientists have started to store as many different kinds of seed known to man in massive refrigerators at the National Seed Bank.

3 Cloning animals

Animal breeders can make clones by a technique called **embryo transplantation**. They do this by fertilising an egg in a Petri dish and allowing the fertilised egg to divide a few times. Before the cells in the embryo start to become different types of cell, they can be separated from each other and placed in the uterus of a number of different females called **host mothers**. These cells then start to develop into new animals and because they've got the same genes they'll end being **clones** of each other (see Figure 3.20)

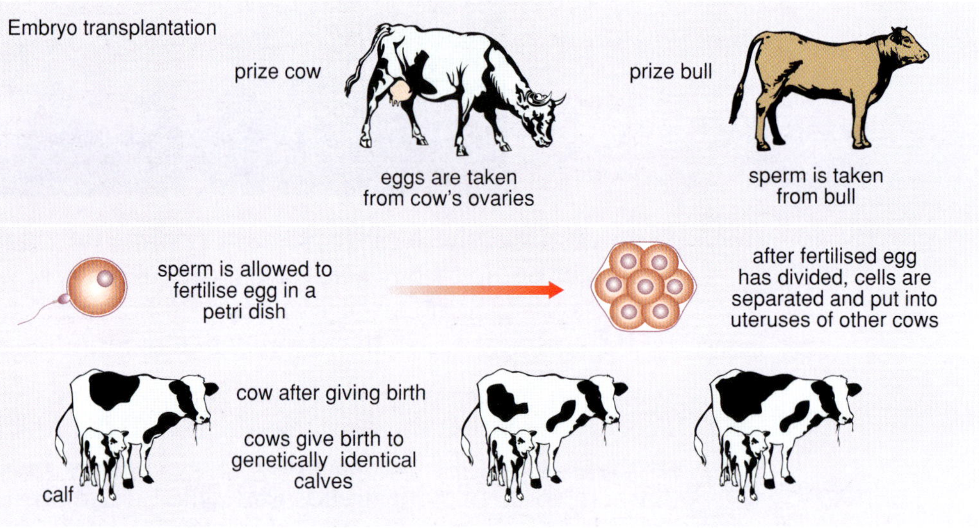
Embryo transplantation
prize cow
prize bull
eggs are taken from cow's ovaries
sperm is taken from bull
sperm is allowed to fertilise egg in a petri dish
after fertilised egg has divided, cells are separated and put into uteruses of other cows
cow after giving birth
cows give birth to genetically identical calves
calf

Figure 3.20 *Cloning cows*

4 Genetic engineering

Changing bacteria

Scientists are able to cut out **genes** from the **chromosomes** of one type of living thing and put them back into the chromosomes of bacteria. Surprisingly, the gene makes the same protein in the bacterial cell as it did in the original organism. This method can be used to make proteins which can be useful as medicines. People suffering from **diabetes** don't make enough **insulin** themselves, but genetic engineering can be used to make huge quantities of this protein. How this is done is explained in Figure 3.21.

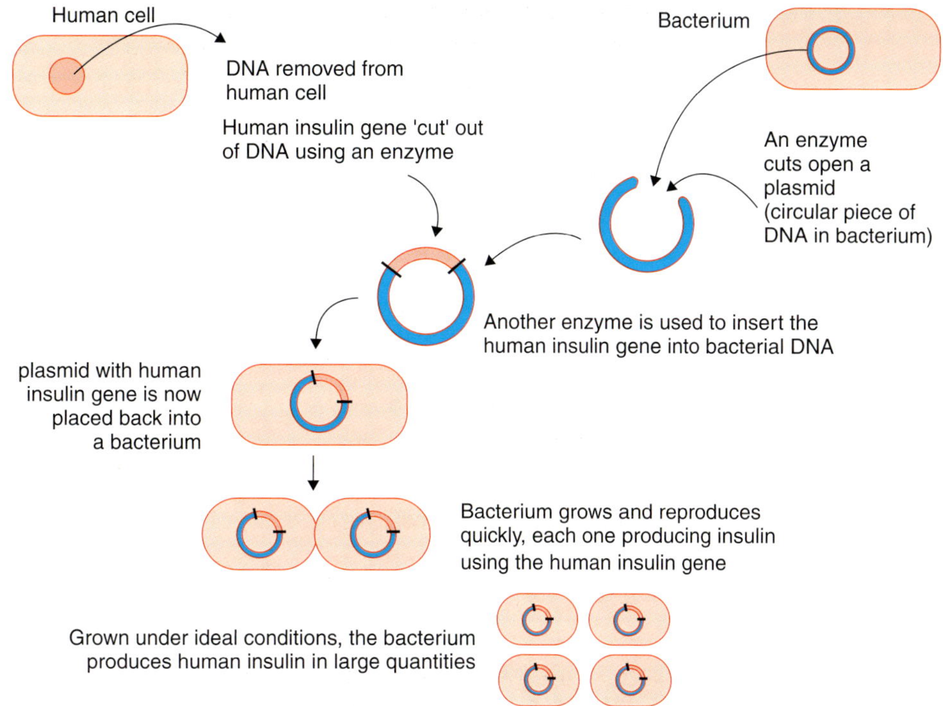
Human cell
DNA removed from human cell
Human insulin gene 'cut' out of DNA using an enzyme
Bacterium
An enzyme cuts open a plasmid (circular piece of DNA in bacterium)
Another enzyme is used to insert the human insulin gene into bacterial DNA
plasmid with human insulin gene is now placed back into a bacterium
Bacterium grows and reproduces quickly, each one producing insulin using the human insulin gene
Grown under ideal conditions, the bacterium produces human insulin in large quantities

Figure 3.21 *Creating insulin for diabetics, through genetic engineering*

Changing animals and plants

It's also possible to put genes from other organisms into the cells of animals and plants. If this is done at an early stage of development, then the gene will be copied as the organism grows and will give the organism the characteristics that the scientists desired.

Is genetic engineering a good thing or a bad thing?

Some people think that genetic engineering is a **positive** thing and give reasons such as:

1 Tomatoes have been genetically engineered so that they stay fresher for longer.

2 Rice has been engineered so that it contains more vitamin A which might help people in developing countries get a better diet.

Some people say that genetic engineering is a **negative** thing because:

1 Genetically engineered plants might in some way be toxic to humans.

2 Some companies might produce plants that are resistant to herbicides (weed killers) so farmers will use more herbicides to get a good crop but the environment will be damaged as a result.

Did you know?

Scientists at the Roslin institute in Scotland have cloned a sheep. They took a cell from the udder of an adult sheep and removed its nucleus. They then removed the nucleus from a sheep egg and replaced it with the nucleus from the udder cell. This cell was then put into the uterus of a sheep where it developed into a lamb. They called the lamb Dolly. She was identical to the sheep that the udder cell was taken from because she contained the same genetic information.

Topic Questions

1 a) Some plants can be grown by taking cuttings. Describe how you would make sure that the cutting gets the best chance of survival.
b) Explain why a new plant grown in this way will look just like the original plant.

2 Plant breeders use a method called tissue culture to produce thousands of clones.
a) Explain what is meant by the word clone.
b) What is the difference between tissue culture and taking cuttings?

3 A farmer grows two varieties of strawberries on his land. One produces small but very sweet strawberries, the other large, tasteless ones. Describe how he could use selective breeding to produce large, sweet strawberries.

4 What problems are associated with selective breeding?

5 Explain simply how scientists carry out genetic engineering.

6 What do you think are the pros and cons of cloning and genetic engineering?

3.5	
Co-ordinated	Modular
DA 10.19	DA 4
SA 10.13	SA 14

Evolution

The Earth is mind-bogglingly old – most scientists agree on a figure of around 4500 000 000 years. There is evidence that simple living things came into existence when the Earth was around 700 million years old. American scientists have found a fossil in a rock from Greenland which they think contains signs of life – they believe this rock is 3850 000 000 years old.

What are fossils?

Fossils are what's left of things that lived millions of years ago – they are normally seen as patterns in rocks. They were created in a number of different ways:

- When things die their bodies are normally decomposed by bacteria. Sometimes, however, the dead thing falls into a **swamp** where the decomposing bacteria can't live because of the **acidity** and **lack of oxygen**. The remains get covered by mud and vegetation which become compressed and eventually form rock.
- **Bones and shells** don't decompose easily. If they're on a beach or a river bed they get covered by layers of silt or sand which eventually turn into sedimentary rock. Sometimes the bones get replaced by **minerals** as they slowly decay in the sediments.
- Occasionally fossils show the outlines of dinosaurs' footprints, or the remains of burrows created by animals millions of years ago – even fossilised dinosaur dung has been found!

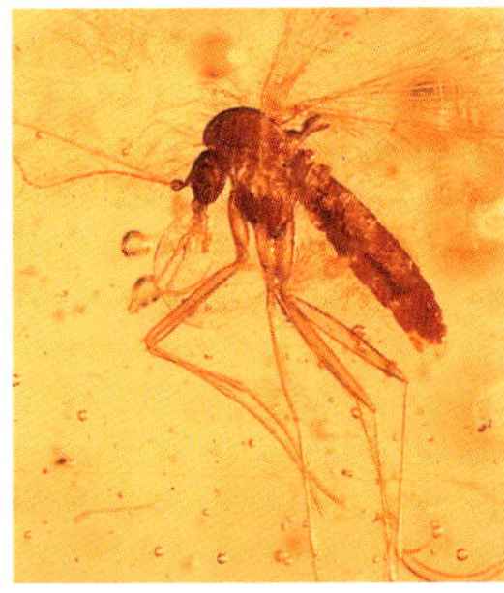

Figure 3.22
This insect was trapped in amber millions of years ago

Did you know?

Resin produced by trees millions of years ago sometimes trapped insects. This resin eventually hardened to become amber. When this happens the insect can remain perfectly preserved for millions of years.

The fossil record

Geologists looking at sedimentary rocks know that the **bottom layers** contain the **oldest fossils** because they were laid down first. Geologists have estimated that the lowest layers of rock in the Grand Canyon (see Figure 3.23) were formed 500 million years ago.

Figure 3.23
The oldest fossils found here are around 500 million years old

Did you know?

Dinosaurs probably became extinct because a large meteor hit the Earth with the same force as a powerful nuclear explosion. Some geologists claim to have found the dust thrown up by the collision in layers of sedimentary rocks.

Geologists have noticed that the oldest sedimentary rocks contain different types of fossils to the newer rocks. It's also clear that some organisms such as dinosaurs have not survived to the present day because dinosaur fossils are present in certain layers only.

Using the fossil record it has been possible to piece together the order in which organisms have come into existence upon our planet. This is shown in the list in Figure 3.24.

Figure 3.24
The fossil record can tell us the history of life on Earth

Organisms found in fossils	How many years ago?
Simple bacteria appear in the fossil record closely followed by simple plants known as blue-green algae	3 billion 1 billion
Simple sea creatures such as jellyfish and trilobites, seaweed is the most common plant	600 million
The first vertebrate	480 million
Sharks are found – some seem to have changed very little up to the present day	400 million
Ferns – the first land plant	350 million
Land invertebrates such as woodlice	350 million
Amphibians and reptiles populate the land	230 million
Dinosaurs – huge lizard-like creatures roam the land	220 million
First mammal appeared	210 million
Archaeopteryx – the first bird	140 million
First flowering plants	120 million
Camels appear	35 million
Grass	20 million
The first human-like creature	4 million

Figure 3.25
This is what artists think Archaeopteryx *looked like*

As you can see, the fossil record shows that living things have become more complicated as time has progressed.

Charles Darwin and Alfred Wallace – their theory of constant change

Charles Darwin, an Englishman from Shrewsbury, and Alfred Wallace, a Welshman from Usk, were nineteenth century explorers. They both came up with the idea that living things change gradually over time to suit their environment. This idea is known as the **theory of evolution**.

The basic idea is that in any population of living things there will be some individuals better suited to their environment than others. These will be more likely to survive and therefore pass on their genes. These 'fitter' genes will then spread throughout the population. It is as though the environment is choosing the best adapted individuals in a process that is called **natural selection**.

How does this theory explain the long necks of giraffes?

Giraffes evolved from a creature that looked a bit like a horse. Just like in any population of living things, some of these creatures would have had longer necks than others. In times of food shortage, the longer necked creatures would be able to get food more easily by eating leaves from higher in the trees than those with shorter necks. The animals with longer necks would be most likely to survive, producing offspring with longer necks. This **natural selection** continued for millions of years producing the very long necked giraffe that we see today.

? Did you know?

Giraffes have the same number of bones in their neck as human beings.

Mutation

Sometimes when genes are copied during cell division, errors take place so that the gene becomes different. This process is known as **mutation** and happens naturally in all types of living things. Certain factors, however, can make mutations happen more frequently:

- **Ionising radiation** from radioactive elements such as uranium, X-rays and ultra violet light (the part of sunlight that burns your skin) can create mutations. A higher dose of radiation means that the likelihood of mutation is increased.
- Some **chemicals** also have the ability to create mutations. There are thousands of chemicals like this in tobacco smoke. If mutations happen in body cells (such as the cells inside the lungs), they sometimes develop into cancers.

But . . . mutation is not always a bad thing. Sometimes changes take place that make a living thing even better suited to its environment. These **beneficial mutations** can increase the probability of survival.

Why antibiotics won't always work

Alexander Fleming first noticed in 1928 that a certain fungus called **mould** produced a chemical which prevented the growth of bacteria. He also realised that these chemicals (now known as **antibiotics**) could be used as a drug to treat people with bacterial infections.

Since the 1940s, different kinds of antibiotics have been used to treat different bacterial infections with great success. However, certain bacteria have developed the ability to **resist** the effects of antibiotics. This happens

because some bacteria have mutated to produce an enzyme that can break antibiotics down. These bacteria reproduce very quickly because they are resistant to the effects of antibiotics.

This is a kind of evolution and it happens in bacteria very quickly because they reproduce every 20 minutes. Doctors are becoming concerned that overuse of antibiotics could lead to a situation in which these valuable drugs become useless because so many bacteria will have evolved to become **antibiotic resistant**.

Extinction

Many of the organisms we find in fossils are not alive today – we say that they have become **extinct**. The fossil record shows that the dinosaurs suddenly became extinct around 65 million years ago. Most scientists agree that this happened because a massive meteor hit the Earth creating clouds of dust high in the atmosphere. This reduced the amount of sunlight getting to the Earth so that the plants that the **herbivorous** (plant eating) dinosaurs ate died out. This in turn lead to the extinction of the dinosaurs – no dinosaur fossil has been found that is younger than 65 million years old.

We know that other organisms have become extinct in more recent times. The woolly mammoth is a close relative of the elephants that exist today. Complete bodies of frozen woolly mammoths have been found in Siberia so we know how they were adapted to survive the cold conditions in which they lived. They had:

- two layers of thick fur
- a layer of insulating fat underneath the skin
- small ears and short legs which helped reduce heat loss.

Figure 3.26
A woolly mammoth which became extinct at the time of the last Ice Age

Some people say that the woolly mammoth was hunted to extinction, others think that they died because of a reduction in the vegetation that they relied upon for their food.

In general, extinction can happen because of:

- changes in the environment
- successful new predators moving into a habitat
- new competitors arriving into the habitat
- new diseases.

If a species is unable to evolve quickly and adapt to these new challenges it becomes extinct.

Did you know?

Over fishing in the North Atlantic means that cod is now considered an endangered species.

Topic Questions

1 What is meant by the word fossil?

2 State three ways in which fossils are created.

3 How is it possible to date a fossil?

4 What is meant by the term evolution?

5 Explain how certain organisms can become extinct.

6 What is a mutation?

7 Explain why ionising radiation can be dangerous.

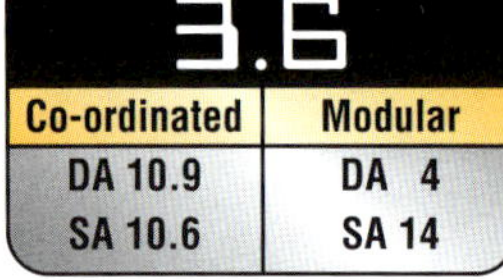

3.6

Co-ordinated	Modular
DA 10.9	DA 4
SA 10.6	SA 14

Hormones and the menstrual cycle

Women are born with all the eggs that they will release in their lifetime. From puberty onwards, one of these eggs will be released each month from the ovaries; this is known as **ovulation**. Ovulation occurs until a woman reaches her late forties or early fifties when this will stop; this is known as the **menopause**.

The **menstrual cycle** is a series of monthly changes that take place inside a woman's reproductive system:

- The uterus lining becomes thicker during the first part of the cycle to receive a fertilised egg.
- An egg is released from an ovary half way through the cycle (**ovulation**).
- If the egg is unfertilised it passes out of the body, along with the lining of the uterus (**menstruation**).

All of the above changes are controlled by chemicals known as **sex hormones**. The pituitary gland (see Figure 3.27) makes a hormone that stimulates the eggs in the ovaries to mature. The ovaries themselves release a hormone (**oestrogen**) that causes the egg to be released.

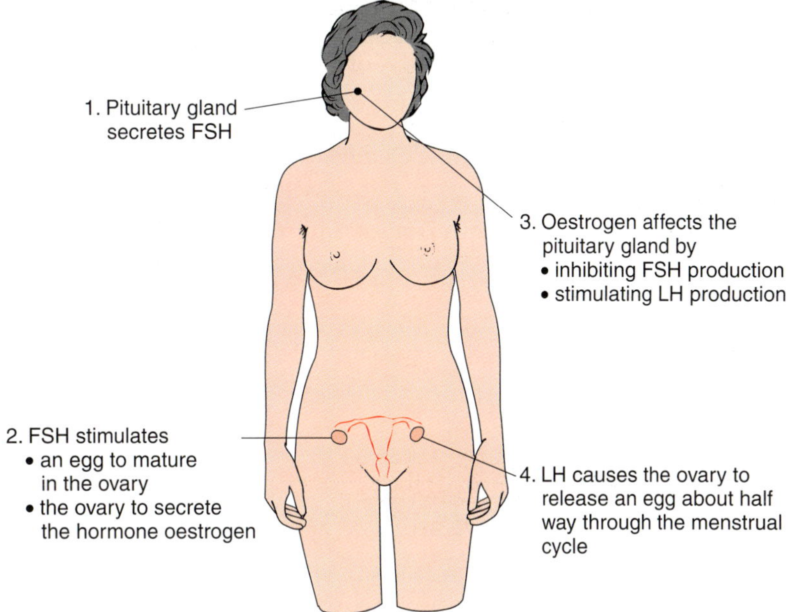
1. Pituitary gland secretes FSH
3. Oestrogen affects the pituitary gland by
• inhibiting FSH production
• stimulating LH production
2. FSH stimulates
• an egg to mature in the ovary
• the ovary to secrete the hormone oestrogen
4. LH causes the ovary to release an egg about half way through the menstrual cycle

Figure 3.27 *Hormones control the menstrual cycle*

The medical use of reproductive hormones

Contraception

The contraceptive pill has hormones that prevent ovulation. If an egg is never released from an ovary, then pregnancy cannot happen. This helps people plan their families, having children when they feel ready for the responsibility.

Problems?

Some studies have shown that taking the contraceptive pill can produce the following side effects:

- breast tenderness
- headaches
- nausea
- changes in weight
- a small increase in the chances of getting breast cancer
- a small increase in the chances of having a stroke (clot in the brain).

Fertility

Some women are unable to have children because their bodies:

- do not make hormones that allow eggs to mature
- do not make hormones that cause ovulation (egg release).

Doctors are able to treat such women with artificial hormones. Many women have been helped to conceive by hormone treatment.

Topic Questions

1 Name two parts of a woman's body that make sex hormones.

2 What does the word ovulation mean?

3 Describe the changes that take place inside a woman's body during the menstrual cycle.

4 Name two medical uses for sex hormones.

5 List the advantages and disadvantages of taking the contraceptive pill.

Summary

- The young of both plants and animals look like their parents.
- Information is passed from parents to children in the sex cells.
- Information is carried by genes.
- Different genes control different characteristics.
- Differences between individuals of the same species may be due to genetic or environmental causes.
- Chromosomes have long molecules of a substance called DNA. A gene is a section of a DNA molecule.
- Chromosomes are found in the nucleus of cells – they're usually found in pairs.
- Chromosomes carry genes that control certain characteristics.
- Genes have different forms called alleles.
- Some alleles are dominant – they overpower the effects of other alleles.
- Some alleles are recessive – two copies are needed before they take effect.
- Cells divide to produce more cells during growth or repair.
- Sexual reproduction involves the joining of male and female sex cells.
- In asexual reproduction, there's only one parent and it produces clones.
- Sexual reproduction produces individuals with a mixture of the genetic information from two parents. These individuals show more variation than clones.
- Gregor Mendel discovered the rules of inheritance.
- Huntington's disease (a disorder of the nervous system) is caused by a dominant allele.
- Cystic fibrosis (a disorder of the cell membranes) is caused by a recessive allele.
- Sickle cell anaemia (a disorder of the red blood cells) reduces how much oxygen the blood can carry. Having just one sickle cell allele can be an advantage in countries with malaria.
- In human cells, one of the 23 pairs of chromosomes carries genes that decide your sex.
- In females, the sex chromosomes are the same (XX). In males, the sex chromosomes are different (XY).
- Taking cuttings from older plants is a simple way of producing more plants. These new plants will be clones because they are genetically identical to the parent plant.
- Artificial selection can be used to produce new types of organisms.
- Selective breeding reduces the number of alleles in a population.
- Modern cloning techniques include:
  - tissue culture – using small groups of cells from part of a plant
  - embryo transplants – splitting apart cells from a developing animal embryo before they become specialised then transplanting the identical embryos into host mothers.

- Genes can be 'cut out' from the chromosomes of one type of organism and transferred to bacterial cells. The transferred gene makes the same protein in the bacterial cell. This process is used to make drugs and hormones, including human insulin.
- Genes can also be transferred to the cells of animals or plants so that they develop with desired characteristics.
- Fossils can be formed in various ways:
  - from the hard parts of animals which do not decay
  - from parts of animals or plants which have not decayed
  - when parts of the plant or animal are replaced by minerals as they decay
  - traces of animals or plants, e.g. footprints, burrows or rootlet traces.
- Fossils show how different organisms have changed over time.
- The theory of evolution states that all species of living things which exist today have evolved from simple life forms which first developed more than 3 billion years ago.
- Evolution occurs through natural selection:
  - organisms within species can show a wide range of variation due to differences in their genes
  - predation, disease and competition cause the death of large numbers of individuals
  - individuals with characteristics most suited to the environment are most likely to survive to breed successfully
  - the genes which helped these individuals survive are passed on to the next generation.
- Hormones are released from the pituitary gland and the ovaries.
- Hormones control the release of an egg from the ovaries and the changes in the thickness of the lining of the uterus.
- Fertility drugs are hormones that stimulate the release of eggs from the ovaries.
- The contraceptive pill is a combination of hormones that prevent the release of eggs from the ovaries.

Examination Questions

1 Choose words from the list in the box to complete the sentences that follow.

alleles	nucleus	genes	pairs	chromosomes

Inside the ____________ of body cells are structures called____________. They are normally found in____________. They have____________ along their length which control the development of certain characteristics. Some genes have different forms called____________. *(5 marks)*

2 The drawings opposite show how horses reproduce. Some of the stages have been labelled.

a) Name the part of cell A that contains genetic information. *(1 mark)*

b) Explain why the foal looks a little like its parents. *(2 marks)*

c) Name cell B. *(1 mark)*

d) What is the name of the process that makes cell C. *(1 mark)*

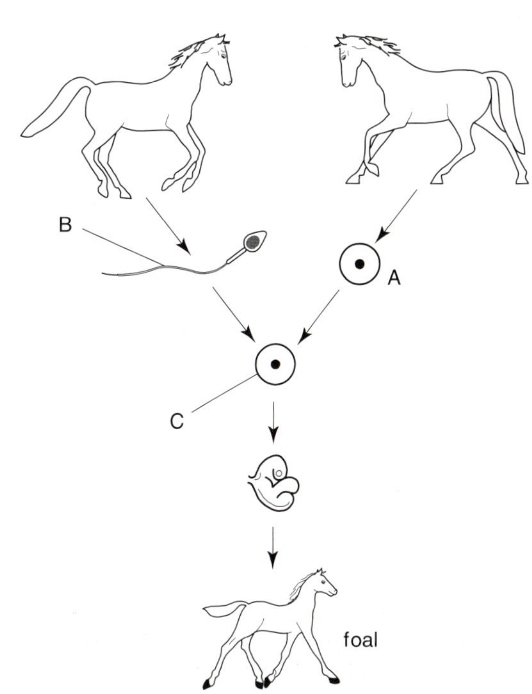

e) Horse breeders use selective breeding to make really fast race horses. Number the stages below to explain how they go about this process.
___ Foals are tested for their speed.
___ Process repeated.
___ Male and female racehorses are selected for their speed.
___ Once they've reached sexual maturity, the fastest foals are bred together.
(4 marks)

3 This couple are expecting their first baby.

Sex chromosomes .... Sex chromosomes ....

a) What are the sex chromosomes in (i) men, and (ii) women? *(2 marks)*
b) What is the probability that their child will be a girl? *(1 mark)*
c) The man has one copy of the allele responsible for the disorder cystic fibrosis. Explain why he doesn't suffer from the disease. *(2 marks)*
d) Name one other type of inherited disorder. *(1 mark)*
e) What was the name of the monk that discovered the laws of genetic inheritance? *(1 mark)*
f) Explain what is meant by the term dominant allele. *(2 marks)*

4 Giraffes live in Africa and feed mainly upon the leaves of trees.

a) Choose words from the list in the box to complete the sentences that follow.

fittest	longer	genes	shorter	population	chance	variation

In a _____ of giraffes, some will have longer necks than others. This is known as _____. The _____ necked giraffes stand a greater _____ of survival because they'll get more food than _____ necked giraffes. Long necked giraffes are more likely to pass on their _____. This is known as the 'survival of the_____'. *(6 marks)*
b) Over millions of years, the neck length of giraffes has changed. Explain how scientists know this. *(2 marks)*
c) What is the name given to changes in species over time? *(1 mark)*
d) What is meant by the term mutation? *(1 mark)*
e) Name two factors that can increase the rate of mutations. *(2 marks)*

5 Sickle cell anaemia is an inherited disorder. It is more common in countries where malaria is common.
a) In individuals with sickle cell anaemia, what type of cells are affected? *(1 mark)*
b) Describe the symptoms of sickle cell anaemia. *(2 marks)*
c) Explain why individuals with only one sickle cell allele are more likely to survive in countries where malaria is common. *(2 marks)*

Chapter 4

The environment

Key terms

population • adaptation • camouflage • photosynthesis • respiration • producers • primary consumers • secondary consumers • biomass • decomposition

4.1 Adaptation and competition

Co-ordinated	Modular
DA 10.20	DA 3
SA 10.14	SA 14

The Earth is really a large rock floating in space. It orbits the Sun at just the correct distance so that the average temperature is about right to support life. As far as we know it is the only planet in our solar system that supports life.

The environment is a mixture of physical factors (things such as temperature, rainfall and acidity of the water) and biological factors (other living things). Organisms are **adapted** to survive in different environments, which means that they have special features to help them cope with the physical and biological factors in that environment. Living things are found on every part of the planet, even at the extremes.

The Arctic

Figure 4.1
Short and stocky, this shape helps conserve heat

Anything that lives here has to be **camouflaged** and capable of dealing with extremely low temperatures.

The fox in the photograph has to catch a lot of prey in order to survive.

- The white fur not only helps them go unnoticed but also reduces the amount of heat lost as the colour **reflects heat radiation** back to the body.
- As you'll realise if you get your hair cut short, a lot of heat is lost through the ears. The arctic fox has small ears to **reduce the surface area** that it can lose heat from.
- The arctic fox looks stocky with a big body and short legs. This shape helps it conserve heat because the surface area (where heat is lost) is **small compared to its volume**.
- Underneath its skin, the arctic fox has a thick layer of **fat** which helps conserve heat as it is an insulator.

Many animals that live in the Arctic have similar adaptations to the arctic fox, think of the polar bear, seals or whales – they all have a body with a

small surface area but a large volume, and they all have a thick layer of fat which helps insulate them from the cold.

The desert

Anything that lives here also has to be camouflaged, but here they have to be able to cope with extremely high temperatures and the lack of water.

Did you know?

A blue whale's heart is as big as a mini and its tongue is the same weight as the average elephant!

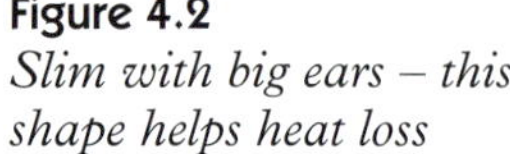

Figure 4.2
Slim with big ears – this shape helps heat loss

Immediately you'll see that the fennec (desert) fox is built differently from the arctic fox.

- It is **slim** so that it has a **large surface area** compared to its volume. The large surface area helps it **lose** heat – elephants increase this surface area again by having folded skin.
- The fox's large ears certainly help it locate its prey but they also provide a large surface area for heat loss.
- The fennec fox has very **low body fat** which ensures that it loses heat easily.
- Its sandy colouring helps **camouflage** it from its prey.

Desert plants

Plants that live in the desert have certain distinctive features which help them survive in arid (dry) conditions.

Figure 4.3
Fat and spiky – this shape helps conserve water

If you look at Figure 4.3 you'll notice that cacti have:

- A **large volume** but only a **small surface area** to lose water from.
- Leaves reduced to spikes to **reduce water loss** and to prevent animals biting its stem.
- Roots that are long but do not go very deep into the ground. Instead they cover a wide surface area for **maximum absorption of water** when it does actually rain.

Populations

In ecology, the word **population** means the **number of organisms** of a certain type in a particular habitat. The population of any species is never constant because some organisms give birth to young and others die. There are many factors that might affect the population of a particular species, most of them revolve around **food**.

If there's lots of food, then newborns of any species stand a good chance of surviving. In times of plenty, populations increase. However, for all types of organisms, predator or prey, the good times never last.

Plants

The most important factors that affect the populations of plants are the amount of light and the availability of water. Plants have adaptations to help them compete for these.

Figure 4.4
The photograph on the right was taken in spring, the photograph on the left was taken in early summer

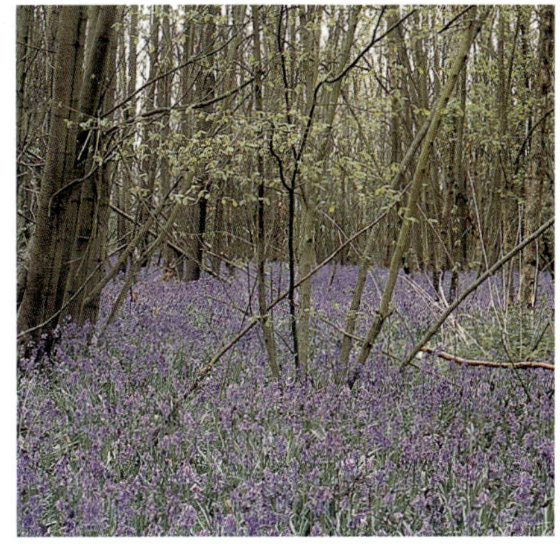

The photograph on the right shows a wood in spring. The bluebells are adapted so that they come out before the leaves on larger trees have grown enough to block out the light. In the summer, the larger trees with big broad leaves have an advantage as they compete well for light. In the autumn, trees reabsorb chlorophyll from their leaves leaving them red and brown. They reuse this chlorophyll in the new leaves they grow the next summer.

Plant eaters

As herbivores (plant eaters) increase in number, they often overgraze the plants that they rely upon. This leads to starvation and to the death of most of the organisms within a population. When the population is high, disease can spread throughout a population also leading to the death of individuals. The graph below (Figure 4.5) shows what normally happens to the population of a herbivore when they're not affected by predators.

Figure 4.5
A typical population curve

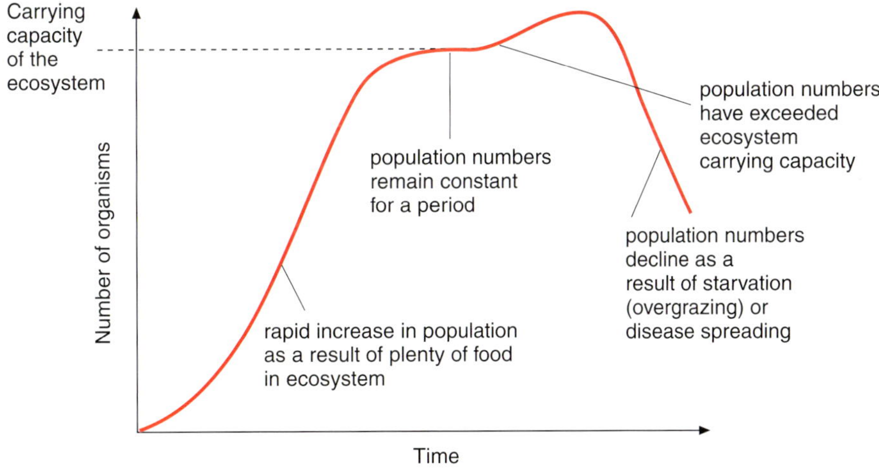
Carrying capacity of the ecosystem
Number of organisms
Time
population numbers remain constant for a period
population numbers have exceeded ecosystem carrying capacity
population numbers decline as a result of starvation (overgrazing) or disease spreading
rapid increase in population as a result of plenty of food in ecosystem

Meat eaters

Predators increase in number until they start to kill too many of their prey. This in turn leads to a decrease in their food and inevitably to a population crash. The number of prey organisms then gradually recovers. A typical predator–prey relationship is shown in Figure 4.6.

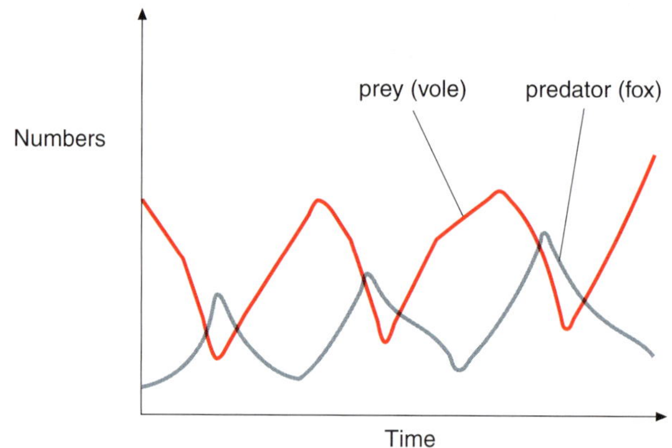

Figure 4.6
Population curves to show the changes in vole and fox populations over time

Topic Questions

1. Explain why arctic foxes:
   a) are white,
   b) have small ears,
   c) have a layer of thick fat underneath their skin.
2. a) What challenges do desert plants like cacti face?
   b) Describe two adaptations of cacti that help them survive in their environment.
3. What does the term population mean?
4. Explain why populations of predators never stay constant.

4.2	
Co-ordinated	Modular
DA 10.21	DA 3
SA 10.15	SA 14

Our effect on the environment

One species has such an impact on the environment that its actions decide the fate of all other species on the planet. This species is us – human beings.

Our population has grown so much during the last two hundred years that we are now destroying habitats of other organisms at a previously unheard of pace. Habitats are being destroyed by farming, quarrying, building new houses and dumping the waste that we produce.

More people own cars nowadays than ever before and so we're using up our reserves of oil and gas. It seems that the higher our living standards, the more the environment is damaged.

Fossil fuels

When we burn fossil fuels (coal or oil), carbon dioxide and water are the main by-products. Other gases such as sulphur dioxide and nitrogen oxides are also produced. All of these gases have some kind of effect on the environment.

Acid rain

Sulphur dioxide and nitrogen oxides dissolve in the droplets of water that make up clouds; this makes the rain acidic. When acid rain falls, it first

has an effect on plants. The trees in the photograph below have been killed by acid rain.

Figure 4.7
Acid rain kills plants

When acid rain falls into streams, it kills fish and other aquatic life – trout die in water below pH 5.5. Acid rain also dissolves limestone and marble, damaging the stonework in some buildings. Recently, fuel companies have recognised the problem and have started to produce 'low sulphur' petrol and diesel which produces less sulphur dioxide when it's burnt.

Waste

Most of the packaging used by companies today is made of plastic which can't be broken down by bacteria. This kind of waste is called 'non-biodegrable' and some environmentalists believe that the best way of dealing with it is to recycle it.

Global warming

The Sun's rays carry light energy and heat energy in the form of infra red radiation. Some of this infra red radiation is reflected back from the Earth and would travel into space if it were not for **greenhouse gases**. Carbon dioxide and methane are known as greenhouse gases because they reflect infra red radiation back to Earth. If it were not for greenhouse gases, the temperature of the Earth would be around $-74^{\circ}C$, much too cold for life.

In one day, the average cow produces enough methane gas to fill a hot air balloon!

The problem is that throughout the twentieth century we have:

- Burnt a lot of fossil fuels, increasing the concentration of carbon dioxide in the atmosphere.
- Bred a lot more cows to keep pace with the demand for meat and milk and cows produce methane when they break wind.
- Produced lots of rice fields, all of which have bacteria living in them that produce methane.

Figure 4.8
A rice field – huge amounts of methane are released from these

Some scientists are concerned that an increase in greenhouse gases will lead to an increase in global temperature leading to changes in the climates of many countries. Some scientists have predicted that:

- the ice cap of the Antarctic will melt
- glaciers around the world will melt
- the seas will expand with heating.

All of these changes mean that sea levels will rise in the future and some low lying countries will

be flooded. Some weather forecasters predict climate changes such as higher wind speeds everywhere, greater rainfall in already rainy parts of the world and less rainfall in drier areas.

How farming affects the environment

Space

Most of the woodland that once covered our country has been cut down to make room for farmland. More farms mean that we can feed more people more easily, but it also means less space for other living things.

Did you know?

At one time bears, wolves and beavers lived in Britain but destruction of their habitat has lead to their extinction.

Pesticides

Farmers use pesticides to kill insects which might otherwise eat some of their crops. The problem is that pesticides are washed into streams and rivers and kill the insects that live there. This then leads to a drop in the fish population.

Herbicides

These are chemicals used by farmers to kill plants that could compete with their crops. Again these chemicals can be washed into rivers, destroying the plants that live there.

Did you know?

An area of rainforest the same size as two football pitches is being removed from rainforests every minute of the day.

Fertilisers

These are chemicals that farmers use to help crops grow better. The problem is that when they're washed into streams and rivers they make the plants in the rivers grow out of control. Eventually, whole rivers become 'choked' by this excessive growth leading to the death of fish.

Deforestation

The whole of Europe was once covered in forests but as the human population grew, more and more of it had to be cut down to make space for farms. The same thing is currently happening in some developing countries such as Borneo and Brazil. The rainforests in these countries are habitats to more kinds of living things than anywhere else on the planet and destroying them means that many of these animals and plants are becoming extinct.

In itself this kind of destruction is depressing but environmental scientists believe that destruction of the rainforests might have unfortunate results for human beings:

- Most medicines are derived from plants. Some of the plants in the rainforest may contain chemicals which could treat diseases such as cancer, so the loss of the rain forests means that these potential cures will now no longer be a possibility.
- Deforestation reduces the amount of carbon dioxide being absorbed during photosynthesis, adding to the greenhouse effect.

Some people believe that the rainforests are so important that other countries should pay a tax to countries such as Brazil so that in return no further deforestation would take place.

Topic Questions

1 Name four substances created when fossil fuels are burnt.

2 Name two greenhouse gases.

3 What is meant by the term global warming?

4 a) Name two substances used by farmers that can damage the environment.
  b) Explain how these substances can damage the environment.

5 Explain how living things are affected by deforestation.

6 How does acid rain affect plants and animals?

4.3 Energy transfers

Co-ordinated	Modular
DA 10.22 SA n/a	DA 3 SA n/a

How do living things get energy?

It's strange to think that the all of the energy that your body gets comes from the Sun. Plants absorb light energy in the process called **photosynthesis** and use it to help them make sugars which they then use in **respiration**. Plants are known as **producers** because the food they make is used by other living things. Animals that eat plants are known as **primary consumers**, animals that eat primary consumers are known as **secondary consumers**. The flow of energy through living things can be represented by a food chain as shown below:

oak tree ⟶ insects ⟶ blue tits ⟶ sparrow hawk

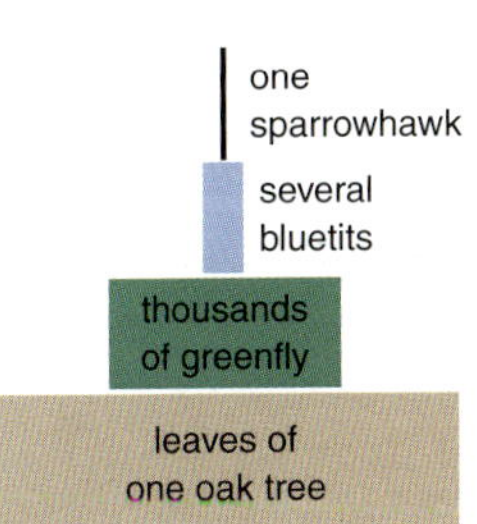

Figure 4.9
This pyramid of biomass shows how the total mass of organisms reduces along a food chain

Food chains always begin with a producer and the arrows represent the energy flow from one organism to the next. The mass of each type of organism is called its **biomass** and food chains can be represented in a pyramid of biomass as shown in Figure 4.9.

It's always exciting to see a rare animal such as a sparrow hawk, red kite or dolphin. There's a good reason that these kind of animals are rare – the energy taken in by living things is mainly used in **moving around**, **heating their bodies** and **helping growth**; only a small part of the energy actually gets stored in their tissues. Figure 4.10 shows how energy is lost in a food chain.

Figure 4.10
A simple food chain showing where the energy is lost

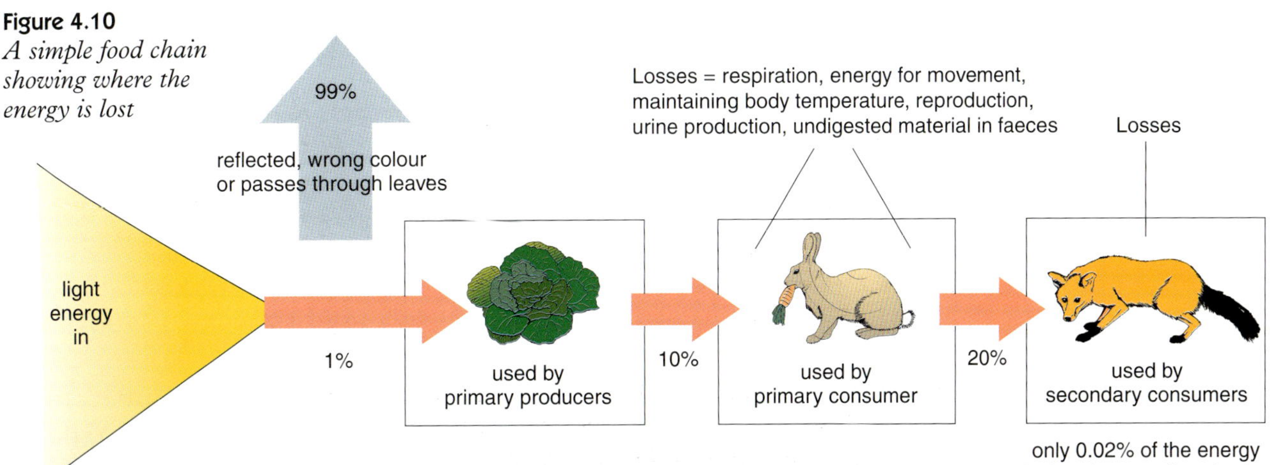

Farmers know that energy is lost in this way and make sure that the animals that they breed for food production are **primary consumers**.

Figure 4.11
In the food chain wheat ⟶ pig ⟶ human, pigs are primary consumers and humans are secondary consumers. Most of the energy consumed by the pig is used for movement, keeping warm or is lost in its waste. The human only receives 20% of the energy consumed by the pig (and that's only if she eats all of it!)

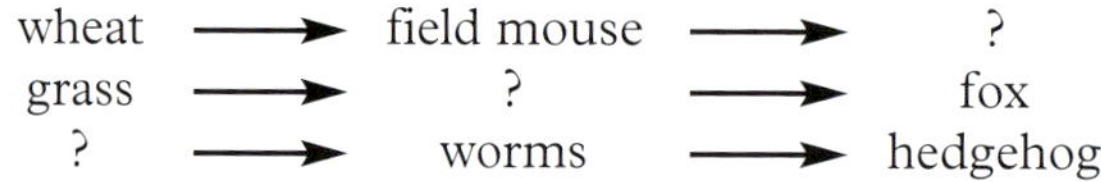

1 Complete the following food chains.

wheat ⟶ field mouse ⟶ ?
grass ⟶ ? ⟶ fox
? ⟶ worms ⟶ hedgehog

2 What is meant by the following terms?

a) producer
b) primary consumer
c) secondary consumer

3 Draw a pyramid of biomass to represent the following food chain.

plankton ⟶ mussel ⟶ dog whelk

4 Explain why farmers breed animals that are herbivores for meat production.

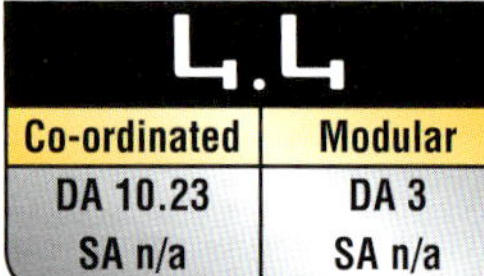

How nature recycles living things

The carbon cycle

All of the carbon atoms in your body have at one time been in other living things; they will even have been in the bodies of other human beings! This is because nature recycles everything. Carbon is **recycled** by the following processes:

- Green plants absorb carbon dioxide during **photosynthesis** – they also release it into the air when they **respire**.
- When plants are **eaten** by animals, the carbon from the plant becomes part of the animal. Respiration then releases carbon dioxide into the atmosphere.

- When animals and plants die, bacteria and fungi release digestive enzymes onto them, breaking their tissues down in a process called **decomposition**. Carbon dioxide is released into the atmosphere when microorganisms respire.
- Human beings add to the carbon cycle when they **burn** fossil fuels.

The whole process is summarised in Figure 4.12.

Figure 4.12
The carbon cycle

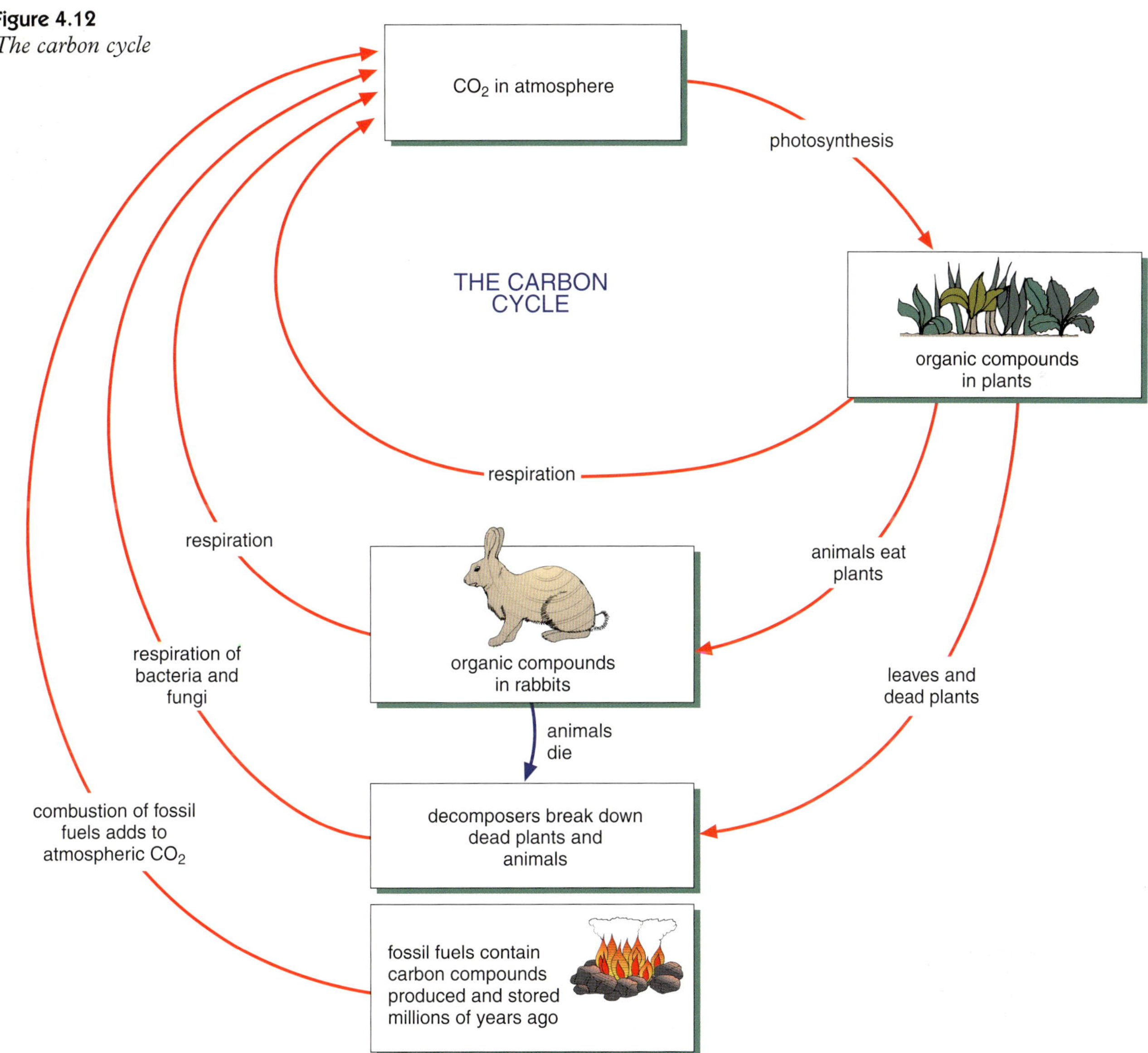

Decomposers

Fungi and bacteria are really important in the recycling of nutrients. Gardeners use them in compost heaps which are really mounds of dead plant material. Compost heaps work best when they're kept **damp** and turned over by a spade every now and then to ensure that the microorganisms get plenty of **oxygen**.

As well as releasing carbon dioxide into the atmosphere, the microorganisms in compost heaps release **nitrates** and other substances that help plants grow better; they're really a free source of fertiliser!

Did you know?

Sewage plants use microorganisms to break down the waste that human beings flush down the toilet.

Topic Questions

1 a) Explain how plants remove carbon dioxide from the atmosphere.

b) In which process is carbon dioxide released into the atmosphere by living things?

c) What are microorganisms?

2 Explain how carbon atoms from the stems or leaves of certain plants can become part of our body.

3 Why are decomposers important to (i) gardeners, and (ii) sewage plants?

Summary

- Animals that live in the desert have a large surface area and low body fat in order to lose heat. They're often sand coloured which helps with camouflage.
- Plants that live in dry conditions have extremely well developed root systems for rapid water absorption. Their leaves are often reduced to spines, they have a thick stem and a thick waxy cuticle to prevent water loss.
- Animals that live in arctic conditions have a low surface area and high body fat to help conserve heat. They're often white which helps with camouflage.
- A population (the number of animals) is limited by the amount of food available.
- The number of predators in a community is affected by the number of prey animals. The number of prey animals is affected by the number of predators.
- Animals compete for resources such as food and space.
- Populations may be affected by:
  - competition for nutrients
  - competition for light
  - predation or grazing
  - disease.
- Humans reduce space for animals and plants by:
  - building
  - quarrying
  - farming
  - dumping waste.
- Humans pollute:
  - water – with sewage, fertiliser or toxic chemicals
  - air – with smoke and gases such as sulphur dioxide
  - land – with toxic chemicals, such as pesticides and herbicides, which may be washed from the land into water.
- Burning fossil fuels makes carbon dioxide, sulphur dioxide and nitrogen oxides. These dissolve in rain making it acidic. Acid rain damages plants and animals that live in water.
- The human population is expanding. Together with an increase in the standard of living this leads to:
  - raw materials, including coal and oil, are being used up
  - more waste is produced
  - more pollution unless waste is properly handled.

- Deforestation in tropical areas has:
  - increased the release of carbon dioxide into the atmosphere
  - reduced the amount of carbon dioxide being removed from the atmosphere.
- Cattle and rice fields release methane into the atmosphere.
- Increasing levels of carbon dioxide and methane may be causing a **greenhouse effect**.
- An increase in the Earth's temperature of only a few degrees Celsius may cause changes in the Earth's climate and a rise in sea levels.
- Green plants capture some of the energy from the Sun.
- This energy is stored in the substances which make up plant cells.
- Biomass (the mass of organisms) at each stage in a food chain is less than it was in the one before.
- You should be able to draw a pyramid of biomass to scale.
- The efficiency of food production is improved by reducing the number of stages in food chains, keeping animals warm and restricting their movement.
- Microorganisms break down materials best in warm, moist conditions. Many microorganisms also work best when there is plenty of oxygen.
- Sewage works use microorganisms to break down human waste.
- Microorganisms break down waste plant materials in compost heaps – this releases substances which plants need to grow.
- In the carbon cycle:
  - carbon dioxide is removed from the air by photosynthesis, and the carbon is used to make carbohydrates, fats and proteins
  - carbon dioxide is returned to the air when green plants respire
  - green plants are eaten by animals that are themselves eaten by other animals. In this way, carbon moves from plants to herbivores and finally to carnivores
  - respiration by animals releases carbon back into the atmosphere as carbon dioxide.
- Dead things are eaten by certain animals and microorganisms – carbon is released into the atmosphere as carbon dioxide when these organisms respire.

Examination Questions

1 The diagram below shows some of the processes that take place in the carbon cycle.

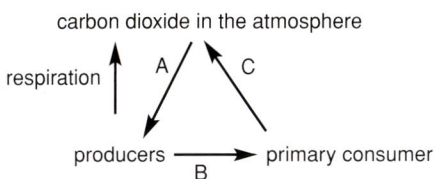

carbon dioxide in the atmosphere
respiration
A
C
producers
B
primary consumer

a) Name the processes labelled A – C. *(3 marks)*

b) Name the organisms responsible for decomposing dead materials or waste. *(1 mark)*

2 Students set up an experiment to see which conditions helped a leaf to decompose most quickly.

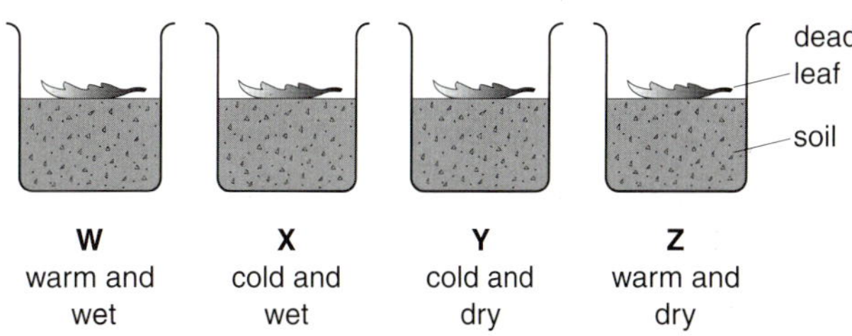

a) In which beaker would you expect the leaf to decay: (i) quickest, and (ii) most slowly? *(2 marks)*

b) Which gas is released into the atmosphere by decomposition? *(1 mark)*

c) Explain why compost heaps are important to gardeners. *(2 marks)*

3 A population of rabbits lived on an island, they had no natural predators. The graph below shows their population over a period of 50 years.

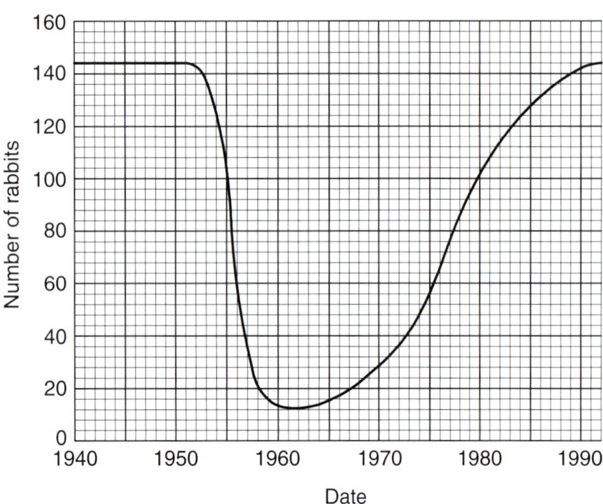

a) What is the population of rabbits in 1961? *(1 mark)*

b) Give one reason for the decline in rabbit numbers between 1952 and 1961. *(1 mark)*

c) What would happen to the population of rabbits if a family of foxes were introduced to the island? *(1 mark)*

d) Write a food chain to show how energy flows through the ecosystem to the foxes. *(3 marks)*

4

a) The mammoth in the photo lived in a very cold climate. Describe how the following features helped it survive:
   (i) small ears *(2 marks)*
   (ii) two layers of thick fur, one short and one long. *(2 marks)*

b) Explain how the mammoth's overall shape helped it survive in its environment. *(2 marks)*

Chapter 5

Metals

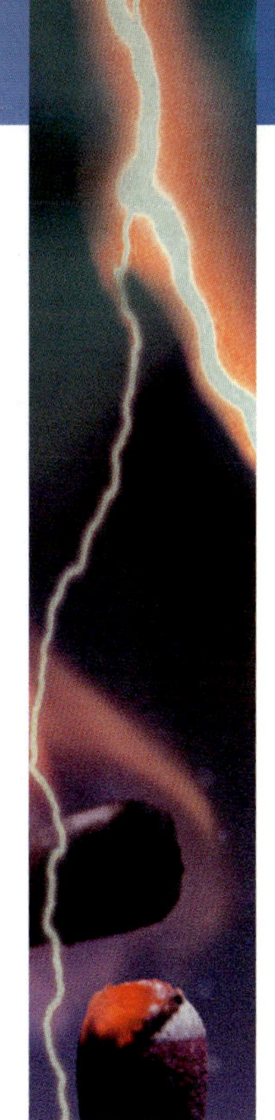

Key terms

elements • atoms • relative atomic mass • Periodic Table • groups • density • hydroxide • alkali • neutral • alloy • catalysts • compounds • ore • smelting • blast furnace • reduction • slag • oxidation • electrolysis • ions • electrodes • anode • cathode • bauxite • cryolite • electrolyte • precipitate • corrode • rusting • sacrificial protection • indicator

5.1 The Periodic Table

Co-ordinated	Modular
SA 11.3	SA 15
DA 11.11	DA 05

Different chemical **elements** have **atoms** with different masses. These masses are very small and are not measured in grams. Instead, the mass of each atom is compared to the mass of a hydrogen atom. A carbon atom is 12 times heavier than a hydrogen atom. So the **relative atomic mass** of carbon is 12. The symbol A_r is used for relative atomic mass. (Don't confuse A_r with Ar, the chemical symbol of the element argon.)

The **Periodic Table** (Figure 5.1) is an arrangement of the elements. The elements are put in rows in order of their relative atomic masses. The vertical columns contain elements with similar chemical properties. These columns are called **groups**. The arrangement isn't perfect. To get a better 'fit' some elements are switched round. Argon and potassium have been switched to make them fit better.

The diagram of the Periodic Table has been coloured to show elements of different types. It is obvious that most of the elements (more than ¾ of them) are metals. Most of the metals are in the two left hand columns (Groups 1 and 2) and the large block of transition elements in the centre.

Tasks

1 Can you find another example in the Periodic Table where two elements have been reversed to make them fit better?

2 Give an example of a period in the Periodic Table.

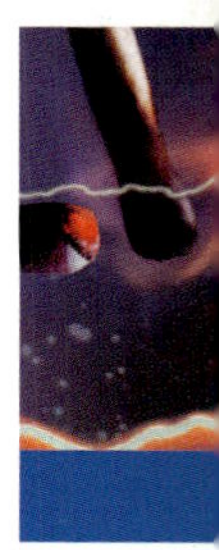

Figure 5.1
The Periodic Table

Mass number A / Atomic number (Proton number) Z: 1 H hydrogen 1

1 (Alkali metals)	2	Transition metals										3	4	5	6	7 (Halogens)	0 (Noble gases)
																	4 He helium 2
7 Li lithium 3	9 Be beryllium 4											11 B boron 5	12 C carbon 6	14 N nitrogen 7	16 O oxygen 8	19 F fluorine 9	20 Ne neon 10
23 Na sodium 11	24 Mg magnesium 12											27 Al aluminium 13	28 Si silicon 14	31 P phosphorus 15	32 S sulphur 16	35 Cl chlorine 17	40 Ar argon 18
39 K potassium 19	40 Ca calcium 20	45 Sc scandium 21	48 Ti titanium 22	51 V vanadium 23	52 Cr chromium 24	55 Mn manganese 25	56 Fe iron 26	59 Co cobalt 27	59 Ni nickel 28	63 Cu copper 29	64 Zn zinc 30	70 Ga gallium 31	73 Ge germanium 32	75 As arsenic 33	79 Se selenium 34	80 Br bromine 35	84 Kr krypton 36
85 Rb rubidium 37	88 Sr strontium 38	89 Y yttrium 39	91 Zr zirconium 40	93 Nb niobium 41	96 Mo molybdenum 42	Tc technetium 43	101 Ru ruthenium 44	103 Rh rhodium 45	106 Pd palladium 46	108 Ag silver 47	112 Cd cadmium 48	115 In indium 49	119 Sn tin 50	122 Sb antimony 51	128 Te tellurium 52	127 I iodine 53	131 Xe xenon 54
133 Cs caesium 55	137 Ba barium 56	139 La lanthanum 57	178 Hf hafnium 72	181 Ta tantalum 73	184 W tungsten 74	186 Re rhenium 75	190 Os osmium 76	192 Ir iridium 77	195 Pt platinum 78	197 Au gold 79	201 Hg mercury 80	204 Tl thallium 81	207 Pb lead 82	209 Bi bismuth 83	Po polonium 84	At astatine 85	Rn radon 86
Fr francium 87	226 Ra radium 88	227 Ac actinium 89															

Elements 58–71 and 90–103 have been omitted.

Key: metal very reactive; metal reactive; metal fairly unreactive; liquid metal or non-metal; non-metal solid; non-metal gas; non-metal unreactive gas

The value used for mass number is normally that of the commonest isotope, eg ^{35}Cl not ^{37}Cl
Bromine is approximately equal proportions of ^{79}Br and ^{81}Br

Topic Questions

1 In which groups are the most reactive metals found?

2 In which groups are the least reactive elements found?

3 a) Which group has only two gases in it?
b) Which groups have only one gas in them?
c) Which elements are liquid at room temperature?

4 Approximately what percentage of the elements are non-metals?

5 Copy the following table and complete it.

Element	Symbol	A_r	Group
	N		
iron			transition
	Li		
platinum			
		127	

5.2

Co-ordinated	Modular
SA 11.3	SA 16
DA 11.11	DA 05

The Group 1 metals

The Group 1 metals are not like most other metals.

- They have very low **density**. In fact lithium, sodium and potassium float on water.
- They are very reactive. If the metals are put in water they react vigorously (see Figure 5.2). The gas hydrogen is given off and the metal **hydroxide** is produced. The metal hydroxide is soluble in water and forms a strongly **alkaline** solution. For lithium the reaction is:

$$\text{lithium} + \text{water} \longrightarrow \text{lithium hydroxide} + \text{hydrogen}$$
$$2Li + 2H_2O \longrightarrow 2LiOH + H_2$$

Because of this reaction the Group 1 metals are often called the alkali metals.

Did you know?

Most of the Group 1 metals were discovered by Sir Humphry Davy between 1807 and 1808. He used electrolysis to obtain them. Sir Humphry Davy also invented the coal miner's safety lamp.

Figure 5.2
a) lithium, b) sodium and c) potassium reacting with water

The alkali metals don't just react with water. They are so reactive that they are kept in oil (see Figure 5.3). But even in oil they react with the oxygen in the air and are soon covered with a layer of metal oxide. For potassium the reaction is:

$$\text{potassium} + \text{oxygen} \longrightarrow \text{potassium oxide}$$
$$4K + O_2 \longrightarrow 2K_2O$$

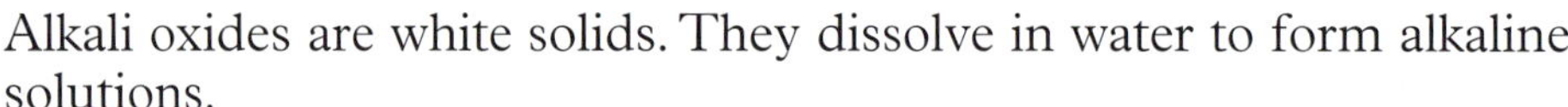

Alkali oxides are white solids. They dissolve in water to form alkaline solutions.

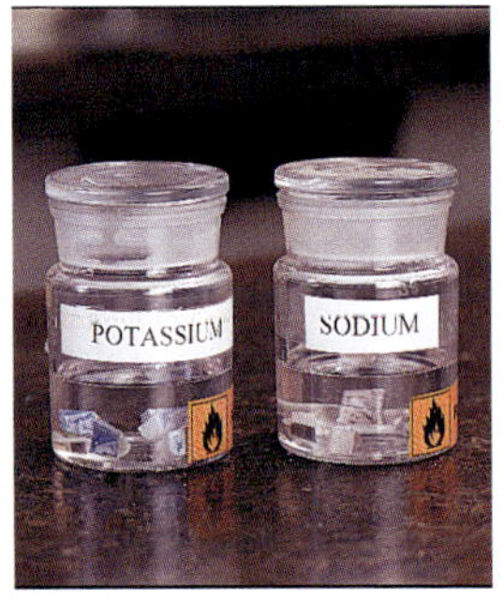

Figure 5.3
Alkali metals are so reactive they have to be stored in oil

Alkali metals will react with many non-metals. For example, they react with chlorine to form the metal chloride. For sodium the reaction is:

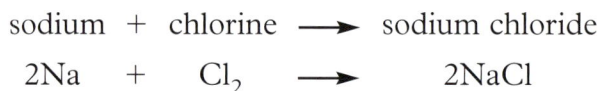

$$\text{sodium} + \text{chlorine} \longrightarrow \text{sodium chloride}$$
$$2Na + Cl_2 \longrightarrow 2NaCl$$

The alkali metal chlorides are colourless crystalline solids. They dissolve in water to form a **neutral** solution.

Tasks

1 Compounds which contain oxygen are called oxides. Compounds which contain chlorine are called chlorides. Find out the names of compounds which contain a) bromine, b) nitrogen and c) sulphur.

2 Find out some other properties of the alkali metals. (For example, which metal has the lowest density? What are their melting points?)

Topic Questions

1 Lithium hydroxide is an alkaline solution. Is the pH of this solution less than 7, exactly 7, or more than 7?

2 Sodium chloride is a neutral solution. What is the pH of sodium chloride solution?

3 What is the common name for sodium chloride?

4 For each of the following state if the solutions are alkaline or neutral:

a) lithium chloride
b) sodium oxide
c) potassium hydroxide

5 Write word equations for the reactions of lithium, sodium and potassium with water, oxygen and chlorine. (Three of these have already been done for you in this section.)

5.3	
Co-ordinated	Modular
SA n/a	SA n/a
DA 11.11	DA 05

Transition elements

The transition elements are in a block in the middle of the Periodic Table. They are all metals and have typical metallic properties. Like the alkali metals, transition metals are good conductors of heat and electricity. They can also be hammered or bent into shape. But in other ways they are not like the alkali metals.

Figure 5.4 *Comparing the properties of alkali metals and transition metals*

Property	Alkali metal	Transition metal
Melting point	Low (potassium 63°C)	High (iron 1539°C)
Strength	Very soft, can be cut with a knife like plasticine	Hard and strong
Reactivity	Very reactive, corrodes easily	Fairly unreactive, some corrode slowly in the air

Did you know?

Some transition metals are needed by plants and animals. They are called 'trace elements' because only a very small amount of them is needed. Iron is needed by animals to produce haemoglobin, which is in red blood cells. A lack of iron causes anaemia.

Transition metals have many uses because they are strong and not very reactive. Steel is an **alloy** of the transition metal iron. Steel is used to make cars, ships and girders for large buildings. Copper is also a transition metal. It is a very good conductor of electricity and is used to make electricity cables.

Figure 5.5 *Rusting and weathered transition metals*

Metals

Most transition metals have a shiny, silvery appearance, but the compounds of transition metals are often coloured. When iron rusts it forms a reddish brown coloured compound. When copper weathers it gets a green coloured coating (see Figure 5.5). The coloured compounds of transition metals are used as glazes in pottery and to colour glass (Bristol blue glass is coloured with cobalt compounds).

Figure 5.6 *Glazed pottery and Bristol blue glass are products of coloured compounds of transition metals*

Many transition metals are used as **catalysts**. Air, water and natural gas are used to make some artificial fertilisers. The transition metals iron and platinum are used as catalysts in this process (see Chapter 7).

Tasks

1 Most transition metals are silvery. Find out the names of some that are not.

2 Find out which transition metals have the lowest and the highest melting points.

Topic Questions

1 Which of the following are transition metals: aluminium, chromium, lead, potassium, tin?

2 The table below gives some details of a number of compounds. Fill in the blanks in the table. The first two have been done for you.

Compound	Colour	Name of the transition metal in the compound
sodium chloride	colourless	none
potassium permanganate	purple	manganese
nickel sulphate	green	
calcium carbonate	colourless	
copper oxide	black	
magnesium sulphate		none
lithium oxide		
vanadium sulphate	violet	
potassium chromate	yellow	

5.4 Ores

Co-ordinated	Modular
SA 11.4	SA 15
DA 11.4	DA 05

The rocks in the Earth's crust contain useful metals. These are usually found as metal **compounds** mixed with other substances. The metal can be extracted from the rock by chemical methods. If it is possible to do this economically then the rock is a useful **ore** of the metal. The ores of some very unreactive metals, like gold, contain the uncombined metal. These ores do not need chemical methods to extract the ores.

Figure 5.7
Iron ore, bauxite and native gold

Task

Find the names of some ores of iron and aluminium.

Topic Questions

1 Name one metal that does not need to be obtained by chemical methods.

2 Clay contains a lot of aluminium, but clay is not an ore of aluminium. Why is this?

5.5 The reactivity series

Co-ordinated	Modular
SA 11.4	SA 15
DA 11.4	DA 05

You should already know that some metals are more reactive than others. The reactivity of metals can be compared by seeing how they react with:

- air (or oxygen) to produce metal oxides
- water or steam to produce metal oxides or hydroxides and hydrogen gas
- dilute acids to produce metal salts and hydrogen gas.

Figure 5.8
A summary of metal reactivity

Metal	Reaction with:			Relative reactivity
	air	water/steam	dilute acids[1]	
potassium	very rapid	very rapid with water	very rapid	most reactive
sodium				
calcium	fairly slow	rapid with water	rapid	
magnesium		fairly slow with water, rapid with steam		
aluminium[2]		slow	fairly rapid	
zinc	slow	very slow even with steam		
iron			slow	
tin				
lead			very slow	
copper	very slow	no reaction even with steam	no reaction	
silver	none			
gold				
platinum				least reactive

[1] Hydrochloric and sulphuric acids behave like this. Weak acids, like ethanoic acid, are much less reactive so all the reactions are slower. Nitric acid behaves in totally different way.
[2] In air aluminium rapidly gets an oxide layer. This seals the surface and stops air, water or acids reaching the metal. So aluminium seems to be less reactive than it really is.

The table in Figure 5.8 summarises the reactions. It is clear that the relative reactivity of the metals is the same whatever they react with.

If a mixture of magnesium and copper is heated in air the magnesium will react but the copper won't. The more reactive metal 'wins the battle' for the oxygen.

$$\text{magnesium} + \text{oxygen} \longrightarrow \text{magnesium oxide}$$
$$2Mg + O_2 \longrightarrow 2MgO$$

The same idea applies if the oxygen is actually combined with the copper. If magnesium is heated with copper oxide the magnesium takes the oxygen.

$$\text{magnesium} + \text{copper oxide} \longrightarrow \text{magnesium oxide} + \text{copper}$$
$$Mg + CuO \longrightarrow MgO + Cu$$

The rule is that a **more reactive metal will displace a less reactive metal from its compound.**

This principle is used in the Thermit process. In this process aluminium reacts with iron oxide. The reaction gets hot enough to melt the iron. The Thermit process is used to weld railway tracks together.

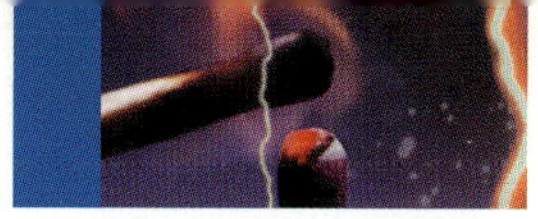

Figure 5.9
The Thermit process

The rule applies to other metal compounds besides oxides. If iron powder is dropped into a solution of copper sulphate the iron displaces the copper. The blue colour of the copper sulphate solution gradually fades and the greyish coloured iron powder slowly gets covered with a brown layer of copper.

$$\text{iron} + \text{copper sulphate} \longrightarrow \text{iron sulphate} + \text{copper}$$
$$Fe + CuSO_4 \longrightarrow FeSO_4 + Cu$$

Figure 5.10

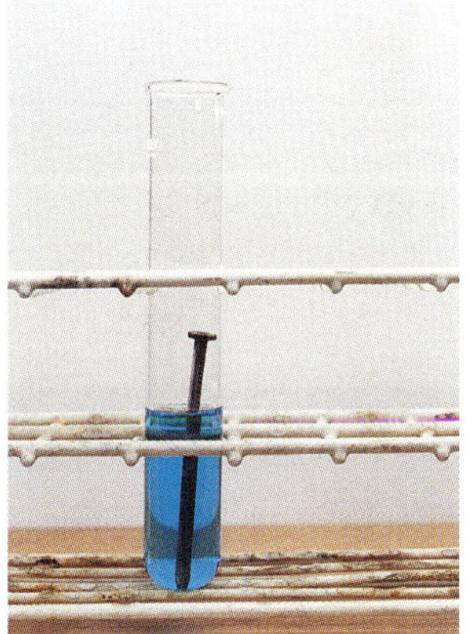

The non-metals carbon and hydrogen are often included in the reactivity series.

Metals above hydrogen will react with water/steam and dilute acids to give off hydrogen. Hydrogen gas will react with the oxides of the metals below hydrogen. For example, with copper oxide:

$$\text{hydrogen} + \text{copper oxide} \longrightarrow \text{water} + \text{copper}$$
$$H_2 + CuO \longrightarrow H_2O + Cu$$

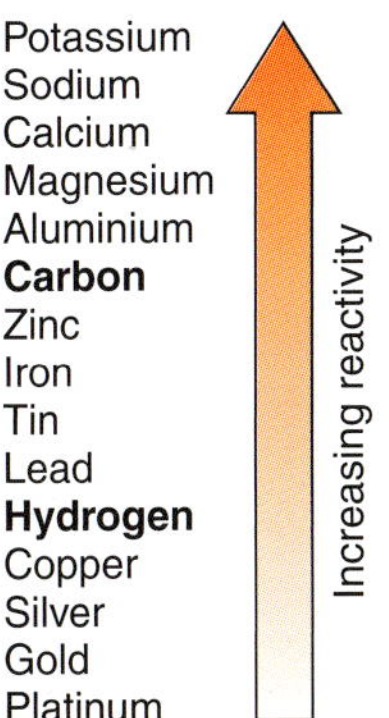

Figure 5.11
The reactivity series showing the non-metals carbon and hydrogen

Carbon also reacts with the oxides of the metals lower in the reactivity series. Carbon dioxide and the metal are formed. This method is used to extract the less reactive materials from their ores. The process is called **smelting**. Most metal ores contain the metal combined with oxygen. In these cases, the smelting process is quite straightforward. Some metal ores are not oxides but can usually be turned into oxides quite easily.

For metals above carbon in the reactivity series a different method of extraction must be used.

Tasks

1 Mercury and nickel are both metals. Pick one of these metals and find out some of its properties. Try to work out where it would go in the reactivity series.

2 Try to find out which of the metallic elements is the most reactive.

Topic Questions

1 For each of the following reactions use the reactivity series to work out what would happen. If there is a reaction, write down the products. If there is not a reaction, write 'No Reaction'.

a) copper and magnesium oxide
b) copper and silver nitrate solution
c) tin and copper sulphate solution
d) iron and potassium nitrate solution
e) iron and water (or steam)
f) silver and water (or steam)

2 Carbon dioxide is used in fire extinguishers because things don't burn in carbon dioxide. But magnesium metal will burn in carbon dioxide. Why is this? What are the products of this reaction?

3 Water gas is made by passing steam over red-hot carbon. What gases are present in water gas?

5.6 Getting iron by smelting

Co-ordinated	Modular
SA 11.4	SA 15
DA 11.4	DA 05

Iron is extracted from iron ore in a **blast furnace**. The commonest iron ore is called haematite. It contains iron oxide (Fe_2O_3). To get the iron from the iron ore the oxygen has to be removed. Removing oxygen from a compound is called **reduction**.

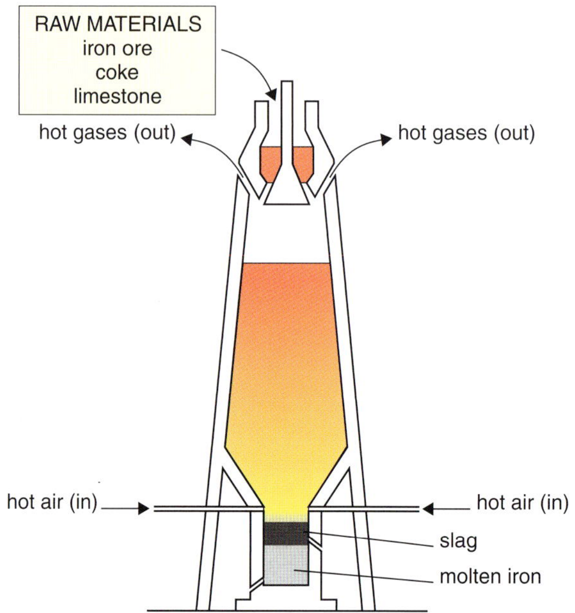

Figure 5.12
The inside of a blast furnace

Carbon is above iron in the reactivity series so carbon can be used to remove the oxygen. Carbon can be got quite easily from coal so it is quite cheap. This helps to make the iron inexpensive. The process of getting metals from their ores by heating them with carbon is called **smelting**.

Coal contains a lot of impurities. If coal is heated with no air present many of the impurities come off as gases. What is left is called **coke**. Coke is almost pure carbon. Iron ore contains a lot of impurities. One of these is sand. Sand is removed in the blast furnace by reacting it with limestone. This produces a substance called **slag**.

The iron ore, coke and limestone are put in the top of the blast furnace. Hot air is pumped in at the bottom. (The hot air blast is why it's called a 'blast furnace'.)

The carbon (coke) burns in the hot air making carbon dioxide.

$$C + O_2 \longrightarrow CO_2$$

This reaction gives out a lot of heat energy and keeps the furnace very hot.

The carbon dioxide reacts with more carbon to produce carbon monoxide.

$$CO_2 + C \longrightarrow 2CO$$

The carbon monoxide then reacts with the iron ore. Iron and carbon dioxide are produced

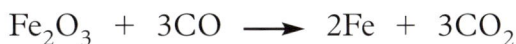

$$Fe_2O_3 + 3CO \longrightarrow 2Fe + 3CO_2$$

Some of the iron ore reacts directly with the coke. This reaction also produces iron and carbon dioxide.

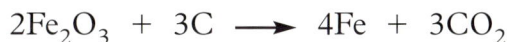

$$2Fe_2O_3 + 3C \longrightarrow 4Fe + 3CO_2$$

In these reactions the carbon monoxide (or the carbon) gain oxygen. This is called **oxidation**. So when iron oxide is reduced the carbon monoxide (or the carbon) is oxidised.

The temperature in the blast furnace is hot enough for the iron to melt. It collects at the bottom of the furnace as molten iron. The slag also melts. It collects on top of the molten iron.

Cast iron contains about 4% carbon. To make steel, most of the carbon has to be removed. Wrought iron contains almost no carbon, mild steel has about 0.5% carbon and hard steel has about 1% carbon.

Tasks

1 A lot of gas comes out of the top of the blast furnace. Why does this gas contain a lot of nitrogen? What other gases are present?

2 Find out how the air pumped into the blast furnace is heated.

Topic Questions

1 Name the three substances that are put into the top of the blast furnace.

2 Besides these three substances, what else is put into the blast furnace?

3 Write word equations for the chemical equations that take place in a blast furnace.

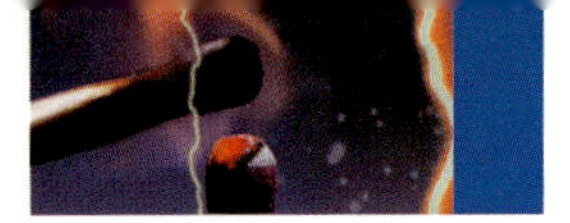

5.7	
Co-ordinated	Modular
SA n/a	SA n/a
DA 11.4	DA 05

Getting aluminium by electrolysis

Aluminium is more reactive than carbon, so smelting can't be used to make aluminium. **Electrolysis** is used to get aluminium from its ore.

Electrolysis only works with ionic compounds. All **ions** have an electric charge. Metals, like aluminium, have ions with a positive charge; non-metals, like oxygen, have ions with a negative charge.

In ionic solids the ions are held firmly in place (see Figure 5.13). When the ionic compounds are molten or dissolved in water the ions are free to move. If **electrodes** connected in a circuit are placed in the liquid the ions move. Positive ions are attracted to the negative electrode (called the **cathode**) and negative ions are attracted to the positive electrode (called the **anode**).

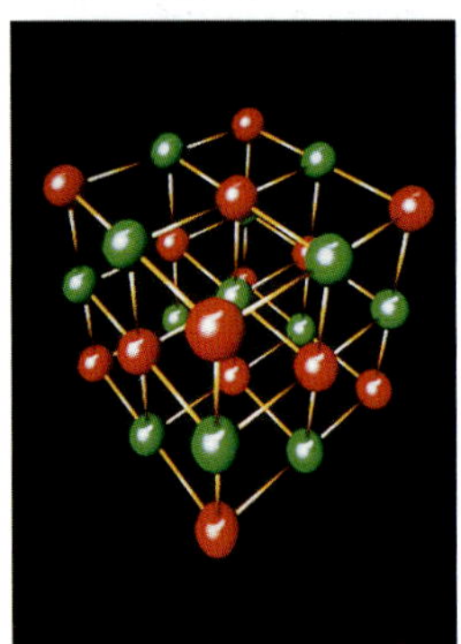

Figure 5.13
The structure of the ionic solid sodium chloride

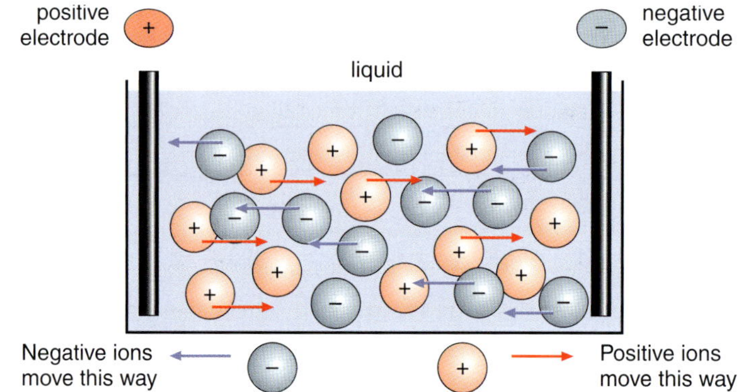

Figure 5.14
The movement of ions during electrolysis of a liquid

The ore of aluminium is called **bauxite**. It is aluminium oxide. It doesn't dissolve in water and has a very high melting point (2050°C) so it is very hard to electrolyse it. Bauxite docs dissolve in **cryolite**. Cryolite is also a compound of aluminium. It has a fairly low melting point. So aluminium is made by passing an electric current through a solution of bauxite dissolved in molten cryolite. Carbon is used to make the electrodes.

During electrolysis, positive aluminium ions go to the cathode. At the cathode the aluminium ions gain electrons and become aluminium metal. At the same time negative oxide ions go to the anode where they lose electrons and become oxygen gas. The oxygen gas attacks the carbon anode producing carbon dioxide. The anodes burn away and have to be replaced frequently.

Sir Humphry Davy discovered aluminium. He called it 'aluminum' – which is what Americans still call it.

Making pure copper by electrolysis

Electrolysis is also used to make pure copper. Impure copper is used as the anode. The **electrolyte** is a solution containing copper ions. The copper in the anode loses electrons to form copper ions. These ions dissolve in the electrolyte. At the cathode the opposite reaction takes place and copper ions from the electrolyte gain electrons to make copper metal. This copper sticks to the cathode.

The impurities are left behind. Some impurities dissolve in the electrolyte and stay there, other impurities don't dissolve and settle out as a **precipitate**.

Did you know?

Electroplating is used in industry to put a thin layer of metal onto the surface of another metal. The thin coating of metal may be used to protect the other metal from corrosion. It can also be used to put a layer of an expensive metal like gold over a cheap metal. The method is very similar to the one used to purify copper.

Tasks

1 Three ionic substances were electrolysed. Complete the table to show what substances were produced.

Electrolyte	Products produced at the anode	Products produced at the cathode
molten sodium chloride		
molten lead bromide		
molten potassium iodide		

2 Very pure copper is used to make electrical wires. Try to find out why it is important to use very pure copper.

3 Why is smelting, not electrolysis, used to make iron?

Topic Questions

1 Which two substances mentioned in this section are compounds of aluminium?

2 Why can't aluminium be made by smelting?

3 Why don't ionic compounds conduct electricity when they are solids?

4 Some molten lithium chloride was electrolysed. Chlorine was given off at the anode and lithium metal was produced at the cathode. If a solution of lithium chloride in water is electrolysed, the anode still gives off chlorine but the cathode gives off hydrogen. Why is this? (Clue: look back to Section 5.2)

5.8 Preventing metals from corroding

Co-ordinated	Modular
SA 11.4	SA 15
DA 11.4	DA 05

Most metals **corrode**. Corrosion occurs when a metal reacts with the oxygen and water in the air. The metal is oxidised to form the metal oxide. The higher a metal is in the reactivity series, the more rapidly it corrodes. Iron corrodes faster than most other transition metals.

Steel is an alloy of iron. Most types of steel corrode quite quickly. The corrosion of iron and steel is called **rusting**.

Iron and steel can be protected from rusting in several ways. Painting the metal is one way – the paint stops air and water reaching the metal. Stainless steel is an alloy. It contains the transition metal chromium and does not rust easily.

Did you know?

Some large steel structures like bridges have to be painted continuously. As soon as the painters reach the end of the structure they start at the beginning again.

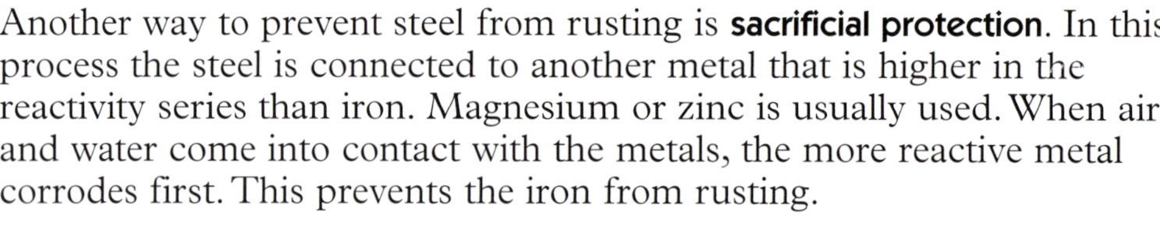

Another way to prevent steel from rusting is **sacrificial protection**. In this process the steel is connected to another metal that is higher in the reactivity series than iron. Magnesium or zinc is usually used. When air and water come into contact with the metals, the more reactive metal corrodes first. This prevents the iron from rusting.

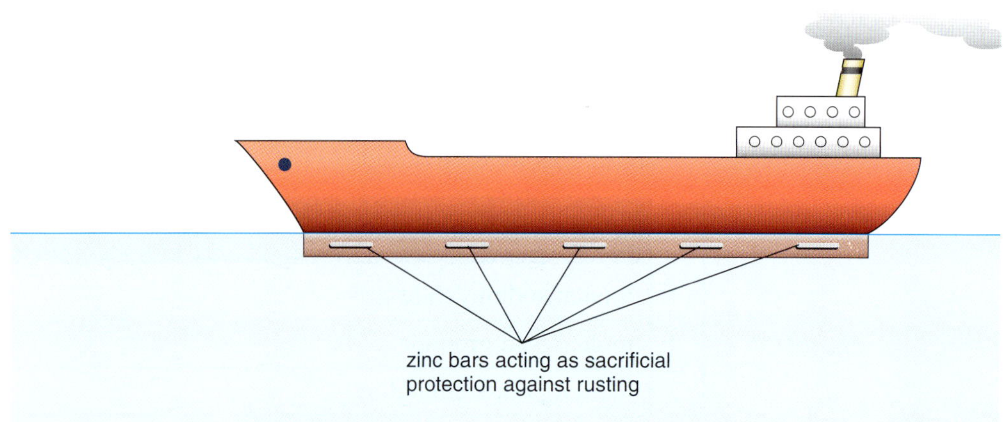

Figure 5.15
Sacrificial protection of a ship's hull

Figure 5.16
When the oxide coating has been removed, aluminium reacts rapidly with water

Figure 5.17
Alloy wheels are made from an aluminium/magnesium alloy

Aluminium does not corrode easily. This seems strange as aluminium is high in the reactivity series. The reason aluminium doesn't corrode easily is that it is self-protecting. When aluminium starts to corrode an oxide layer forms. This layer seals the surface of the metal. Oxygen and water can't get through this oxide layer so no more corrosion takes place. If this protective layer is removed, aluminium will react quite rapidly with water.

Like iron, aluminium can be made into alloys by adding small amounts of other metals. Magnesium is often used in aluminium alloys. These alloys are harder, stronger and stiffer than pure aluminium.

Tasks

1 Name some metals that do not corrode.

2 Aluminium can be protected by a process called anodising. Find out how aluminium is anodised.

Topic Questions

1 Give two uses for magnesium mentioned in the text.

2 There are several ways in which iron and steel can be protected against rusting. Name as many as you can.

3 The following sentences all contain errors. Correct the errors.

a) Iron can be protected from rusting by connecting it to a block of copper.
b) Aluminium does not rust very easily.
c) Stainless steel is an alloy of iron and magnesium.

4 Why is aluminium not very reactive?

5.9 Neutralisation

Co-ordinated	Modular
SA 11.4	SA 15
DA 11.12	DA 05

Many substances dissolve in water. The solutions produced may be acidic, alkaline or neutral. The pH scale is used to measure how acidic or alkaline a solution is. Acidic solutions have a pH less than 7, alkaline solutions have a pH greater than 7 and neutral solutions have a pH of exactly 7. The pH of pure water is 7, so water is neutral.

Figure 5.18 *Universal indicator colours at different pH values*

pH	1	2	3	4	5	6	7	8	9	10	11	12	13	14
Colour	← red →			orange	yellow		green	blue		← purple →				
	← ACID →							← ALKALI →						

When an acid reacts with an alkali they neutralise each other. The products of these reactions are always a salt and water.

acid + alkali ⟶ salt + water

For example:

hydrochloric acid + sodium hydroxide ⟶ sodium chloride + water

$HCl + NaOH \longrightarrow NaCl + H_2O$

Sodium chloride is the substance we call 'salt'. It should really be called 'common salt' because there are hundreds of different substances called salts. The reactions of some alkalis and acids are listed in the table below.

Figure 5.19
Table showing the reactions of some acids and alkalis

Acid	+	Alkali		Salt	+	Water
nitric acid	+	potassium hydroxide	⟶	potassium nitrate	+	water
HNO_3	+	KOH	⟶	KNO_3	+	H_2O
hydrochloric acid	+	lithium hydroxide	⟶	lithium chloride	+	water
HCl	+	LiOH	⟶	LiCl	+	H_2O
sulphuric acid	+	sodium hydroxide	⟶	sodium sulphate	+	water
H_2SO_4	+	2NaOH	⟶	Na_2SO_4	+	$2H_2O$
nitric acid	+	lithium hydroxide	⟶	lithium nitrate	+	water
HNO_3	+	LiOH	⟶	$LiNO_3$	+	H_2O
sulphuric acid	+	potassium hydroxide	⟶	potassium sulphate	+	water
H_2SO_4	+	2KOH	⟶	K_2SO_4	+	$2H_2O$

In each reaction the product is a salt and water. Notice that salts have a 'metal bit' and an 'acid bit'. The name of the 'metal bit' is the same as the name of the metal. But the name of the 'acid bit' is not the same as the name of the acid.

Did you know?

Splashes of alkali in your eye are often more dangerous than splashes of acid.

Hydro**chlori**c acid makes salts called **chlori**des.

Nitric acid makes salts called **nitr**ates.

Sulphuric acid makes salts called **sulph**ates.

Indigestion is caused by excess acid in the stomach. Indigestion tablets contain weak alkalis. These neutralise the acid.

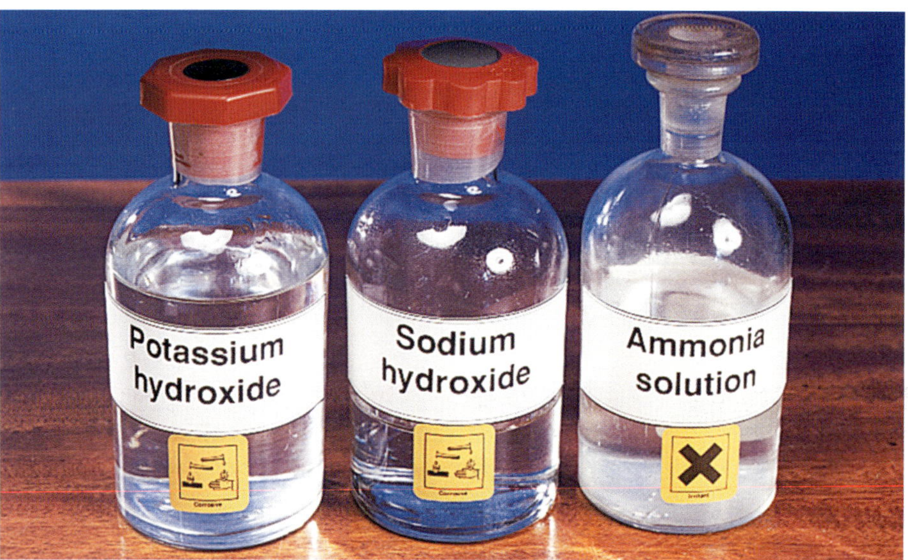

Figure 5.20
Pictures of bottles of common alkalis

Ammonia

Ammonia is not a metal, it is a gas. Ammonia is not even an element, it is a compound with the formula NH_3. But ammonia dissolves in water to form an alkaline solution. Ammonia solution can be neutralised by acids to produce ammonium salts. (Notice how the ending of ammonia is changed in its salts to make it sound like a metal.) If ammonia is neutralised by hydrochloric acid the salt formed is ammonium chloride.

Tasks

1 There are many different acids. Try to find out the name of the acid that:

a) is in oranges and lemons.
b) is in vinegar
c) makes rhubarb leaves poisonous
d) is in stinging nettles

2 Four common alkalis are mentioned in the text. Try to find out the names of some other alkaline solutions.

Topic Questions

1 The pH of some solutions were tested. The results are in the table.

Which solution is:

a) neutral?
b) strongly acidic?
c) the most alkaline?

Solution	pH
A	2.5
B	6.8
C	7.0
D	8.3
E	10.5

2 Name the acid used to make the following salts:

a) lithium sulphate
b) calcium chloride
c) magnesium nitrate

3 What is the name of the salt formed when ammonia is neutralised by nitric acid?

4 Alkalis are used in indigestion medicines. They neutralise the hydrochloric acid in the stomach. 'Milk of Magnesia' contains magnesium hydroxide. What salt is formed in your stomach if you take 'Milk of Magnesia'?

5 The word equation for the reaction between ammonia and an acid is:

ammonia + acid ⟶ ammonium salt

How does this reaction differ from the reaction of the common alkalis with an acid?

5.10 Making salts from alkalis

Co-ordinated	Modular
SA 11.4	SA 15
DA 11.2	DA 05

To make a pure salt from an alkali, exactly the right amount of acid must be added. When the right amount has been added the solution will be neutral. To tell when this has happened an **indicator** is used. Indicators have one colour in alkaline solutions and another colour in acidic solutions.

Indicator	Colour in acidic solutions	Colour in alkaline solutions
methyl orange	red	yellow
phenolphthalein	colourless	red
litmus	red	blue

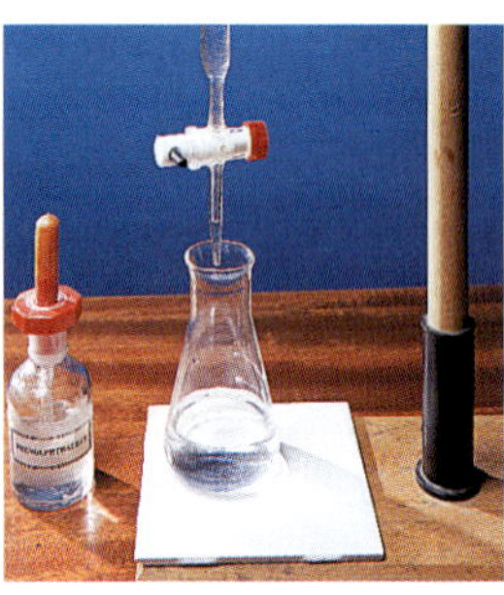

Figure 5.21

To exactly neutralise the alkali, a few drops of indicator are added to it. The acid is added carefully until one drop of acid causes the indicator to change colour. The solution is now neutral. It contains just the salt. There is no excess acid or alkali. But the salt is contaminated with the indicator. To get the pure salt the same volumes of acid and alkali used in the experiment are mixed. This time no indicator is added. The pure salt can be obtained from the solution by evaporating the water.

Task

Universal indicator has many colour changes. Can you work out how this is done?

Topic Questions

1 A pupil accidentally mixes some methyl orange and phenolphthalein indicators. What colour will the mixture be in:

a) an acidic solution?
b) an alkaline solution?

2 Explain why the mixture of indicators from in Question 1 would not be very useful in deciding if something were acid or alkaline.

3 Name the acid and the alkali used to make the following salts:

a) potassium chloride
b) sodium nitrate
c) lithium sulphate
d) ammonium chloride

5.11 What makes a substance acidic or alkaline?

Co-ordinated	Modular
SA 11.4	SA 15
DA 11.12	DA 05

Figure 5.22 gives the names and formulae of several acids.

Figure 5.22 *The names and formulae of some common acids*

Name of acid	Formula
Hydrochloric acid	HCl
Sulphuric acid	H_2SO_4
Nitric acid	HNO_3
Carbonic acid	H_2CO_3
Hydrofluoric acid	HF

Hydrogen is the only element that is in all the acids. So it is the hydrogen that gives acids their special properties. In acids the hydrogen exists as ions. The hydrogen ions have the symbol $H^+(aq)$.

Water contains hydrogen. It is not acidic because the hydrogen in water does not exist as ions. Natural gas (methane, CH_4) is also not an acid because the hydrogen atoms do not exist as ions.

Figure 5.23 *The names and formulae of some common alkalis*

Name of alkali	Formula
Lithium hydroxide	$LiOH$
Sodium hydroxide	$NaOH$
Potassium hydroxide	KOH
Calcium hydroxide	$Ca(OH)_2$

All alkalis contain the OH group. In alkalis this group exists as a hydroxide ion with the formula $OH^-(aq)$. So all alkalis contain hydroxide ions.

Tasks

1 The table lists some compounds that contain hydrogen. Decide from the information given which of these compounds could be acids.

	Formula of compound	Some information about the compound
A	HIO_3	pH of solution = 5
B	HCN	pH of solution = 6.5
C	N_2H_4	pH of solution = 7.8
D	C_2H_4	a gas that is insoluble in water

2 Most compounds that contain hydrogen are not acids. Water and methane are examples mentioned in the text. Most compounds with an OH group in them are not alkalis. Try to find some compounds that contain the OH group that are not alkalis.

Topic Questions

1 Copy the following table into your book and tick the correct boxes.

Name of substance	Could it contain hydrogen ions?	Is the pH less than 7?	Is the pH equal to 7?	Is the pH greater than 7?
Lithium hydroxide				
Citric acid				
Sodium chloride				
Ammonia solution				
Nitric acid				

2 What elements are present in the hydroxide ion?

3 What is the charge on the hydroxide ion?

4 The hydrogen ion can be written $H^+(aq)$.

a) What does the 'plus' sign (+) mean?
b) What does (aq) stand for?

5.12

Co-ordinated	Modular
SA n/a	SA n/a
DA 11.12	DA 05

Making salts of transition metals

Transition metals can also form salts. If the oxide or hydroxide of a transition metal reacts with an acid, a salt is produced. The only difference is that transition metal oxides and hydroxides are insoluble in water.

Metal oxides and hydroxides are called bases. Alkalis are bases that dissolve in water. The general reaction for an acid with a base is:

$$\text{acid} + \text{base} \longrightarrow \text{salt} + \text{water}$$

To make salts of transition metals the metal oxide (or hydroxide) is added to the acid. The mixture is stirred. As the metal oxide reacts it seems to dissolve. Sometimes it is necessary to heat the mixture to get the metal oxide to react. When the metal oxide has reacted, more is added. Once all the acid is neutralised no more metal oxide will 'dissolve'. The excess metal oxide is now filtered off. The solution contains just the salt of the transition metal.

Figure 5.24 *Making copper(II) sulphate*

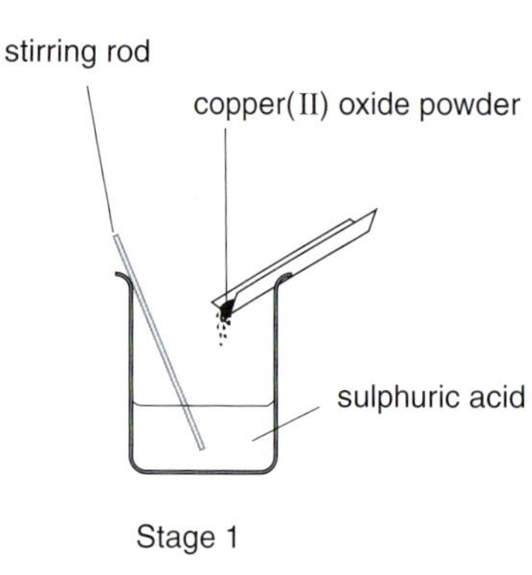

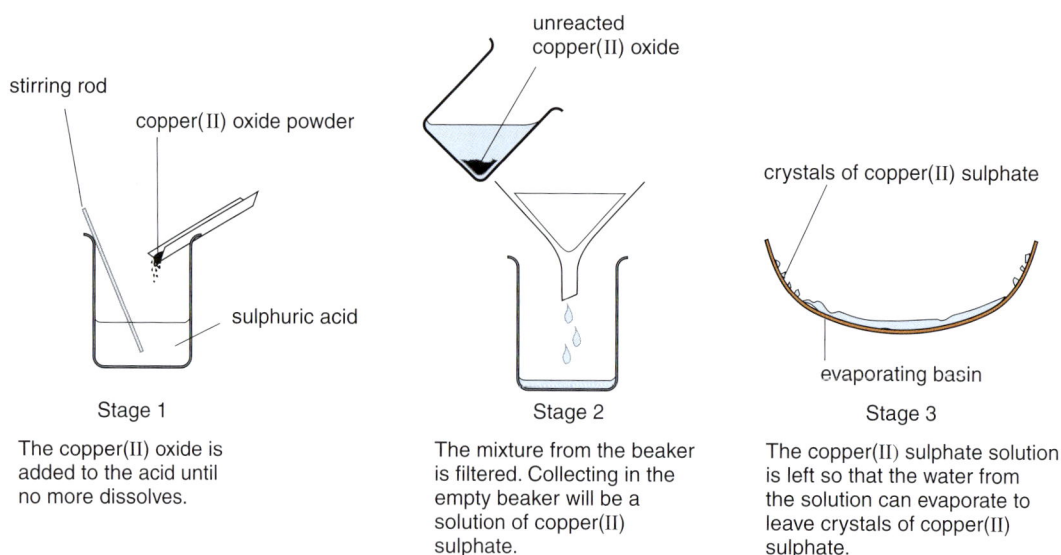

The copper(II) oxide is added to the acid until no more dissolves.

The mixture from the beaker is filtered. Collecting in the empty beaker will be a solution of copper(II) sulphate.

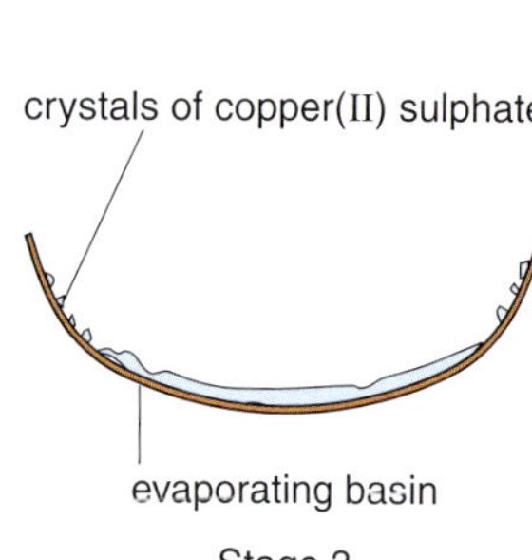

The copper(II) sulphate solution is left so that the water from the solution can evaporate to leave crystals of copper(II) sulphate.

Task

When making the salts of transition metals indicators are not needed. But even if they were needed they wouldn't be very useful in most cases. Why is this?
(Clue: look at Section 5.3)

Topic Questions

1 Look at the Periodic Table and pick four or five transition metals. Write down the names of the salts formed by the oxides of each of these metals with:

a) hydrochloric acid
b) nitric acid
c) sulphuric acid

2 Why is it not necessary to use an indicator when making salts of transition metals?

3 Draw a diagram of the apparatus you would use to filter the excess metal oxide from the solution.

Summary

- Relative atomic mass is the mass of an atom of an element compared to a hydrogen atom.
- The metals in Group 1 of the Periodic Table are very reactive. They react with water to form alkaline solutions. They react rapidly with air and are stored in oil.
- Group 1 metals are called alkali metals because they form alkaline solutions.
- Transition elements are found as a block in the middle of the Periodic Table.
- Transition elements are all metals. They are less reactive, have higher melting points and are stronger than the alkali metals.
- Transition metals have coloured compounds.
- Transition metals are used as catalysts in many industrial reactions.
- Useful metals like iron and aluminium are obtained from ores found in the earth.
- Some unreactive metals, like gold, occur in the earth as the element.
- The reactivity series is an arrangement of metals in order of their reactivity. (Carbon and hydrogen are not metals but are included in the table.)
- A metal (or carbon, or hydrogen) can displace a less reactive metal from its compounds.
- Smelting is the process of heating metal ores with carbon to extract the metal.
- Iron is extracted from iron ore by smelting in a blast furnace.
- Aluminium is too reactive to be extracted by smelting. Electrolysis is used instead.
- Electrolysis is used to purify copper.
- Most metals corrode in air.
- Iron can be protected against rusting (corroding) by sacrificial protection using either zinc or magnesium.
- Iron alloys (like stainless steel) also help protect iron from corrosion.
- Aluminium is protected from corrosion by a naturally forming layer of aluminium oxide on its surface.
- Acids (with a pH less than 7) will neutralise alkalis (with a pH greater than 7) to form a salt and water.
- All acids contain hydrogen ions ($H^+(aq)$) and all alkalis contain hydroxide ions ($OH^-(aq)$).
- Salts can be made by reacting a metal hydroxide or oxide with an acid.

Examination Questions

1 Put the following metals in order of their reactivity. Put the most reactive metal first.

aluminium	gold	iron	potassium

(4 marks)

2 In the Periodic Table which two of the following elements have been switched to make them fit better – argon, iron, neon, potassium, sodium? *(2 marks)*

3 Which two of the following elements are transition metals – aluminium, copper, hydrogen, iron, sodium? *(2 marks)*

4 Alkalis will react with acids. Their products always include a salt.

Which two of the following word equations are correct?

ammonia + hydrochloric acid → ammonium chloride + water

lithium hydroxide + sulphuric acid → lithium sulphate

potassium hydroxide + hydrochloric acid → potassium chloride + water

potassium hydroxide + nitric acid → potassium sulphate + water

sodium hydroxide + sulphuric acid → sodium sulphate + water

(2 marks)

5 Which two of the following statements are true about the method of manufacturing aluminium?

- The method used is electrolysis.
- The method used is smelting.
- The ore used is bauxite.
- The ore used is clay.
- The ore used is haematite. *(2 marks)*

6 Iron can be extracted from its ore in a blast furnace.

a) Iron ore is mainly:
A aluminium oxide
B iron metal mixed with aluminium oxide
C iron oxide
D iron sulphide *(1 mark)*

b) The substances put into the top of the blast furnace are:
A iron ore and coal
B iron ore, coke and coal
C iron ore, limestone and coal
D iron ore, limestone and coke *(1 mark)*

c) In the blast furnace iron ore is reduced. Which gas reduces the iron ore?
A carbon
B carbon dioxide
C carbon monoxide
D oxygen *(1 mark)*

d) The substances that collect at the base of the blast furnace are:
A iron and sand
B iron and slag
C iron ore and sand
D iron ore and water *(1 mark)*

Chapter 6

Earth materials

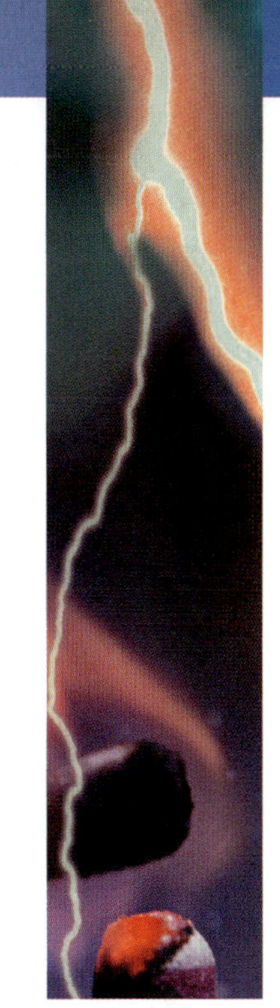

Key terms thermal decomposition • fossil fuels • mixture • evaporated • condense • molecules • volatile • viscosity • catalyst • oxidation • polymerisation • biodegradeable • recycled • organism • sedimentary rocks • microorganisms • magma • lava • weathering • seismic activity • lithosphere • tectonic plates

6.1 The uses of limestone

Co-ordinated	Modular
SA 11.4	SA 15
DA 11.5	DA 06

Limestone is found in many parts of the United Kingdom. It is used for building. In limestone areas you will find many houses are built of it.

Figure 6.1
A limestone cottage in the Cotswolds

Metal carbonates, like metal oxides and hydroxides, will neutralise acids. Limestone is mainly calcium carbonate ($CaCO_3$). So powdered limestone can be used to neutralise acidity in lakes or in the soil.

Most metal carbonates decompose when heated. This is an example of a **thermal decomposition** reaction:

$$\text{metal carbonate} \xrightarrow{\text{heat}} \text{metal oxide} + \text{carbon dioxide}$$

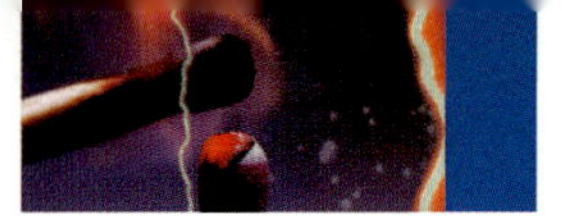

For limestone, the reaction is:

$$\underset{\text{limestone}}{\text{calcium carbonate}} \xrightarrow{\text{heat}} \underset{\text{quicklime}}{\text{calcium oxide}} + \text{carbon dioxide}$$

$$CaCO_3 \longrightarrow CaO + CO_2$$

The quicklime (calcium oxide) produced will react with water to produce slaked lime.

$$\underset{\text{(quicklime)}}{\text{calcium oxide}} + \text{water} \longrightarrow \underset{\text{(slaked lime)}}{\text{calcium hydroxide}}$$

$$CaO + H_2O \longrightarrow Ca(OH)_2$$

The reaction is very vigorous and gives out a lot of heat. Slaked lime is used to reduce the acidity of soil.

Did you know?

Different crops need soil with different pH values. Plants in the cabbage family do best on slightly alkaline soils but potatoes prefer a more acidic soil. Most soils have a pH between 6 and 8.

Figure 6.2
A rotary kiln for the manufacture of cement

Did you know?

Concrete will set hard even under water.

Cement is made by roasting powdered limestone with powdered clay in a rotary kiln.

Concrete is formed from a mixture of cement, sand, crushed rock and water. The cement powder reacts slowly with the water. As it sets it binds the other substances into a hard, solid mass.

Limestone is also used to make glass. To do this a mixture of limestone, sand and 'soda' (sodium carbonate) is heated strongly.

The different uses of limestone are summarised in Figure 6.3.

Figure 6.3
Some uses of limestone

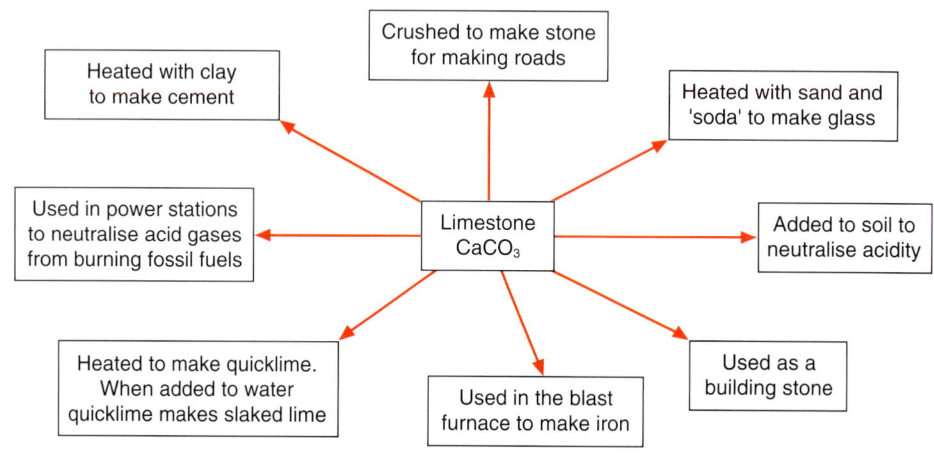

Did you know?

Glass beads have been found which were made in Egypt in about 2500 BC.

Tasks

1 List as many uses of limestone as you can. Try to find some that are not given in the text.

2 What name is given to a solution of slaked lime in water? What is this solution used for in the laboratory?

3 Powdered limestone can be used to neutralise acidity in lakes. How might the lake become acidic to start with?

Topic Questions

1 What are the chemical names and formulae of the following?

a) limestone
b) quicklime
c) slaked lime
d) 'soda' (used in making glass)

2 What is meant by thermal decomposition? Give some examples of thermal decomposition reactions.

3 As concrete sets people sometimes say the concrete is drying. Read the text carefully and explain why this is wrong.

6.2 Crude oil and hydrocarbons

Co-ordinated	Modular
SA 11.2	SA 15
DA 11.3	DA 06

Crude oil, coal and natural gas are found in the Earth's crust. They are **fossil fuels** and are the remains of plants and small animals that died more than 100 000 000 years ago. After they died, the plants and animals were covered with layers of sediment. The sediment stopped air getting to the dead organisms. Heat and pressure slowly changed these organisms to coal, oil and natural gas.

Crude oil is a **mixture** of a large number of compounds. Most of these compounds contain just carbon and hydrogen. For this reason they are called hydrocarbons. (Be careful not to confuse hydrocarbons with carbohydrates – they are quite different.)

Did you know?

About 2 tonnes of North Sea crude oil are used each year for every person in the UK.

The different hydrocarbons in crude oil can be separated by fractional distillation (see Figure 6.4). In this process the crude oil is **evaporated**. The vapour is passed into a tower that is hot at the bottom and cool at the top. As the vapour rises up the tower it cools. As it cools the hydrocarbons **condense**. Those with the highest boiling points condense first, near the bottom of the tower.

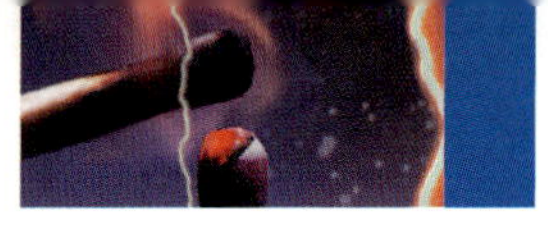

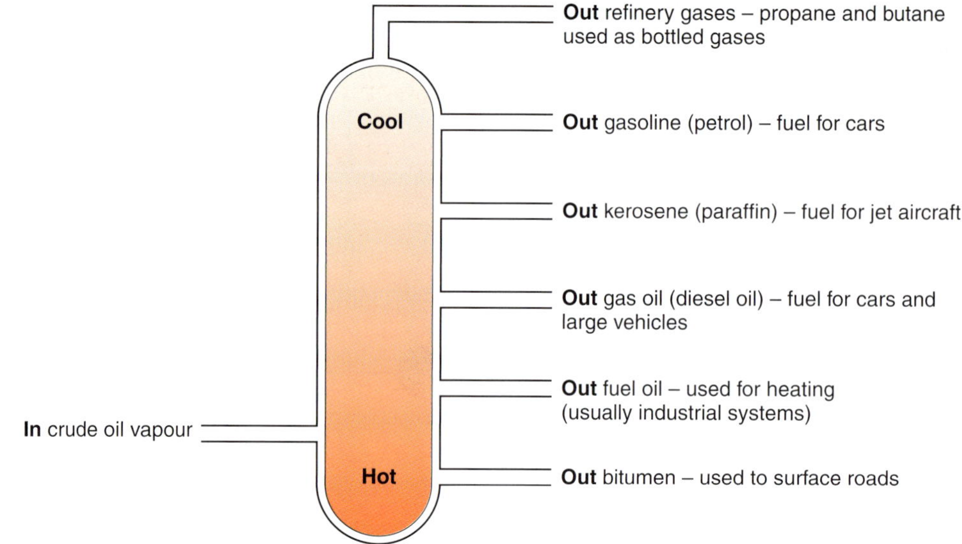

Figure 6.4 *Fractional distillation of crude oil*

Did you know?

North Sea crude oil contains about 2% refinery gas, 35% gasoline, 11% kerosene, 30% gas oil, 20% fuel oil and lubricating oil and 2% bitumen.

The hydrocarbons whose **molecules** have the most carbon atoms have the highest boiling points. So these hydrocarbons are the least **volatile**. Figure 6.5 compares the properties of some hydrocarbon fractions.

Figure 6.5 *Comparing the properties of some hydrocarbon fractions*

Number of carbon atoms	Boiling point / °C	Appearance	What happens when it burns
4 – 10	less than 100	Colourless liquid with low **viscosity**	Burns easily with almost colourless, smokeless flame
15 – 20	about 200	Yellow liquid, higher viscosity	Burns with yellow, smoky flame
over 50	over 350	Black, very viscous	Very difficult to burn

Figure 6.4 shows that the different hydrocarbon fractions have different uses. Those that burn well are used as fuels. So hydrocarbons with large molecules (many carbon atoms) are not very useful as fuels.

Tasks

1 Find out what the difference is between hydrocarbons and carbohydrates.

2 Find out the names of some of the hydrocarbons in crude oil. See what patterns you can find in their names.

Topic Questions

1 List the following crude oil fractions in order of their boiling points (lowest boiling point first) – bitumen, fuel oil, gas oil, gasoline, kerosene, refinery gas.

2 What two chemical elements are present in all hydrocarbons?

3 Which fossil fuel is not a hydrocarbon? Explain your answer.

4 Crude oil is a **mixture** of hydrocarbons. Hydrocarbons are **compounds** containing the **element** carbon.

Use these sentences to explain how **elements**, **compounds** and **mixtures** differ from each other.

6.3 Cracking hydrocarbons

Co-ordinated	Modular
SA 11.2	SA 15
DA 11.3	DA 06

The different fractions in crude oil are not in the proportion that we need them. To overcome this problem, a process called cracking is used. When large hydrocarbon molecules are 'cracked', they are broken down into smaller molecules. To do this the hydrocarbon is vaporised and passed over a **catalyst**. Thermal decomposition of the large hydrocarbon molecules takes place. The smaller molecules are more useful. Some of them can be used as fuels. They can also be used to make plastics.

Did you know?

Some Venezuelan crude oil contains over 85% fuel oil and only about 6% gasoline and kerosene.

Task

Try to find out what different types of crude oil contain.

Topic Question

A hydrocarbon molecule contains 6 carbon atoms. When it is cracked how many carbon atoms might there be in the products?

6.4 Fossil fuels

Co-ordinated	Modular
SA 11.2	SA 15
DA 11.3	DA 06

All fossil fuels contain carbon. Except for coal, fossil fuels also contain hydrogen. Other fuels, like wood, also contain carbon and hydrogen. None of these fuels are pure. The impurities include sulphur compounds. When fuels are burned they are **oxidised**. Figure 6.6 shows an experiment to show that burning hydrocarbons produce carbon dioxide and water vapour.

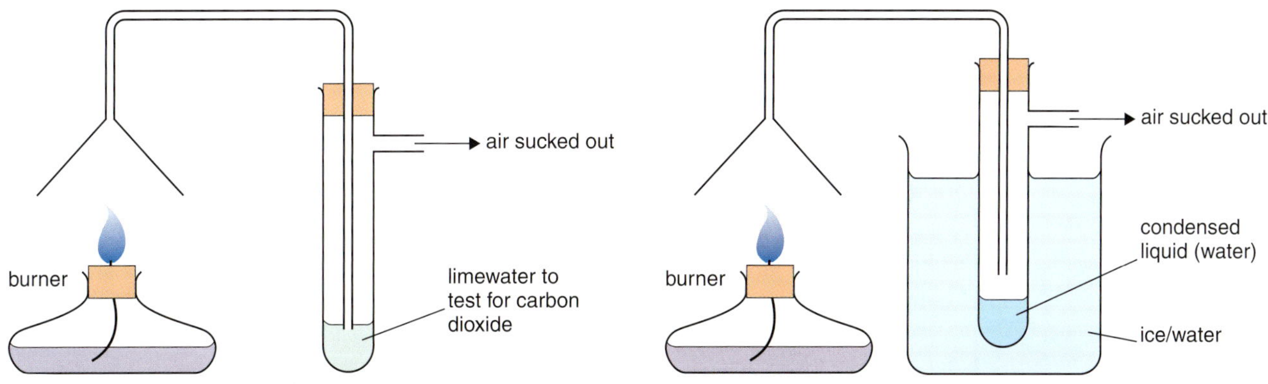

Figure 6.6
Apparatus to show that burning hydrocarbons produce carbon dioxide and water vapour

The carbon in the fuel is oxidised to carbon dioxide.

$$C \text{ (in fuel)} + O_2(g) \longrightarrow CO_2(g)$$

The hydrogen is oxidised to water (usually water vapour).

$$H \text{ (in fuel)} + O_2(g) \longrightarrow H_2O(g)$$

The sulphur in the impurities is oxidised to sulphur dioxide.

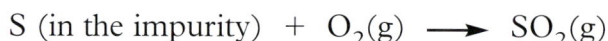

$$S \text{ (in the impurity)} + O_2(g) \longrightarrow SO_2(g)$$

Sulphur dioxide dissolves in water. If there is a lot of sulphur dioxide in the air it can make rain very acidic. Acid rain can contaminate lakes and rivers. (In some parts of Europe there are lakes with a pH of 2 or 3.) Very acidic water can kill fish and other aquatic creatures. It can also contaminate the land causing trees and other plants to die.

Figure 6.7
Trees damaged by acid rain

Carbon dioxide is also an acidic gas, but it is not as acidic as sulphur dioxide so is not such a big problem. Carbon dioxide in the air traps the Sun's energy. This can cause the atmosphere to warm up (the greenhouse effect). It may lead to global warming. Global warming is where the whole Earth gets warmer. If the Earth warms up even slightly, big changes to the weather may occur. One of these changes is that the polar ice caps get smaller.

Tasks

1 Try to find out how the amount of carbon dioxide in the atmosphere has changed in the past 50 years.

2 Now do the same for sulphur dioxide.

Topic Questions

1 Why do all fossil fuels contain carbon? (Clue: go back and read Section 6.2)

2 Many power stations burn fossil fuels. In doing so they produce sulphur dioxide. The sulphur dioxide produced will go into the atmosphere. How can the amount of sulphur dioxide coming from a power station be reduced? (Clue: go back and read Section 6.1)

6.5	
Co-ordinated	Modular
SA 11.2	SA 15
DA 11.3	DA 06

Plastics (polymers)

Ethene and propene are gases. They are made during the cracking process (Section 6.3). Each of these gases can be made to **polymerise**. In this process the small molecules join together to form very long molecules. The products are called polymers. They are generally known as plastics. Ethene polymerises to make poly(ethene). This plastic is better known as polythene. Propene polymerises to form poly(propene), which is often called polypropylene.

Poly(ethene) is used to make plastic bags and bottles. Poly(propene) is a stronger plastic. It is used to make ropes and milk crates.

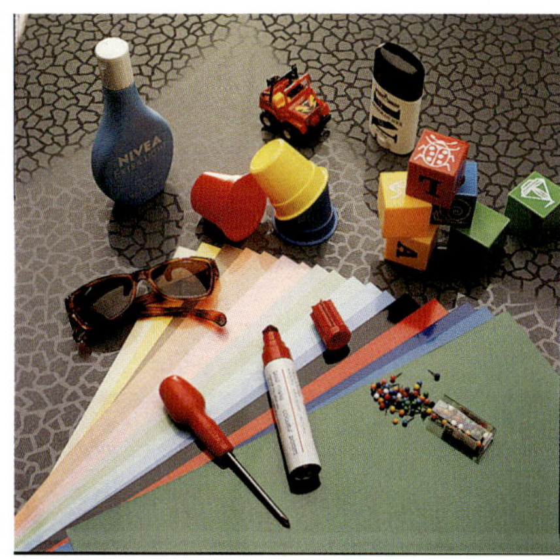

Figure 6.8
Some common plastic products

A lot of the rubbish we throw away is **biodegradable**. Paper, cardboard, wool, cotton and wood will all decompose in the ground. Even metals will gradually corrode. Plastics are not biodegradable and do not corrode. If they are buried in the ground they stay there. So it is not a good idea to throw plastics in the dustbin.

Many plastics can be **recycled**. Others can be burned in incinerators. Incinerators produce fairly cheap energy. But they can also produce toxic fumes.

Figure 6.9
A landfill site

Figure 6.10
An incinerator

Figure 6.11
Collecting skip for plastic bottles

Tasks

1 The gas ethene can be made into the plastic poly(ethene). See if you can find out what 'poly' means. (Clue: What is the general name for shapes like hexagons and octagons?)

2 Try to find out how much plastic waste people in your class throw in the dustbin each week. How much do they put out for recycling?

Topic Questions

1 a) Name two gases that can be used to make plastics.
b) Which two plastics are made from these gases?

2 Why is putting plastic in the dustbin bad for the environment?

3 What are the advantages and disadvantages of:

a) recycling plastic waste?
b) incinerating plastic waste?

6.6	
Co-ordinated	Modular
SA n/a	SA n/a
DA 11.9	DA 06

The Earth's atmosphere

Although air is a mixture its composition doesn't change very much.

Figure 6.12 *The composition of the air*

Gas	Amount
nitrogen	79%
oxygen	20%
other gases (mainly argon)	about 1%

The other gases include carbon dioxide, water vapour and the noble gases.

Did you know?

Water vapour is the only gas in the air where the amount depends on where you are. There is a lot more water vapour in the air in a tropical rain forest than in the middle of the Sahara desert.

Did you know?

The amount of argon in the air is about 1%. Argon is used in electric light bulbs. There is enough argon in a normal school laboratory for about 15 000 light bulbs.

In fact, the composition of the air hasn't changed much in the past 200 000 000 years. But before that it was very different.

In the first billion years of the Earth's existence there was a lot of volcanic activity. These volcanoes put gases into the atmosphere. Most of the gas was probably carbon dioxide but water vapour, ammonia (NH_3) and methane (CH_4) were also present. In those days there was very little oxygen in the atmosphere. In fact the atmosphere was similar to that found on Mars and Venus today.

As the Earth cooled the water vapour condensed to form oceans. A lot of the carbon dioxide dissolved in the oceans. Some marine **organisms** used this carbon dioxide to make calcium carbonate and form shells. As these organisms died the shells formed **sedimentary** rocks.

Other organisms died and were buried under layers of sediment. These organisms slowly turned into fossil fuels.

Once plants were established on the Earth they began to convert carbon dioxide to oxygen. Many of the **microorganisms** on the Earth were 'oxygen haters'. As the oxygen level went up they died.

The oxygen in the atmosphere also reacted with the ammonia and the methane.

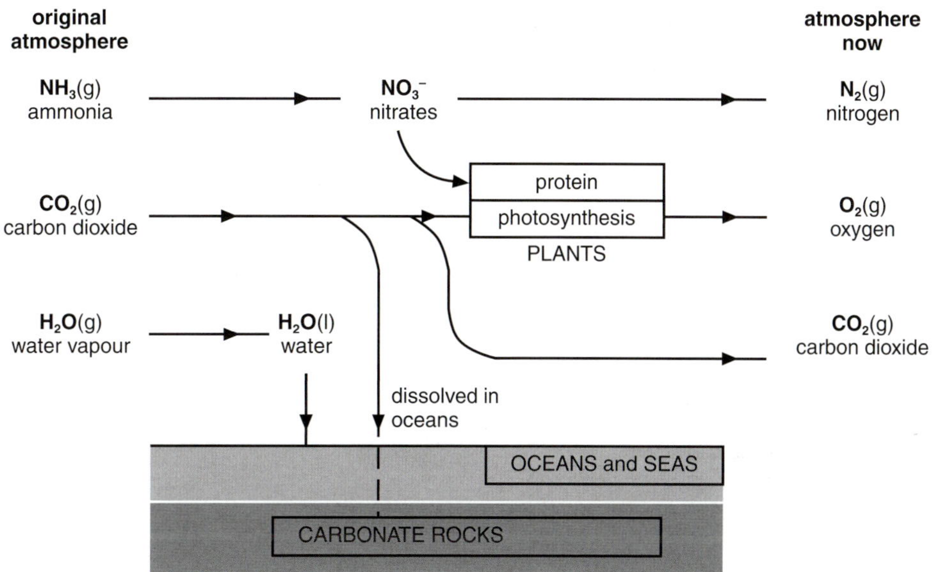

Figure 6.13 *The evolution of the atmosphere*

Tasks

1 Try to find out the amount of a) carbon dioxide, and b) argon in the atmosphere.

2 Find out how plants and animals keep the carbon dioxide levels in the atmosphere fairly constant. What could happen if the carbon dioxide content in the air increased? What changes could cause the carbon dioxide level to increase?

3 Try to draw a time line sketching the changes to the Earth's atmosphere.

Topic Questions

1 List the main gases in the atmosphere. Put them in order, starting with the one there is most of.

2 Write a paragraph explaining why the carbon dioxide content of the original atmosphere decreased.

3 The atmosphere now contains a lot of nitrogen. Which gas mentioned in the text could have been the source of this nitrogen?

4 Much of the original atmosphere was carbon dioxide. As the Earth cooled and oceans formed a lot of the carbon dioxide dissolved in the water. Use the data in the table to draw a graph of the way the solubility of carbon dioxide in water changes with temperature.

Temperature/°C	0	10	15	20	30	40	50	60
Vol of CO_2 (in cm^3) that dissolves in 1 cm^3 of water	1.71	1.19	1.02	0.88	0.66	0.53	0.44	0.36

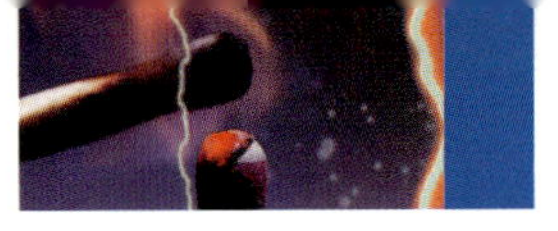

6.7	
Co-ordinated	Modular
SA n/a	SA n/a
DA 12.12	DA 06

The structure of the Earth

The Earth is nearly spherical. It is made up of layers rather like an onion.

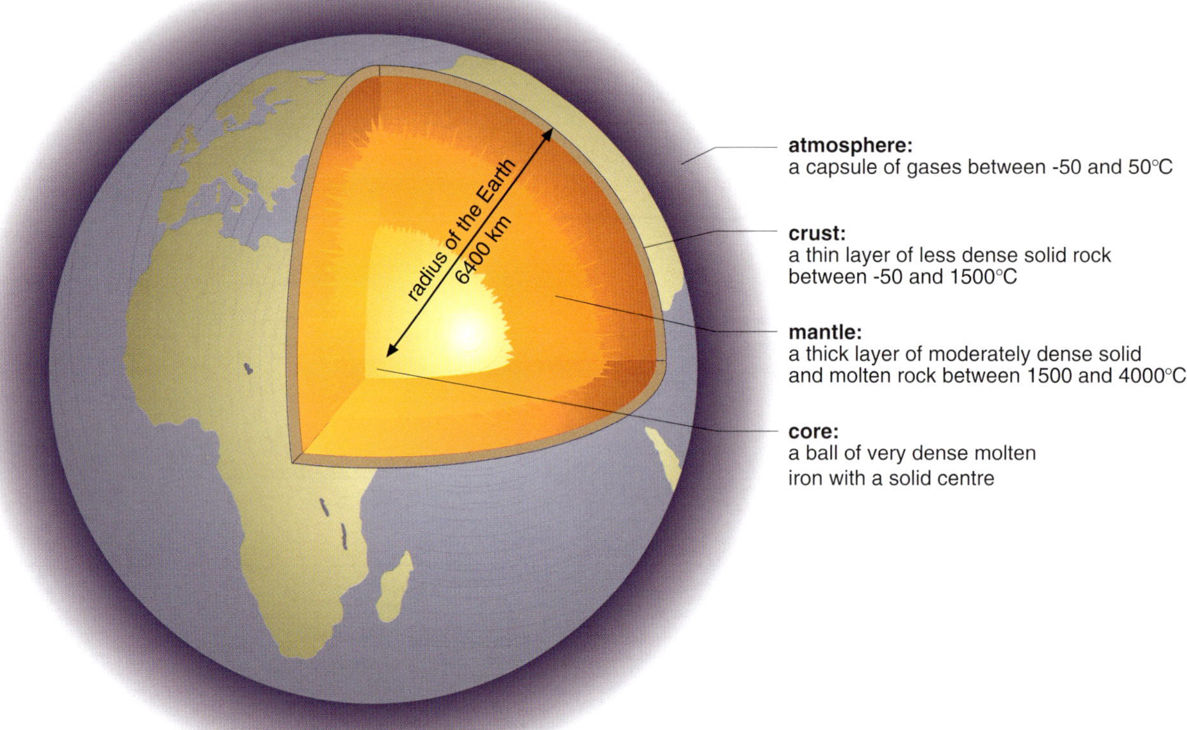

Figure 6.14

The outer layer is a thin crust. Underneath the crust is the mantle. It extends almost half way to the Earth's centre. The mantle is unusual – it behaves like a solid in most ways but it flows very slowly like a very viscous liquid.

In the centre of the Earth is the core. The core is made mainly of iron and nickel. Most of the core is molten but the central part is solid. The density of the core is high. Scientists know this because they know the overall density of the Earth and the density of the rocks in the crust. Because the crust has a much lower density than the whole Earth it means that the density of the core must be high.

Did you know?

The radius of the Earth at the equator is 6378 km, but at the poles it is 6357 km.

Task

Try to find out how scientists worked out the structure of the Earth. (Clue: look in Chapter 12)

Topic Questions

1 Draw a diagram of a section through the Earth. Label the different layers.

2 What evidence do we have to show that the Earth's mantle is molten?

3 The core of the Earth is mainly iron. On Earth the density of iron is 7.8 g/cm^3. The density of the Earth's core is over 10 g/cm^3. Explain why the core density is so high.

6.8	
Co-ordinated	Modular
SA n/a	SA n/a
DA 11.10	DA 06

The rock cycle and the rock record

Igneous rocks

Igneous rocks are formed when molten **magma** cools and solidifies. If this cooling takes place inside the Earth's crust the rocks are called **intrusive** rocks. Granite is an example of an intrusive rock. If the magma comes to the surface as **lava** from a volcano and cools on top of the Earth the rocks formed are called **extrusive** rocks. Basalt is an extrusive rock.

(a)

(b)

Figure 6.15 *The different sized crystals in a) granite and b) basalt*

Sedimentary rocks

Sedimentary rocks are formed in the ocean. They are made from mud, pieces of **weathered** rock or the shells of dead sea creatures which settle on the ocean floor. As the layer gets thicker the lower layers are compressed into rock. Mud becomes mudstone, sand forms sandstone and shells are compressed into limestone.

Figure 6.16 *Examples of sandstone and limestone*

Metamorphic rocks

Metamorphic rocks are formed when sedimentary rocks are changed by heat and pressure. This can occur if the sedimentary rock gets buried deep in the Earth. Slate is formed when mudstone is metamorphosed and limestone changes into marble.

Figure 6.17 *Examples of metamorphic rock*

The changes that produce these different types of rocks are summarised in Figure 6.18.

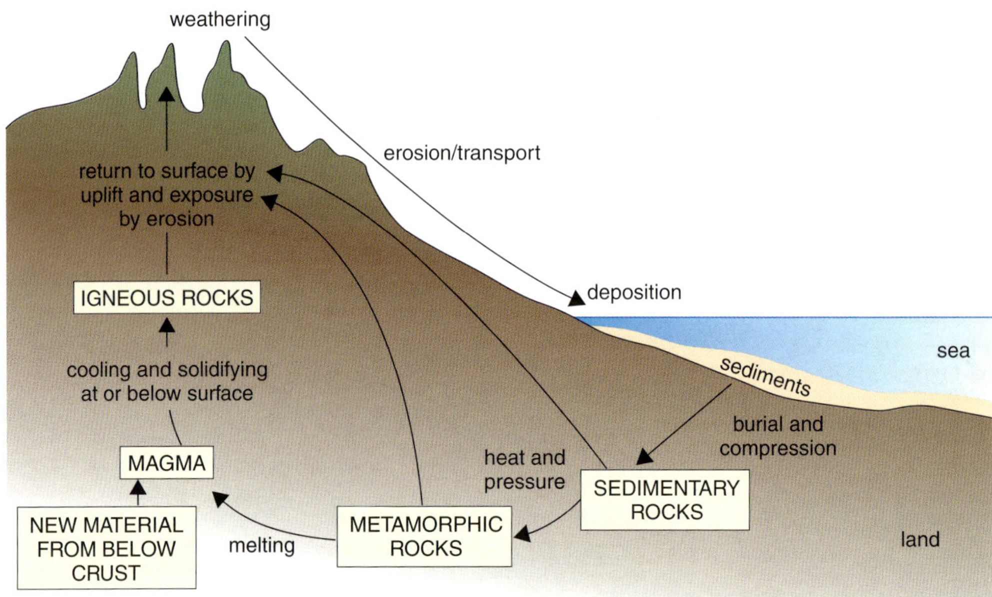

Figure 6.18
The rock cycle

The Earth's crust is unstable. It is continually moving. Over millions of years this movement can force rocks into mountain ranges like the Alps or the Himalayas. The sedimentary rocks formed beneath the ocean are raised out of the water.

Usually younger sedimentary rocks are on top of older ones but this is not always the case. Sometimes the huge forces can tilt the rock layers – even turn them upside down! Figure 6.19 shows some of the ways that layers of sedimentary rocks can be distorted. Sometimes it is possible to see ripple marks caused by waves in the ocean millions of years ago. The fossilised remains of animals and plants can be found in some sedimentary rocks.

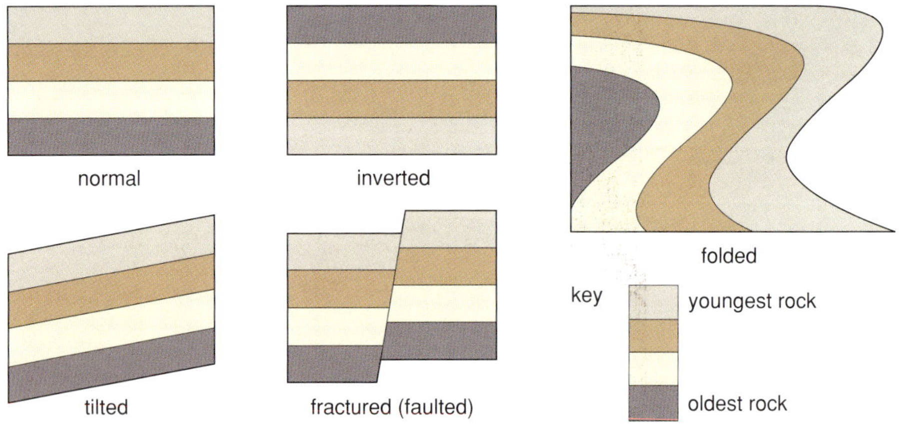

Figure 6.19
Sedimentary rock structures

Rocks that are exposed to the air will be gradually broken down. The processes that do this are weathering and erosion.

Figure 6.20
Examples of weathering and erosion

There are three main types of weathering:

- **Biological weathering** is where rocks are gradually broken down when the roots of plants grow down into cracks in the rocks.
- **Chemical weathering** occurs when water with dissolved carbon dioxide reacts with rocks. This type of weathering attacks limestone.
- **Physical weathering** is when frost and ice cause rocks to shatter.

Rocks can also be worn away by sand particles blown about by strong winds or by waves crashing against cliffs. This is called erosion.

Streams and rivers transport the particles of weathered or eroded rock. Eventually the particles reach the oceans where they settle out and new sedimentary rocks are slowly formed.

Predicting earthquakes

It is almost impossible to predict an earthquake. Scientists know where the most severe earthquakes are likely to occur but there is no way of predicting *when* they will happen. Earthquakes happen when stresses in the rocks build up and are suddenly released. It's a bit like blowing up a balloon. As you blow it up you know that eventually it will burst – but you can't tell exactly when.

Predicting volcanic eruptions

Volcanoes usually give warning signs before they erupt. The warning signs include:

- Increased **seismic activity**.
- Bulges in the ground caused by a build up of magma below ground.
- An increase in the amount of sulphur dioxide in the atmosphere.
- An increase in the temperature in the volcano.
- Smoke, fumes and traces of lava coming out of the volcano.

Sometimes these warning signs occur just a few hours before an eruption. At other times they might go on for years before the volcano erupts.

Tasks

1 Find out about the ways in which intrusive rocks and extrusive rocks differ.

2 Why do scientists find it so difficult to predict when earthquakes and volcanic eruptions are likely to happen?

Topic Questions

1 What is a) magma, and b) lava?

2 Name two igneous, two sedimentary and two metamorphic rocks.

3 Why are fossils only found in sedimentary rocks?

4 Explain how a sedimentary rock can be changed into a metamorphic rock.

5 Why is the process described in this section called the 'rock cycle'?

6.9 Earth movements

Co-ordinated	Modular
SA n/a	SA n/a
DA 11.10 12.13	DA 06

In the last section metamorphic rocks were mentioned. High temperatures and pressures change sedimentary rocks to metamorphic rocks. It is movement of the Earth's crust that causes these high temperatures and pressures. The same movements created the great mountain ranges like the Rockies, Andes and Himalayas.

The idea that the Earth's **lithosphere** moves is not new. Alfred Wegener first suggested it in 1915. He called the process 'continental drift'. He used the idea to explain an observation first made 300 years earlier – that South America and Africa seem to fit together like pieces in a jigsaw puzzle. The idea also explained why rock formations and fossils in South America were similar to those in Africa.

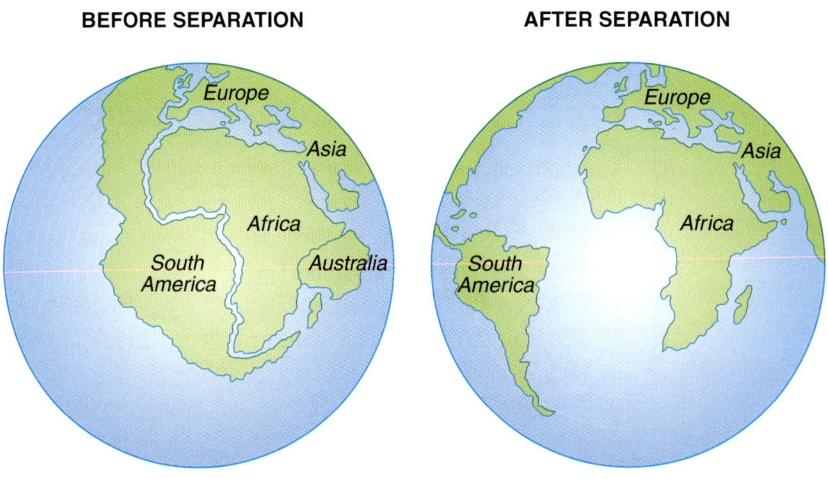

Figure 6.21 *Two maps made in 1858 showing how the continents of South America and Africa may once have fitted together*

Wegener tried to explain how the continents moved. For example, he thought that it was the spin of the Earth that forced the continents to separate. Other scientists easily proved that this could not be the case. The forces caused by the Earth's rotation were too small to move the continents.

Because his explanations were clearly wrong, Wegener's theory was not taken seriously. But in the 1950s, discoveries were made that supported Wegener's ideas. It was discovered that the Earth's surface was made of a number of separate 'plates' which moved. These plates were called **tectonic plates**. The plates move just a few centimetres each year. Most earthquakes and volcanic activity occur where these plates meet. About 50 years after Wegener's theory was suggested, it became accepted as fact.

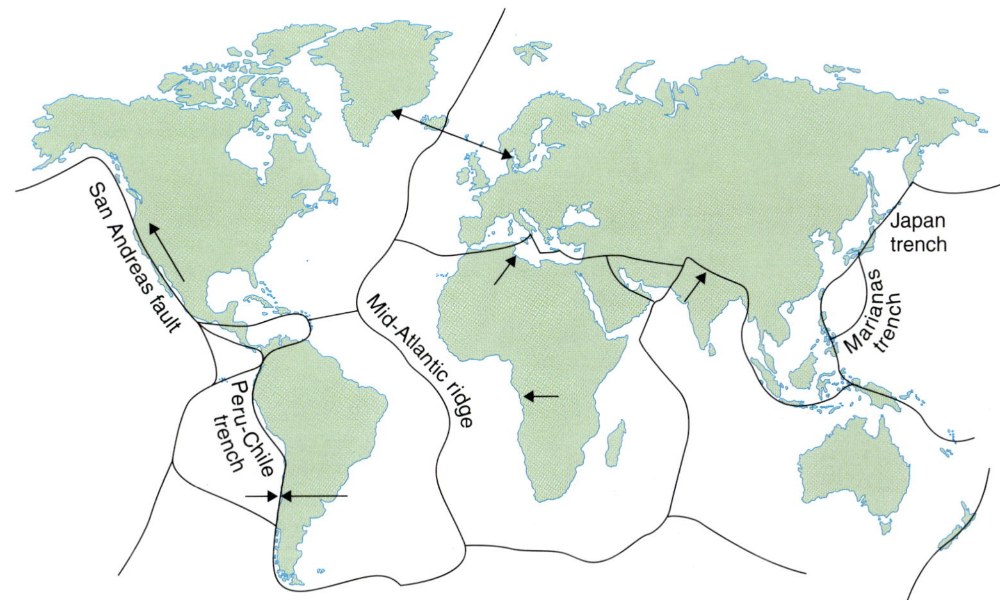

Figure 6.22 *The Earth, showing plate boundaries and direction of movement*

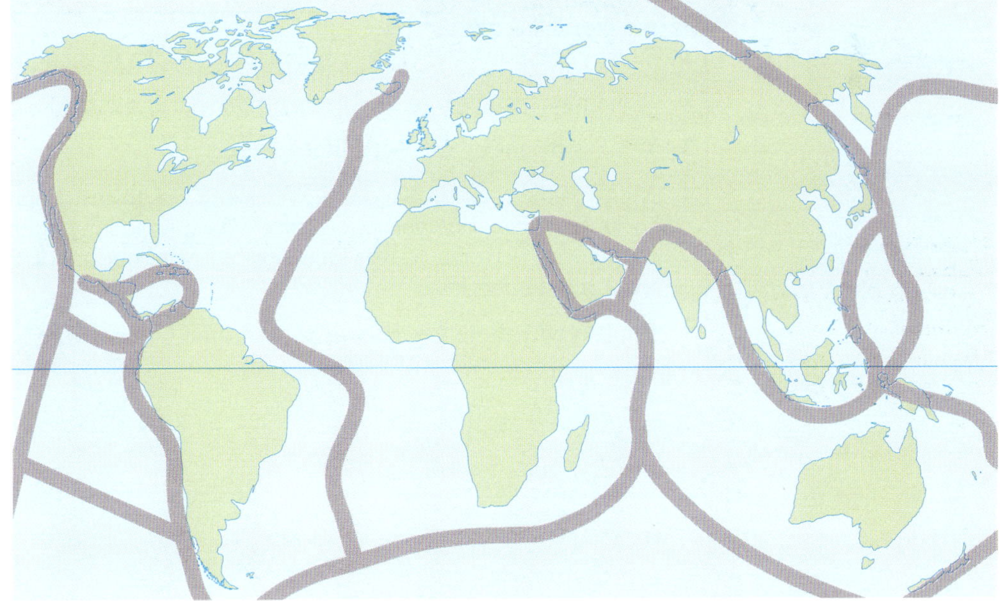

Figure 6.23 *Earthquake zones*

Tasks

Read the extract below then answer the questions.

In support of his presentation of the case for continental drift, Wegener marshalled an imposing collection of facts and opinions. Some of his evidence was undeniably cogent, but so much of his advocacy was based on speculation and special pleading that it raised a storm of adverse criticism. Most geologists, moreover, were reluctant to admit the possibility of continental drift, because no recognized natural process seemed to have the remotest chance of bringing it about. Polar wandering, the 'flight from the poles', and the westerly tidal drift have all been discarded as operative factors. Nevertheless, the really important point is not so much to disprove Wegener's particular views as to decide from the relevant evidence whether or not continental drift is a genuine variety of earth movement.

(Arthur Holmes, *Principles of Physical Geology*, 1944)

1 How many years after Wegner published his theory was this book written?

2 What do the following words mean?

a) cogent
b) advocacy
c) speculation
d) operative

3 a) Why was continental drift not recognised as even a possibility in 1944?
b) Why is continental drift accepted today?
c) What is the 'recognised natural process' that causes continental drift?
d) Is it possible that ideas we believe to be false now might be proven true in the future?

4 Why was there so much opposition to Wegener's views?

Topic Questions

1 What conditions will change sedimentary rocks to metamorphic rocks?

2 What name is given to the plates on the Earth's surface?

3 What is meant by the word lithosphere?

4 Many of the rocks in Scotland, Wales and Southwest England are igneous. Igneous rocks are formed by volcanic action. Volcanic action occurs mainly at the boundaries between plates. There are no plate boundaries within about 1500 km of the United Kingdom.

Explain why there are igneous rocks in the United Kingdom.

Summary

- Limestone has many uses. These include:
  - making roads
  - making buildings
  - making quicklime, glass and cement
  - neutralising acidity in soil and the waste gases from power stations
  - the manufacture of iron in a blast furnace.
- Fossil fuels include coal, natural gas and crude oil.
- When fossil fuels burn the produce carbon, water vapour and carbon dioxide.
- Fossil fuels contain impurities. During burning gases like sulphur dioxide are produced. Sulphur dioxide is the cause of acid rain.
- Crude oil is a mixture of hydrocarbons.
- Hydrocarbons are compounds containing *only* hydrogen and carbon.
- Crude oil can be broken down into different fractions by fractional distillation.
- Crude oil fractions include:
  - refinery gases
  - gasoline (petrol)
  - kerosene
  - gas oil (diesel oil)
  - fuel oil
  - bitumen.
- The different fractions of crude oil contain hydrocarbons with different numbers of carbon atoms in the molecule.
- Low boiling fractions are pale coloured, free-running and highly flammable. They contain hydrocarbons with few carbon atoms.
- High boiling fractions are dark coloured, viscous and hard to burn. They contain hydrocarbons with many carbon atoms.
- Large hydrocarbon molecules can be 'cracked' into smaller molecules, which can make them more useful.
- Cracking can produce substances like ethene and propene. These can be polymerised to make useful plastics.
- Plastics are not biodegradable. Disposing of them can be difficult. It is often better to recycle or incinerate them.
- The Earth's atmosphere contains nitrogen (20%), oxygen (30%) and traces of various other gases.
- The composition of the Earth's atmosphere does not change very much – except for water vapour.
- In the distant past, the Earth's atmosphere was very different in composition from what it is today.
- The Earth has a very thin, solid crust. Beneath the crust is the mantle and, in the centre of the Earth, the core.
- There are three main sorts of rock:
  - igneous
  - sedimentary
  - metamorphic.
- Igneous rocks are formed when molten rock from inside the Earth melts.
- Sedimentary rocks are formed by the deposition of the shells of sea creatures or rock fragments produced by weathering and erosion of existing rocks.
- Metamorphic rocks are formed when other rocks are subjected to high temperatures and/or pressures.
- The Earth's crust is made of tectonic plates. These plates are moving very slowly. At the boudaries between these plates are zones of high seismic activity. These are the areas where earthquakes and volcanoes are most common.

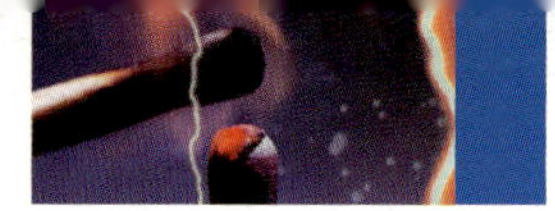

Examination Questions

1 The following substances are produced by the fractional distillation of crude oil. Choose the correct substance from the list in the box to complete the following sentences.

bitumen	gasoline	kerosene	refinery gas

The substance that comes off at the top of the fractionating tower is ______________. The substance used to surface roads is ______________. The substance that includes petrol for use in cars is ______________. The substance that includes diesel oil is ______________.

2 Which two of the following are NOT uses for limestone – making bricks, making cement, making detergents, making glass, making iron in a blast furnace? *(2 marks)*

3 Which two of the following statements are correct?
- Poly(ethene) is made from ethene and is used to make ropes.
- Poly(propene) is made from ethene and is used to make bottles.
- Poly(propene) is made from propene and is used to make ropes.
- Poly(propene) is made from propene and is used to make bottles.
- Poly(ethene) is made from ethene and is used to make bottles. *(2 marks)*

4 Which two of the following statements are true of 'cracked' hydrocarbons?
- They are less flammable.
- They are more useful.
- They have fewer carbon atoms.
- They have higher boiling points.
- They are darker in colour. *(2 marks)*

5 These questions are about rocks.

a) Cliffs can be broken down by the action of waves. This is called:
A biological weathering
B chemical weathering
C physical weathering
D erosion *(1 mark)*

b) The small pieces of rock from the cliff will settle at the bottom of the sea. The first thing they may become is:
A fossils
B igneous rock
C metamorphic rock
D sedimentary rock *(1 mark)*

c) Molten rock beneath the Earth's crust is called:
A igneous rock
B lava
C magma
D the core *(1 mark)*

d) When tectonic plates are in contact with each other, very high temperatures and pressures occur. This can turn:
A metamorphic rocks into sedimentary rocks
B sedimentary rocks into metamorphic rocks
C igneous rocks into sedimentary rocks
D igneous rocks into metamorphic rocks *(1 mark)*

6 When fuels are burned gases are produced.

a) Carbon dioxide causes:
A acid rain which pollutes lakes
B the greenhouse effect
C heavy rainfall
D very poisonous fumes *(1 mark)*

b) Sulphur dioxide causes:
A acid rain which pollutes lakes
B the greenhouse effect
C global warming
D smoke *(1 mark)*

c) Acidic gases can be removed by:
A bubbling the gas through water from a river
B dissolving them in concentrated sulphuric acid
C filters
D passing the gas over limestone *(1 mark)*

d) Water vapour is not produced if the fuel is:
A coal
B diesel oil
C fuel oil
D natural gas *(1 mark)*

7 These questions are about the composition of the atmosphere and how it has changed over many millions of years.

a) In the atmosphere now, which of the following gases can have a variable percentage?
A nitrogen
B oxygen
C water vapour
D noble gases *(1 mark)*

b) Which gas in the early atmosphere was mainly converted into nitrogen?
A ammonia
B carbon dioxide
C methane
D water vapour *(1 mark)*

c) Which gas disappeared from the atmosphere first as the Earth cooled?
A carbon dioxide
B nitrogen
C oxygen
D water vapour *(1 mark)*

d) Which gas was 'trapped' in the rocks as carbonates or fossil fuels?
A ammonia
B carbon dioxide
C methane
D water vapour *(1 mark)*

Chapter 7

Patterns of change

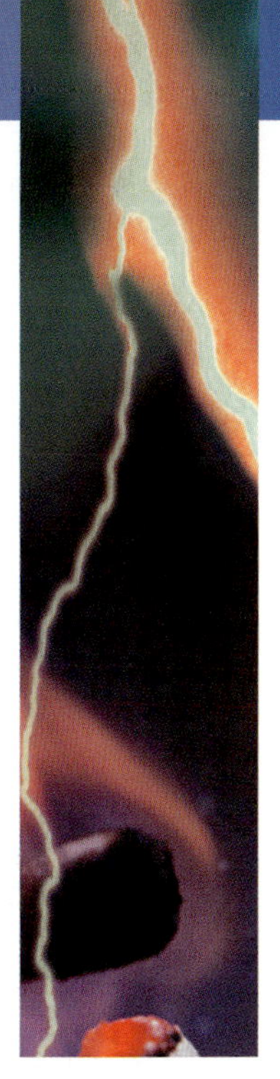

Key terms

rusting • reactants • precipitate • microorganisms • limewater • insoluble • denaturing • exothermic reaction • endothermic reaction • thermal decomposition • anhydrous • hydrated • reversible reactions • fertilisers • oxidation • neutralisation

7.1 Chemical hazard symbols

Co-ordinated	Modular
SA 11 intro	SA 16
DA 11 intro	DA 07

Most chemicals are harmful in some way. But some chemicals are especially hazardous. Containers with hazardous chemicals in have a special symbol fixed to them. The symbol indicates what the danger is with that chemical. The symbols are internationally recognised. Figure 7.1 shows these symbols and what they mean.

Figure 7.1
The hazardous chemical symbols

Oxidising
These substances provide oxygen which allows other materials to burn more fiercely.

Highly flammable
These substances easily catch fire.

Toxic
These substances can cause death. They may have their effects when swallowed or breathed in or absorbed through the skin.

Harmful
These substances are similar to toxic substances but less dangerous.

Corrosive
These substances attack and destroy living tissues, including eyes and skin.

Irritant
These substances are not corrosive but can cause reddening or blistering of the skin.

Tasks

1 For each of the hazardous chemical symbols try to find the name of at least one chemical that has that symbol.

2 Try to find some chemicals that have more than one hazardous chemical symbol on them.

Topic Questions

1 The *Oxidising* and *Highly flammable* symbols are quite similar to look at. In what way are the symbols different?

2 What is the difference between a *Toxic* substance and a *Harmful* substance?

3 How does a chemical with the *Oxidising* symbol differ in its behaviour from one with the *Highly flammable* symbol?

7.2	
Co-ordinated	Modular
SA 11.5	SA 16
DA 11.13	DA 07

How to change the speed of chemical reactions

Chemical reactions do not happen instantly. Some, like an explosion, happen very quickly. Others, like the **rusting** of iron, happen quite slowly.

The speed (sometimes called the rate) of a reaction depends on several factors. By controlling these factors the speed of a chemical reaction can be controlled. This is important in industry. If the speed of an industrial reaction can be increased then more product can be made in the same time. This reduces the cost of the product.

Even in the home it is sometimes useful to speed up chemical reactions. Cooking food causes chemical reactions to take place. Using a microwave or a pressure cooker speeds up these reactions so the food cooks more rapidly.

(a)

(b)

Figure 7.2
(a) Microwave cooker and (b) a pressure cooker

The following factors will increase the speed of a reaction:

- an increase in **temperature**
- an increase in the **concentration of reactants** that are in solution
- an increase in the **pressure** of gases
- an increase in the **surface area** of solid reactants
- the presence of a **catalyst**.

A **catalyst** is a substance that that increases the speed of a reaction but does not get used up in the reaction. Many industrial catalysts are very expensive. This doesn't matter because the catalyst can be used over and over again.

Did you know?

It is wrong to say that a catalyst doesn't take part in the reaction. It does take part – but it doesn't get used up. One way this can happen is if the catalyst reacts with one of the reactants A to form a temporary substance X:

$$A + catalyst \longrightarrow X$$

X now reacts with the other reactant B to produce the product AB and give back the catalyst:

$$X + B \longrightarrow AB + catalyst$$

The overall reaction is:

$$A + B \longrightarrow AB$$

Different reactions need different catalysts. There are many reactions for which no catalyst is available – so these reactions can't be speeded up with a catalyst.

Figure 7.3
A large industrial 'mesh' catalyst

Did you know?

Catalytic converters in cars change pollutants like the oxides of nitrogen and carbon monoxide into nitrogen and carbon dioxide. To do this expensive transition metals like platinum, palladium and rhodium are used as the catalysts. These catalysts are 'poisoned' by lead, so unleaded petrol must be used.

As a reaction takes place the amount of each reactant drops. At the same time the amount of the products formed increases. Figure 7.4 shows how these amounts change during a reaction.

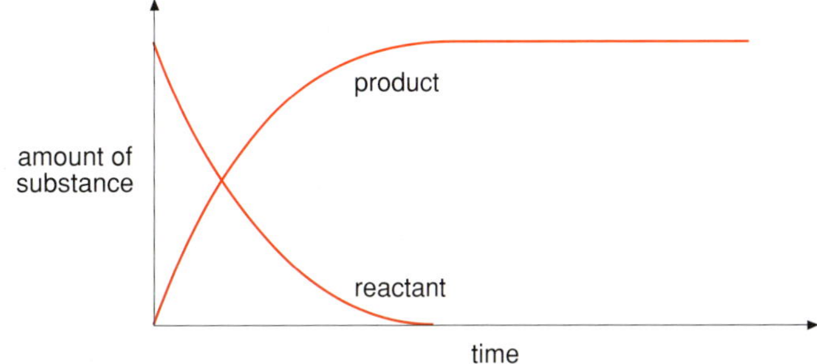

Figure 7.4
The change in the amounts of reactants and products during a reaction

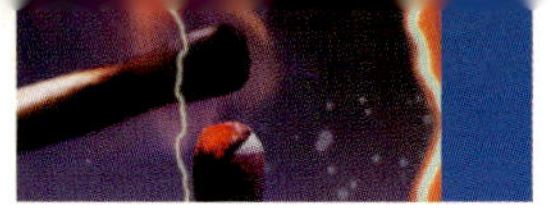

Notice the shape of the lines. There is a gradual change in their steepness during the reaction. When the reactants are used up the reaction is complete. When this happens the line is horizontal. This shows that no more reactant is being used. It also shows that no more product is being formed.

Graphs of this type can be used to see how the speed of a reaction changes.

Tasks

1 Find some examples of reactions that are a) very fast, and b) very slow.

2 Why does a pressure cooker allow food to be cooked more rapidly?

3 Why does breaking a piece of marble into smaller pieces increase the surface area of the marble?

4 Look at Figure 7.4. Why does the steepness of the graph get less as the reaction proceeds?

5 What are the catalysts used in the following reactions:

a) Decomposition of hydrogen peroxide ($2H_2O_2 \longrightarrow 2H_2O + O_2$)?
b) Manufacture of ammonia by the Haber process ($N_2 + 3H_2 \longrightarrow 2NH_3$)?
c) Manufacture of nitric acid from ammonia in the Ostwald process?
d) Manufacture of sulphuric acid using the Contact process?
e) Inside the catalytic converter in a car?

What do you notice about each of these catalysts?

Topic Questions

1 What is a catalyst?

2 What factors can speed up a chemical reaction? How could you slow down a chemical reaction?

3 The table below shows some results from an experiment to find out how fast a reaction went.

Time (mins)	0	1	2	3	4	5	6	7	8	9	10	11	12
Amount of product (cm^3)	0.0	3.5	7.0	10.0	13.0	15.5	18.0	20.1	21.9	23.5	25.0	26.0	26.5

a) Draw a graph of these results.
b) What volume of product would you expect to get after 15 minutes?
c) Sketch on the same axes what the graph would look like if the reactants had been more concentrated.

4 Why does an increase in the surface area of a solid reactant cause the reaction to go faster?

5 Catalysts help to reduce the costs of industrial processes. Why is it important for industry to keep its production costs as low as possible?

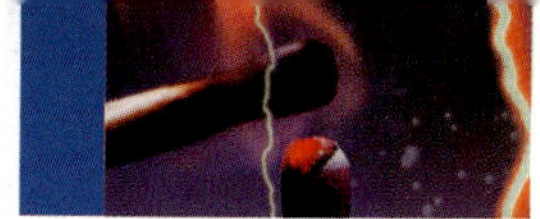

7.3	
Co-ordinated	Modular
SA 11.5	SA 16
DA 11.13	DA 07

Why the speed of chemical reactions changes

Changing the temperature to affect the speed of a reaction

For a chemical reaction to take place particles of the reactants have to collide with each other. The collision has to have enough energy to break down the reactant particles and let them make products.

The smallest amount of energy needed for a reaction to take place is called the **activation energy**. Particles that collide with less energy than the activation energy will not react.

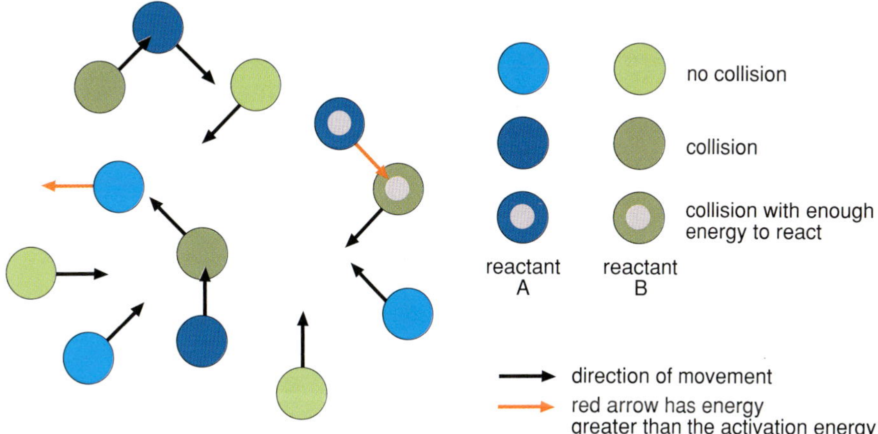

Figure 7.5
Reactants moving and colliding

The hotter a substance is the more energy its particles have and the faster they move. So, at higher temperatures it is more likely that collisions between particles will occur. It is also more likely that the energy of these collisions will be above the activation energy. So a greater percentage of the collisions will result in a reaction. This means that increasing the temperature of a reaction speeds the reaction up.

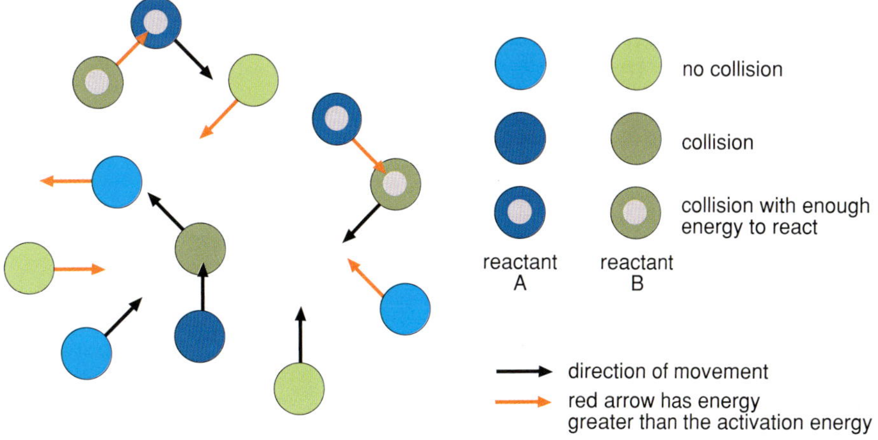

Figure 7.6
Reactants moving and colliding at a higher temperature

The effect of temperature on the speed of a reaction can be investigated using the apparatus shown in Figure 7.7.

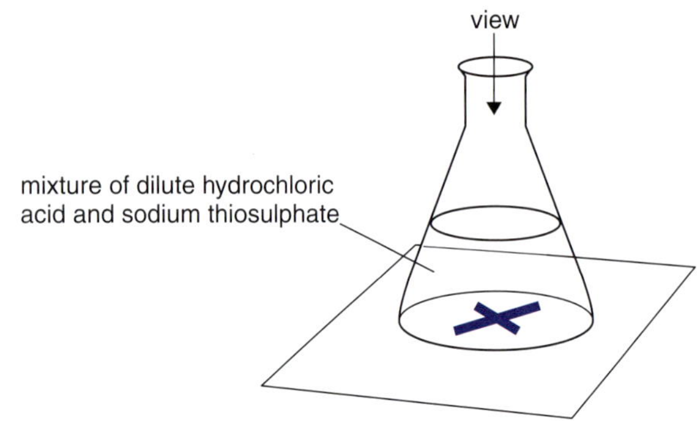

Figure 7.7
Apparatus to show how temperature affects the rate of a reaction

In this method a solution of sodium thiosulphate reacts with dilute hydrochloric acid. The reaction produces a **precipitate** of sulphur and the mixture slowly goes cloudy. The experiment is timed from when the reactants are mixed until the black cross is no longer visible. Figure 7.8 is a graph of some results obtained using this method.

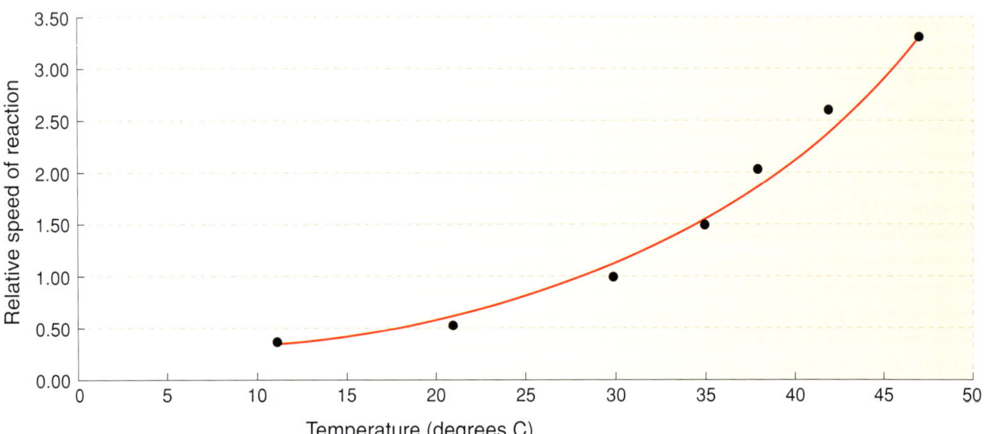

Figure 7.8
How the speed of the reaction between sodium thiosulphate and hydrochloric acid changes with temperature

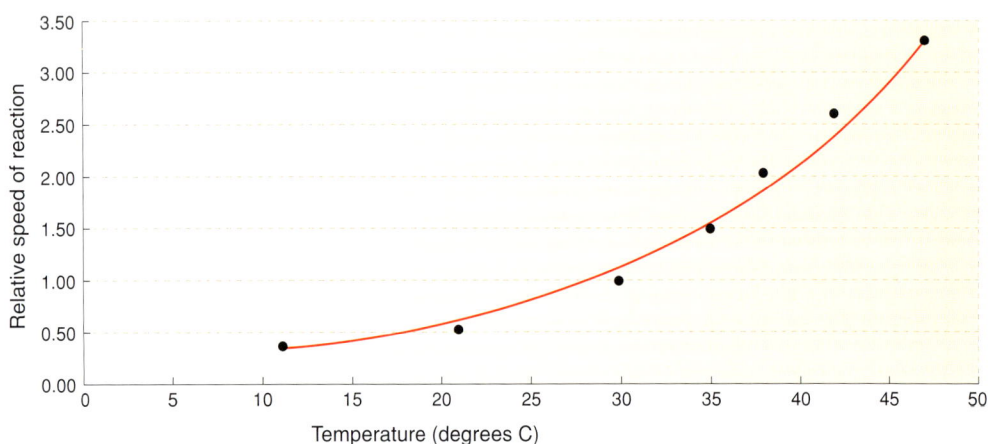

Changing the concentration of reactants to affect the speed of a reaction

Increasing the concentration of the reactants also increases the speed of a reaction.

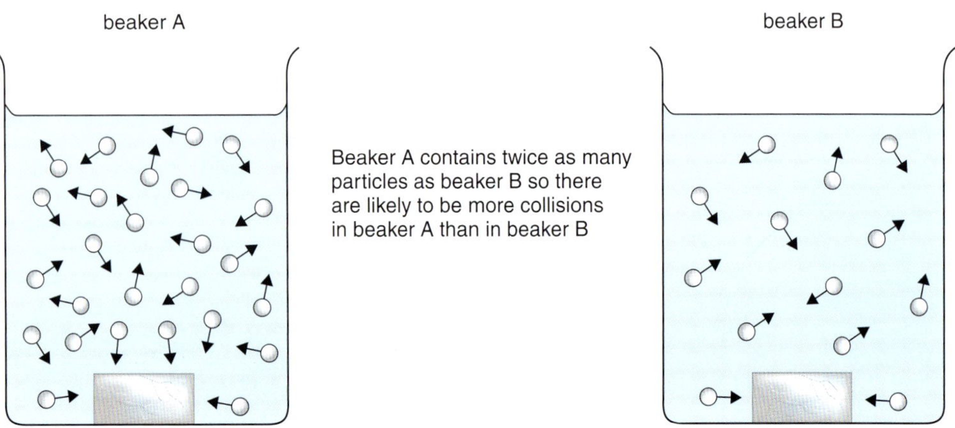

Figure 7.9
How increasing the concentration of a reactant increases the speed of a reaction

In Figure 7.9, particles of acid are shown reacting with a block of marble in a beaker. With more concentrated acid there are more particles. This means there will be more collisions between the acid particles and the block of marble, so the reaction is faster.

This method can be used to investigate the reaction between hydrochloric acid and marble chips. The apparatus used is shown in Figure 7.10.

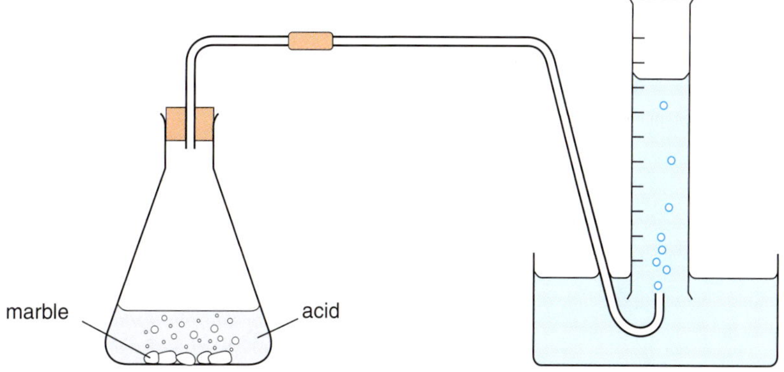

Figure 7.10 *Changing the concentration of a substance in solution*

The equation for the reaction is:

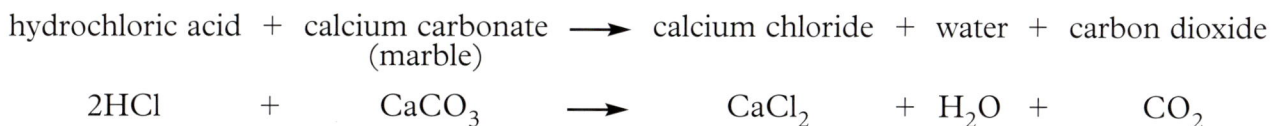

hydrochloric acid + calcium carbonate (marble) ⟶ calcium chloride + water + carbon dioxide

$$2HCl + CaCO_3 \longrightarrow CaCl_2 + H_2O + CO_2$$

In this reaction carbon dioxide gas is given off. The gas is collected in a gas jar. The volume collected is recorded every minute. Figure 7.11 is a graph of the results for this experiment.

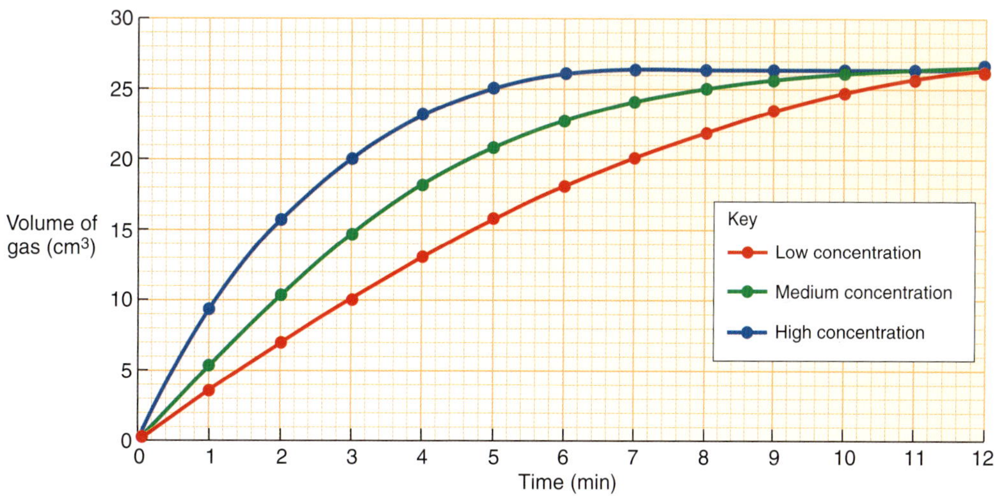

Figure 7.11 *The effect of concentration on reaction rate*

Using more concentrated acid means more carbon dioxide is given off each minute. This means the graph is steeper to start with. A steeper line means a faster reaction.

Changing the pressure of gases to affect the speed of a reaction

Increasing the pressure of reactions with gases pushes the particles closer together. The effect is exactly the same as increasing the concentration of a solution. So, with reactions involving gases, increasing the pressure speeds up the reaction.

Tasks

1 Draw a diagram like those in Figures 7.5 and 7.6 to show what would happen if *all* the particles had energy greater than the activation energy.

2 a) Dilute sulphuric acid solution reacts with zinc metal as follows:

sulphuric acid solution + zinc ⟶ zinc sulphate solution + hydrogen gas

$$H_2SO_4(aq) + Zn(s) \longrightarrow ZnSO_4(aq) + H_2(g)$$

The reaction is fairly slow and can be carried out easily in a school laboratory.

Plan an investigation into the speed of the reaction between sulphuric acid solution and zinc metal. You may choose to investigate any factor that affects the speed of the reaction.

b) The reaction is catalysed by copper sulphate solution. Plan an investigation to find out about the effect of copper sulphate on the reaction.

Topic Questions

1 Hydrochloric acid solution reacts with magnesium metal as follows:

hydrochloric acid + magnesium ⟶ magnesium chloride + hydrogen

$$2HCl + Mg \longrightarrow MgCl_2 + H_2$$

What has to happen to a **particle** of hydrochloric acid before it can react with the magnesium?

2 Why does the concentration of a reactant affect the speed of the reaction?

3 The table shows the results for an investigation into the speed of reaction between two solutions, A and B.

Concentration of A (concentration of B constant)	0.1	0.2	0.3	0.4	0.5	0.6	0.7	0.8
Relative speed of reaction	1.00	1.85	2.90	4.05	5.10	6.20	7.30	8.55

Concentration of B (concentration of A constant)	1.0	1.5	2.0	2.5	3.0	3.5
Relative speed of reaction	1.00	1.60	2.05	2.65	3.10	3.55

a) Plot separate graphs for each of these sets of results.
b) Estimate the relative speed of the reaction for solution B at a concentration of 0.5.
c) How would the rate of the reaction change if the concentrations of **both** A and B were doubled?

4 The table shows the results for an investigation into the speed of a reaction in which the temperature was altered.

Temperature /°C	22	31	43	51	63
Relative speed of reaction	1.0	1.7	3.8	6.8	13.2

a) Plot a graph of these results.
b) What is the relative speed of the reaction at 56°C?
c) Estimate the speed of the reaction at (i) 10°C and (ii) 70°C.

5 Use the idea of activation energy to explain why hydrochloric acid and marble chips start to react as soon as they are mixed, but natural gas and air do not react until they are ignited.

7.4

Co-ordinated	Modular
SA 11.6	SA 16
DA 11.14	DA 07

Fermentation

Chemical reactions take place in the cells of all living things. These reactions produce new substances.

Microorganisms can be used to produce useful substances. Yoghurt is made from milk by using bacteria. These bacteria change lactose (the sugar found in milk) into lactic acid.

Yeast is a microorganism. Yeast cells will change sugar into alcohol and carbon dioxide. The process is called fermentation:

$$\text{sugar} \longrightarrow \text{alcohol} + \text{carbon dioxide}$$

$$C_6H_{12}O_6 \longrightarrow 2C_2H_5OH + 2CO_2$$

Fermentation is used to produce alcoholic drinks like beer and wine. The reaction is quite slow so the mixture has to be left for quite a long time.

Did you know?

To make alcoholic drinks like whisky and brandy, the alcoholic mixture has to be distilled. The process is quite complicated because some of the substances produced by fermentation are toxic. It is both illegal and very dangerous to try to produce spirits by distilling home brewed wine.

Did you know?

Sparkling wines (like champagne) are bottled before the fermentation has stopped. The carbon dioxide is trapped in the bottle and makes the wine fizzy.

Fermentation is also used to make bread rise. In this process it is the carbon dioxide that is used. As bubbles of carbon dioxide are produced they cause the bread to rise. The bread is only left for a short time to rise so very little alcohol is produced. Baking the bread kills the yeast so the reaction stops.

Figure 7.12 *Some products of fermentation*

Testing for carbon dioxide

If carbon dioxide is bubbled through **limewater** the solution goes cloudy. This is because carbon dioxide reacts with calcium hydroxide to produce calcium carbonate. Calcium carbonate is **insoluble** in water so a precipitate forms.

Tasks

1 Find out the alcohol content of some well known drinks.

2 Find out why the fermentation reaction stops when the alcohol content reaches about 15%.

Topic Questions

1 What is the laboratory test for carbon dioxide?

2 What are the two products of the fermentation reaction? Give one use for each of these products.

3 There are many different compounds that can be called alcohols. What is the correct chemical name for the alcohol produced in the fermentation reaction?

7.5 Reactions using enzymes

Co-ordinated	Modular
DA 11.14	DA 07
SA 11.6	SA 16

Most chemical reactions go faster when the temperature rises. Reactions in living cells also speed up as the temperature increases. But once the temperature goes above about 45°C the reactions slow down again. This is because reactions in living cells use catalysts called enzymes. Enzymes are protein molecules. They work by what is sometimes called the 'lock and key' process.

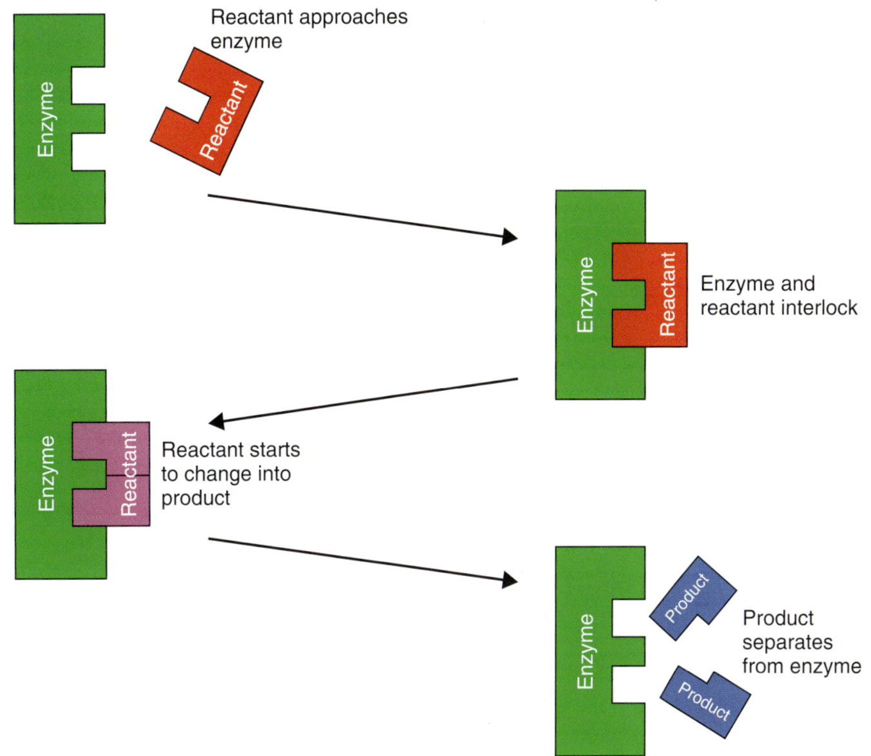

Figure 7.13
How enzymes catalyse reactions

In this process the shape of the enzyme molecule is important. The reactant 'fits' neatly into spaces in the enzyme molecule. The enzyme then breaks the reactant and releases the separate parts. But as an enzyme molecule gets hotter it vibrates more. If the vibrations are too great it can be difficult for the reactant molecule to fit into the spaces in the enzyme. So enzyme catalysts become less effective above about 45°C. If the temperature gets too high the enzyme may be permanently damaged. This is called **denaturing**.

Enzyme catalysts are very efficient. They speed up the reactions inside cells considerably. But in many cases a particular enzyme will only break down molecules of one specific compound.

Enzyme catalysts are also very sensitive to changes in pH. For example, an enzyme that works well in slightly acidic conditions usually won't work well in alkaline conditions.

Task

Try to find out the name of an enzyme in the body that works best in acidic conditions.

Topic Questions

1 Why are enzyme catalysts temperature sensitive?

2 Explain how enzymes speed up a chemical reaction.

Uses of enzymes

Enzymes are used in many processes. In the home they are present in biological detergents. These detergents contain protein-digesting and fat-digesting enzymes.

Many industrial reactions can be carried out more efficiently using enzymes.

Figure 7.14
Some uses of enzymes

Use	Enzyme	Action	Comments
Biological washing powders	• Lipases • Proteases	• Remove fat stains • Remove stains caused by proteins	Will work well at low temperatures
Baby foods	Proteases	Start to digest some of the protein	Makes it easier for the baby to digest the food
Sweeteners	Isomerase	Converts the glucose into a sugar called fructose	Fructose is much sweeter than glucose. This means less sugar needs to be used to make food sweet. It is useful for making special foods for slimmers
Making sugar syrup	Carbohydrases	Changes starch from plants into sugars like glucose	Makes glucose syrup used in in many soft drinks

To be useful in industry enzymes must be:

- very pure
- all the same (there must be no other enzyme present)
- able to be produced in large quantities.

The enzymes are obtained from microorganisms. To produce enzymes that are identical the microorganisms must be identical. If any of the microorganisms are the wrong type then the enzyme produced could be contaminated with other enzymes. If there is any contamination then the product the enzyme helps to produce may be impure.

Enzymes are produced in large fermentation tanks like those in Figure 7.15. In these tanks the pH and temperature are carefully controlled. This helps to ensure that unwanted enzymes are not produced.

Many industrial reactions need expensive equipment to create high pressures. Others need high temperatures – so energy costs are high. Some of these reactions can be made to take place at normal temperatures and pressures using enzymes. This saves industry a lot of money.

Figure 7.15
Fermentation tanks

Task

Try to find out how normal detergents work. In what ways are biological detergents better than normal detergents?

Topic Questions

1 Give some uses of enzymes.

2 Which sort of enzyme will turn glucose into fructose? What advantage does fructose have over glucose?

3 What are the advantages and the disadvantages of using enzymes as catalysts in industrial processes?

7.6 Energy transfer in chemical reactions

Co-ordinated	Modular
SA n/a	SA n/a
DA 11.16	DA 07

Burning is a chemical reaction. Energy is given out as heat (or, more accurately, energy is transferred from the reaction to the surroundings).

Reactions that give out energy are called **exothermic reactions**. Usually the energy is given out as heat.

Did you know?

The word *exothermic* comes from two word stems. *Exo* meaning 'out' (as in *ex*hale or *ex*it) and *therm* meaning 'heat' (as in *therm*ometer or *therm*al underwear).

Did you know?

Chemical reactions don't always transfer energy as heat. In an explosion some of the energy is transferred as light and some as sound. In a battery the energy is transferred as electricity.

Some exothermic reactions need to be heated to get them started. For example, lighting a Bunsen burner. This happens because the energy of the particles of gas and oxygen (in the air) is below the activation energy (see Section 7.3). The flame gives the particles enough energy to get them started.

Some reactions need to be heated all the time. Reactions that need to take in heat are called **endothermic reactions**. **Thermal decomposition** reactions are endothermic.

Figure 7.16
(a) Exothermic reaction
(b) Thermal decomposition of copper carbonate – an endothermic reaction

Task

Go back and review some of the reactions you have met in your science course so far. Try to find out if they are exothermic or endothermic.

Topic Questions

1 What is meant by an exothermic reaction?

2 Lighting a Bunsen burner requires energy. Once the Bunsen burner has been lit, the flame gives out heat. Is the reaction exothermic or endothermic? Explain your answer.

3 Two solutions A and B are mixed. An exothermic reaction takes place. What happens to the temperature of the mixture? Explain your answer.

7.7	
Co-ordinated	Modular
SA n/a	SA n/a
DA 11.15	DA 07

Reversible reactions

Once a reaction has finished you can't usually make it go backwards. For example, if you burn a piece of paper you can't turn the ash back into paper again. But there are some reactions that can be made to go backwards.

If you heat blue hydrated copper sulphate crystals, water is given off and white **anhydrous** copper sulphate is produced:

hydrated (blue) copper sulphate ⟶ anhydrous (white) copper sulphate + water

If you now add water to the anhydrous copper sulphate, it turns back to blue hydrated copper sulphate:

anhydrous (white) copper sulphate + water ⟶ hydrated (blue) copper sulphate

Reactions like this are called **reversible reactions**.

To show that a reaction is reversible a double-headed arrow ($\rightleftharpoons$) is used. So the reaction above can be written as:

hydrated (blue) copper sulphate $\rightleftharpoons$ anhydrous (white) copper sulphate + water

Testing for water

Anhydrous copper sulphate can be used as a test for water. A drop of liquid is added to anhydrous copper sulphate. If the liquid contains water the white powder turns blue. The test does **not** prove that the liquid is pure water.

To make anhydrous copper sulphate the blue crystals have to be heated. This means the reaction below is endothermic:

hydrated (blue) copper sulphate ⟶ anhydrous (white) copper sulphate + water

The reverse reaction must then be exothermic.

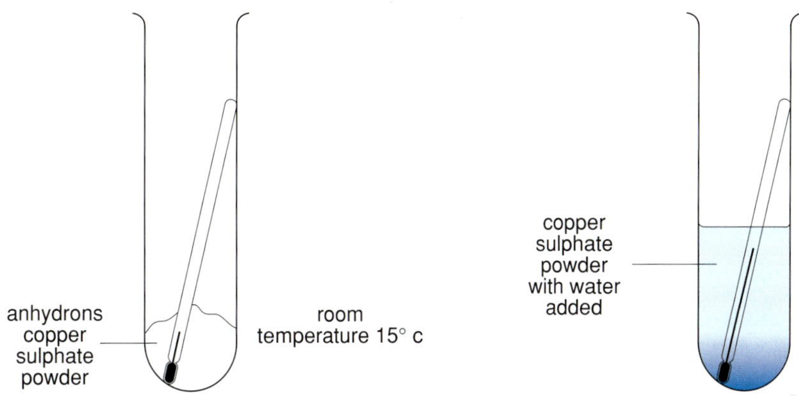

Figure 7.17 *The reaction between water and anhydrous copper sulphate is exothermic*

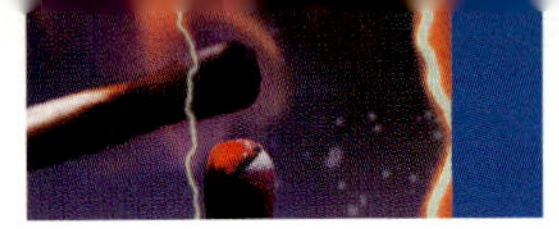

Did you know?

The amount of energy given out in a reversible reaction is exactly the same as the energy needed to make the reaction go in reverse.

If this were not the case it would be possible to get 'free' energy.

Suppose in the reaction:

$$A + B \rightleftharpoons C + D$$

more energy was given out than was used up in the reverse reaction:

$$C + D \rightleftharpoons A + B$$

If A and B were reacted, only some of the energy would be needed to make the reaction go backwards. The extra energy could then be used for something else. If the process were repeated over and over there would be an endless source of 'free' energy.

Another example of a reversible reaction is the one between ammonia gas and hydrogen chloride gas:

$$\text{ammonia gas} + \text{hydrogen chloride gas} \rightleftharpoons \text{ammonium chloride}$$

$$NH_3(g) + HCl(g) \rightleftharpoons NH_4Cl(s)$$

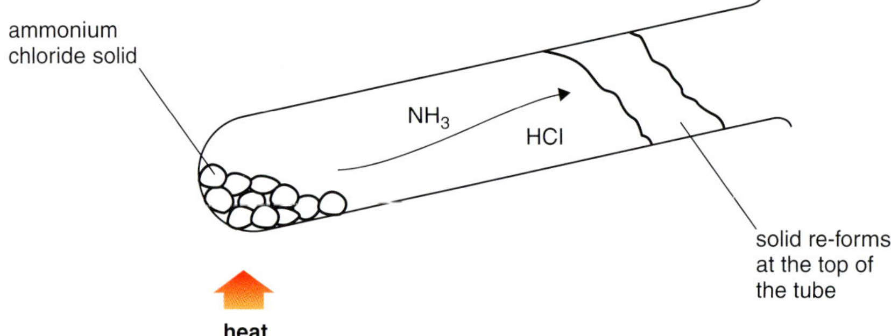

Figure 7.18
The action of heat on solid ammonium chloride

Tasks

1 Find some reactions that are reversible. (Clue: look in this book to start with)

2 Which of the following processes are **not** reversible?

- baking a cake
- boiling water
- making a cup of instant coffee
- defrosting a refrigerator or deep-freeze.

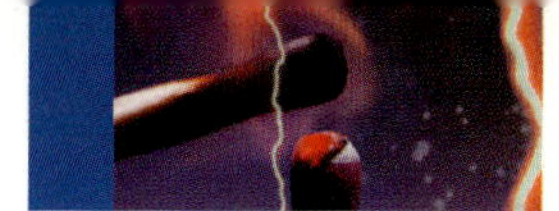

Topic Questions

1 What is the symbol used for a reversible reaction?

2 What do the words hydrated and anhydrous mean?

3 Anhydrous cobalt chloride is blue. It reacts with water to form pink, hydrated cobalt chloride. The reaction is exothermic and reversible.

a) Give one possible use of blue cobalt chloride.

b) How could you make anhydrous cobalt chloride from the hydrated substance?

4 If ammonium chloride is heated it breaks down into hydrogen chloride and ammonia gases:

$$NH_4Cl(s) \rightleftharpoons NH_3(g) + HCl(g)$$

endothermic	exothermic	neutralisation
oxidation	reduction	thermal decomposition

a) Which of the words in the box describes the forward reaction?

b) Which of the words in the box describes the reverse reaction?

7.8 The Haber process and nitrate fertilisers

Co-ordinated	Modular
SA n/a	SA n/a
DA 11.6	DA 07

Plants need nitrogen to help them grow. Air is almost 80% nitrogen but most plants are not able to make use of this nitrogen. They have to get their nitrogen from nitrogen-based **fertilisers**. Farmers use fertilisers to increase the yield of crops. The Haber process is a way of converting nitrogen from the air into ammonia. Ammonia can then be used to make other useful substances – including fertilisers.

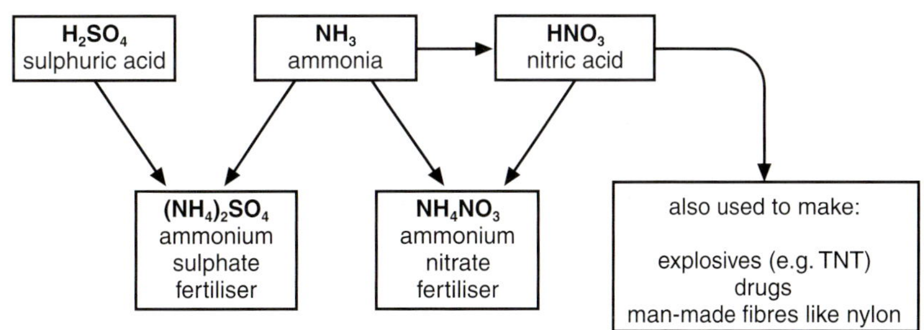

Figure 7.19
The uses of ammonia

Did you know?

There are many organic fertilisers. These do not use artificial chemicals. They include 'dried blood' (about 12% nitrogen), 'fish meal' (6–10% nitrogen) and 'hoof and horn meal' (about 14% nitrogen). Animal manure is also a supplier of nitrogen to the soil.

The Haber process makes ammonia (NH_3) by combining nitrogen and hydrogen. Nitrogen is obtained from the air; hydrogen from natural gas (CH_4).

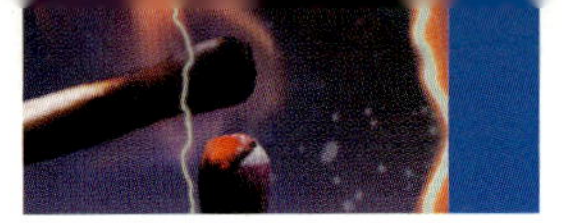

Did you know?

Modern chemical factories are very expensive to build. Using cheap raw materials (like air and natural gas) is important. It makes it unlikely that a cheaper method of manufacture will be discovered. This means the factory can be used for a long time and pay back the cost of building it.

Figure 7.20
An industrial plant producing ammonia by the Haber process

The nitrogen and hydrogen are purified and passed over a catalyst. The catalyst used is iron. The reaction needs a high temperature (about 450°C) and high pressure (about 200 atmospheres).

$$\text{nitrogen} + \text{hydrogen} \rightleftharpoons \text{ammonia}$$
$$N_2 + 3H_2 \rightleftharpoons 2NH_3$$

The reaction is reversible. This means that some of the ammonia produced will decompose back to nitrogen and hydrogen. The effect of this is that only some of the nitrogen and hydrogen are changed into ammonia. The high temperature and pressure ensure that a reasonable quantity of ammonia is produced fairly quickly.

The ammonia is removed by cooling the gases. The ammonia gas becomes a liquid and is drained off. The nitrogen and hydrogen that have not reacted are recycled.

Some of the ammonia is used to make nitric acid. There are two main stages in the process.

Stage 1

Ammonia and air are passed over a platinum catalyst. The ammonia is **oxidised** by the oxygen in the air:

$$\text{ammonia} + \text{oxygen} \longrightarrow \text{nitrogen monoxide} + \text{water}$$
$$4NH_3 + 5O_2 \longrightarrow 4NO + 6H_2O$$

The reaction is very exothermic and helps to keep the catalyst at 900°C – the temperature needed for the reaction to take place quickly.

Stage 2

The nitrogen monoxide is cooled and reacted with water and more oxygen to produce nitric acid:

$$\text{nitrogen monoxide} + \text{water} + \text{oxygen} \longrightarrow \text{nitric acid}$$

$$4NO + 2H_2O + 3O_2 \longrightarrow 4HNO_3$$

Ammonium nitrate fertiliser is made by reacting ammonia with nitric acid:

$$\text{ammonia} + \text{nitric acid} \longrightarrow \text{ammonium nitrate}$$

$$NH_3 + HNO_3 \longrightarrow NH_4NO_3$$

Ammonia is an alkaline substance, so this reaction is a **neutralisation** reaction.

Nitrate fertilisers can increase the yield of crops. But because the fertilisers are fairly cheap the farmer might use too much. And even if he is very careful, heavy rain can wash the fertiliser past the plant's roots before the plant can use it. The nitrates can then be washed into rivers and eventually get into drinking water. Nitrates in drinking water are harmful because they can cause certain types of cancer.

Tasks

1 What part of a plant benefits most from nitrogen-based fertilisers? What **natural** products are a source of nitrogen-based fertilisers?

2 Find out how hydrogen is obtained from natural gas.

3 Nitrogen-based fertilisers increase the yield of crops but they can contaminate drinking water. Do you think that using artificial nitrogen-based fertilisers is beneficial? Consider carefully what would happen if they were banned.

Topic Questions

1 What is the percentage of nitrogen in the air? What other gases are in the air?

2 What is the formula of ammonia? Write an equation for the formation of ammonia from its elements.

3 What is the catalyst used in the Haber process?

4 Write an equation for a) an oxidising reaction, and b) a neutralising reaction mentioned in this section.

5 The Haber process needs high temperature and pressure. Why are they needed? What temperature and pressure are used?

6 At room temperature and pressure ammonia is a gas. Yet when the mixture of ammonia, nitrogen and hydrogen from the Haber process is cooled the ammonia collects as a liquid. Explain why this happens.

7.9	
Co-ordinated	Modular
SA n/a	SA n/a
DA 11.8	DA 07

Chemical calculations

Although atoms are very small they do have mass. But because the masses of atoms are so small they are not measured in grams. Instead the mass of an atom is compared to the mass of a hydrogen atom. An atom of carbon is 12 times heavier than a hydrogen atom so its mass is 12. These masses are called relative atomic masses (given the symbol A_r). So the relative atomic mass of carbon is 12. The relative atomic masses of some common elements are given in Figure 7.21. Knowing the relative atomic masses allows scientists to work out exactly what is happening in a chemical reaction.

? Did you know?

1 gram of hydrogen contains 600 000 000 000 000 000 000 000 (6 hundred thousand million milllion million) hydrogen atoms. If hydrogen atoms were laid out in a row there would be about 300 million of them in 1 cm. If 1 gram of hydrogen had its atoms laid out in a straight line the line would be about 20 000 000 000 km long and it would take a ray of light about 18 hours to travel that distance!

Figure 7.21 *The relative atomic masses of some elements*

Name	Symbol	A_r	Name	Symbol	A_r
Aluminium	Al	27	Lithium	Li	7
Calcium	Ca	40	Magnesium	Mg	24
Carbon	C	12	Nitrogen	N	14
Chlorine	Cl	35.5	Oxygen	O	16
Copper	Cu	63.5	Potassium	K	39
Hydrogen	H	1	Sodium	Na	23
Iron	Fe	56	Sulphur	S	32

Calculating relative formula masses

Relative atomic masses can be used to calculate the relative formula mass (M_r) for compounds. The method is to add together the relative atomic masses of *every* atom in the compound. To do this the correct formula of the compound must be used. The examples below show how relative formula mass is calculated.

Example 1 Carbon monoxide

The formula of carbon monoxide is CO. Carbon monoxide has 1 carbon atom and 1 oxygen atom. So the relative formula mass of carbon monoxide is obtained by adding the relative atomic masses of these *two* atoms.

$$\text{number of atoms in CO} = 1 \times \text{C} + 1 \times \text{O}$$
$$\text{number of atoms} \times A_r = 1 \times 12 + 1 \times 16$$
$$12 + 16$$
$$M_r = 28$$

Example 2 Water

The formula of water is H_2O. Water has 2 hydrogen atoms and 1 oxygen atom. So the relative formula mass of water is obtained by adding the relative atomic masses of *all three* atoms.

$$\text{number of atoms in } H_2O = 2 \times \text{H} + 1 \times \text{O}$$
$$\text{number of atoms} \times A_r = 2 \times 1 + 1 \times 16$$
$$2 + 16$$
$$M_r = 18$$

Example 3 Aluminium chloride

The formula of aluminium chloride is $AlCl_3$. Aluminium chloride has 1 aluminium atom and 3 chlorine atoms. So the relative formula mass of aluminium chloride is obtained by adding the relative atomic masses of *all four* atoms.

$$\text{number of atoms in } AlCl_3 = 1 \times \text{Al} + 3 \times \text{Cl}$$
$$\text{number of atoms} \times A_r = 1 \times 27 + 3 \times 35.5$$
$$27 + 106.5$$
$$M_r = 133.5$$

Example 4 Sodium sulphate

The formula of sodium sulphate is Na_2SO_4. Sodium sulphate has 2 sodium atoms, 1 sulphur atom and 4 oxygen atoms. So the relative formula mass of sodium sulphate is obtained by adding the relative atomic masses of *all seven* atoms.

Relative formula masses can be very large. For polymers like poly(ethene) and PVC they can be several thousand.

$$\text{number of atoms in } Na_2SO_4 = 2 \times \text{Na} + 1 \times \text{S} + 4 \times \text{O}$$
$$\text{number of atoms} \times A_r = 2 \times 23 + 1 \times 32 + 4 \times 16$$
$$46 + 32 + 64$$
$$M_r = 142$$

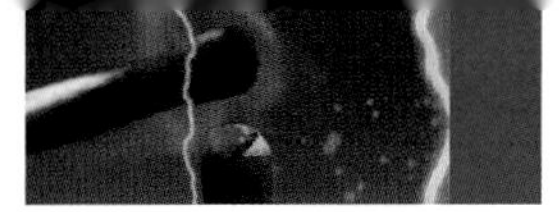

Calculating the percentage of an element in a compound

If the relative formula mass of a compound is known, then the percentages of each element present can be calculated.

Example 1 What is the percentage of hydrogen in water?

The relative formula mass of water is 18. There are 2 atoms of hydrogen. Each hydrogen atom has a relative atomic mass of 1. So the fraction of the formula mass that is hydrogen is $^{2}/_{18}$.

$$\text{percentage of hydrogen in water} = \frac{2}{18} \times 100 = 11.11\%$$

Example 2 What is the percentage of chlorine in aluminium chloride?

The relative formula mass of aluminium chloride is 133.5. There are 3 atoms of chlorine. Each chlorine atom has a relative atomic mass of 35.5. So the fraction of the formula mass that is chlorine is $^{3 \times 35.5}/_{133.5} = {}^{105.5}/_{133.5}$.

$$\text{percentage of chlorine in } AlCl_3 = \frac{105.5}{133.5} \times 100 = 78.65\%$$

Example 3 What is the percentage of oxygen in sodium sulphate?

The relative formula mass of sodium sulphate is 142. There are 4 atoms of oxygen. Each oxygen atom has a relative atomic mass of 16. So the fraction of the formula mass that is oxygen is $^{4 \times 16}/_{142} = {}^{64}/_{142}$.

$$\text{percentage of oxygen in } Na_2SO_4 = \frac{64}{142} \times 100 = 45.07\%$$

Example 4 What is the percentage of nitrogen in ammonium nitrate (NH_4NO_3)?

This is a little bit more difficult because nitrogen occurs *twice* in the formula.

First calculate the relative formula mass. Notice that both nitrogen atoms have to be included.

$$\begin{aligned} \text{number of atoms in } NH_4NO_3 &= 2 \times N + 4 \times H + 3 \times O \\ \text{number of atoms} \times A_r &= 2 \times 14 + 4 \times 1 + 3 \times 16 \\ & \quad\; 28 \quad + \quad 4 \quad + \quad 48 \\ M_r &= 80 \end{aligned}$$

Second calculate the percentage. The relative formula mass of ammonium nitrate is 80. There are 2 atoms of nitrogen. Each nitrogen atom has a relative atomic mass of 14. So the fraction of the formula mass that is nitrogen is $^{2 \times 14}/_{80} = {}^{28}/_{80}$.

$$\text{percentage of nitrogen in } NH_4NO_3 = \frac{28}{80} \times 100 = 35\%$$

The mathematical formula for calculating the percentage of an element in a compound is:

$$\text{percentage of element Z} = \frac{(A_r \text{ of Z}) \times \text{number of atoms of Z in compound}}{M_r \text{ of compound}} \times 100$$

Task

Select some of the balanced equations in this chapter. Calculate the relative formula (or atomic) masses of all the substances in the equation. What do you notice about the total mass of all the reactants and the total mass of all the products?

Topic Questions

1 The following compounds are mentioned in this chapter. Calculate the relative formula mass for each of these. (Use the values for relative atomic masses in Figure 7.21.)

a) ammonia (NH_3)
b) hydrochloric acid
c) hydrogen peroxide (H_2O_2)
d) calcium carbonate
e) glucose

2 Calculate the percentage of:

a) nitrogen in ammonia
b) chlorine in hydrochloric acid
c) oxygen in calcium carbonate
d) carbon in glucose
e) hydrogen in ethanol (C_2H_5OH)

3 Explain why it is useful for scientists to be able to calculate relative formula masses.

Summary

- Chemicals that are particularly hazardous display a symbol warning of the danger.
- Chemical reactions do not happen instantaneously.
- Different reactions take place at different speeds (rates). Some, like rusting, are very slow; others, like an explosion, are very fast.
- Reactions can be speeded up by:
  - increasing the temperature
  - increasing the concentration of the reactants that are in solution
  - increasing the pressure of the reactants that are gases
  - increasing the surface area of reactants that are solids
  - using a catalyst (in some cases).
- A catalyst is a substance that speeds up a reaction but is not used up by the reaction.
- Catalysts are widely used in industry. Many of them are transition metals or their compounds.

- Different reactions use different catalysts.
- Some reactions do not have catalysts.
- The speed of a reaction can be followed by monitoring the speed at which a reactant is used up or the speed at which a product is formed.
- For a reaction to occur the reactant particles must collide with enough energy.
- The lowest amount of energy that will cause a collision to result in a reaction is called the activation energy.
- At a higher temperature, particles move faster and are more likely to collide so the reaction goes faster.
- At a higher temperature, more collisions will have energy greater than the activation energy so the reaction goes faster.
- At a higher concentration (or for gases pressure), the particles are closer together so the reaction goes faster.
- The chemical reactions in the cells of living things are catalysed by enzymes.
- Enzymes work because their shape allows certain molecules to 'fit' inside them. This makes enzymes very specific – an enzyme will normally only catalyse one specific reaction.
- Fermentation is an example of an enzyme catalysed reaction.
- Enzymes work best at about 40–45°C. Above that temperature their structure is damaged and they don't work.
- Carbon dioxide will turn a solution of calcium hydroxide in water (called limewater) cloudy. This is the laboratory test for carbon dioxide.
- Exothermic reactions give out energy.
- Endothermic reactions take in energy.
- Thermal decomposition reactions are endothermic.
- Reversible reactions can be made to go in either direction.
- A liquid that contains water will turn white (anhydrous) copper sulphate blue. This is the laboratory test for the presence of water.
- The Haber process is used to manufacture ammonia (NH_3) from atmospheric nitrogen and natural gas.
- Ammonia is used to make chemicals like nitric acid and fertilisers.
- Nitrate fertilisers increase the yield of crops. If overused they can get into rivers and cause contamination of drinking water.
- Atoms of different elements have different masses.
- The relative atomic mass (A_r) of an element is the mass of its atom compared to the mass of a hydrogen atom.
- The relative formula mass (M_r) of a compound is found by adding together the relative atomic masses of every atom in the compound.
- The percentage of an element in a compound can be calculated using the formula:

$$\text{percentage of element Z} = \frac{(A_r \text{ of Z}) \times \text{number of Z atoms in compound}}{M_r \text{ of compound}} \times 100$$

Examination Questions

1 Draw a line to match the following chemical hazard symbols with the correct meaning.

Chemical hazard symbol	Meaning of the symbol
	Flammable
	Harmful
	Oxidising
	Toxic

(3 marks)

2 One way to speed up a chemical reaction is to use a catalyst.

a) Explain what is meant by the term catalyst. *(1 mark)*

b) Name **two** other ways in which a reaction can be speeded up. *(2 marks)*

3 Hydrochloric acid reacts with marble chips.

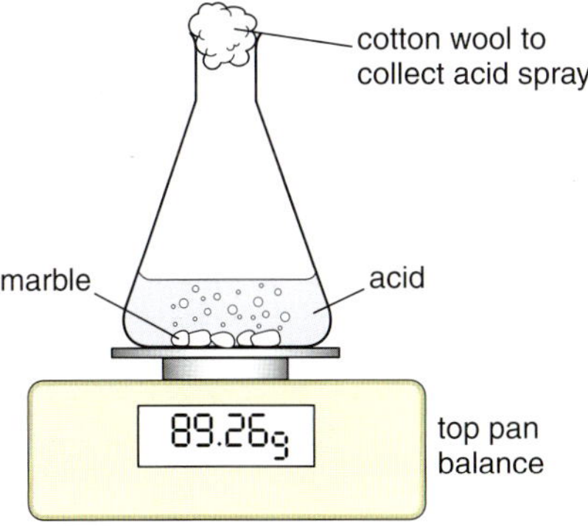

The equation for the reaction is:

$$2HCl(aq) + \underset{\text{marble}}{CaCO_3(s)} \longrightarrow CaCl_2(aq) + H_2O(l) + CO_2(g)$$

During the reaction there is a loss in mass. The table shows how much mass is lost as the reaction proceeds.

Time/min	0	1	2	3	4	5	6	7	8	9
Mass lost/g	0	1.5	2.5	3.2	3.6	3.7	3.8	3.9	4.0	4.0

a) Plot a graph of these results *(3 marks)*

b) On the same graph sketch the graph you would expect if the acid had a higher concentration. *(2 marks)*

c) Explain why there is a loss in mass during the reaction. *(2 marks)*

4 The speed of a reaction was studied by measuring how much of one of the reactants was left.

Time/min	0	1	2	3	4	5	6	7
% of reactant left	100	55	25	12	6	3	0.8	0

a) Plot a graph of these results. *(3 marks)*

b) How long did it take for half the reactant to be used up? *(1 mark)*

c) What percentage of the reactant was left after 2½ minutes? *(1 mark)*

5 The effect of concentration and temperature on the speed of a reaction was studied. The tables show the results obtained.

Concentration mol/dm^3	0	0.25	0.5	0.75	1.0	1.5	2.0
Relative speed of reaction	0	0.1	0.23	0.34	0.48	0.70	1.0

Temperature °C	20	30	40	50	60	70
Relative speed of reaction	1.0	1.8	3.8	7.0	13.5	26.0

a) Plot the graphs for both sets of results. *(3 marks)*

b) Using the idea of particles, explain why:

i) the concentration graph is a straight line. *(2 marks)*

ii) the temperature graph is not a straight line. *(3 marks)*

6 Sugar can be converted into alcohol and carbon dioxide. The reaction uses a microorganism to provide a catalyst. The process is used to make alcoholic drinks.

a) What is the name of this process? *(1 mark)*

b) What other use is made of this process? *(1 mark)*

c) What is the name of the microorganism that provides the catalyst for this reaction? *(1 mark)*

d) Many reactions can be catalysed by enzymes. Explain how temperature affects the speed of a reaction catalysed by an enzyme. *(3 marks)*

7 In the table below are three reactions. For each reaction tick **two** of the boxes that correctly describe the reaction.

Reaction	Endothermic	Exothermic	Neutralisation	Oxidation	Thermal decomposition
$2Mg + O_2 \longrightarrow 2MgO$	☐	☐	☐	☐	☐
$CaCO_3 \longrightarrow CaO + CO_2$	☐	☐	☐	☐	☐
$HCl + NaOH \longrightarrow NaCl + H_2O$	☐	☐	☐	☐	☐

(6 marks)

8 This question is about the Haber process and what it produces.

a) The Haber process is used to make ammonia from hydrogen and nitrogen. The equation for the reaction that takes place is:

$$N_2 + 3H_2 \rightleftharpoons \underset{\text{ammonia}}{2NH_3}$$

i) What does the symbol $\rightleftharpoons$ mean? *(1 mark)*

ii) From where is the nitrogen used in the Haber process obtained? *(1 mark)*

b) Some of the ammonia is converted into nitric acid (HNO_3). The nitric acid can be reacted with more ammonia to produce ammonium nitrate (NH_4NO_3).

i) Write a word equation for this reaction. *(2 marks)*

ii) What type of reaction is this? *(1 mark)*

iii) Give one use of ammonium nitrate. *(1 mark)*

c) Nitrates can get into drinking water.

i) Explain how nitrates get into drinking water. *(3 marks)*

ii) Why are nitrates in drinking water harmful? *(2 marks)*

Chapter 8

Structure and bonding

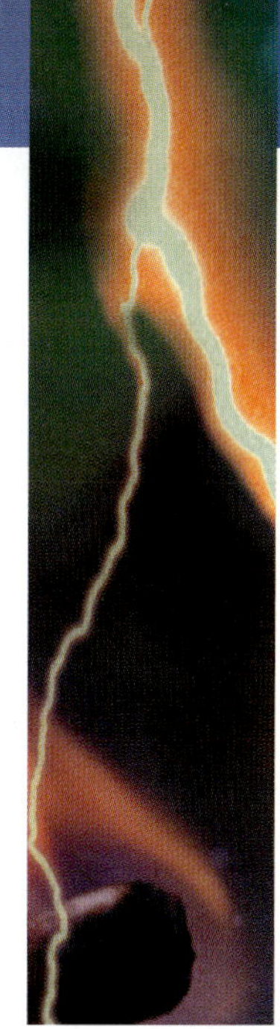

Key terms

element • **compound** • **bonds** • **ions** • **relative atomic mass** • **period** • **group** • **reactants** • **products**

8.1 Changes of state

Co-ordinated	Modular
SA 11 intro	omitted
DA 11 intro	DA 08

Particles in a solid are held together by attractive forces. The particles are not stationary, they vibrate. As a solid gets hotter, its particles vibrate more. The increased vibration will eventually overcome the forces holding the particles together. The particles become separate and are free to move about. This process is melting. Melting takes place at a specific temperature called the melting point.

If a liquid is heated the particles move more quickly. If a particle gains enough energy to overcome the attractive forces of the other particles around it, it can escape from the liquid. The particle has evaporated from the liquid. As the liquid is heated more and more particles have enough energy to evaporate. So, as the temperature goes up the speed of evaporation increases. If the temperature is high enough the liquid will boil. The temperature at which this happens is called the boiling point.

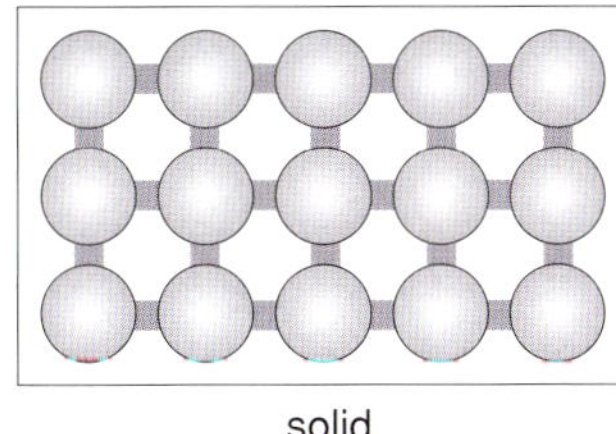

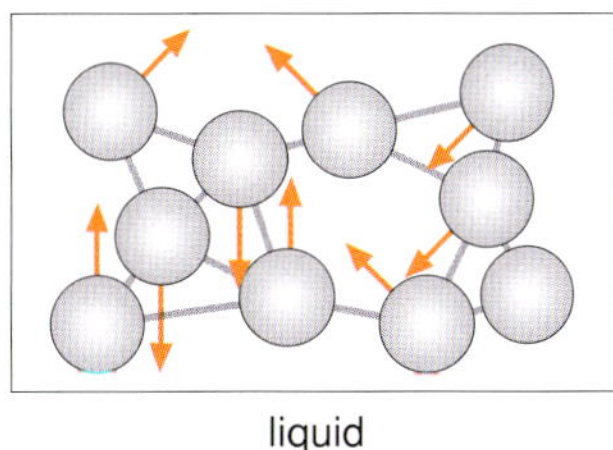

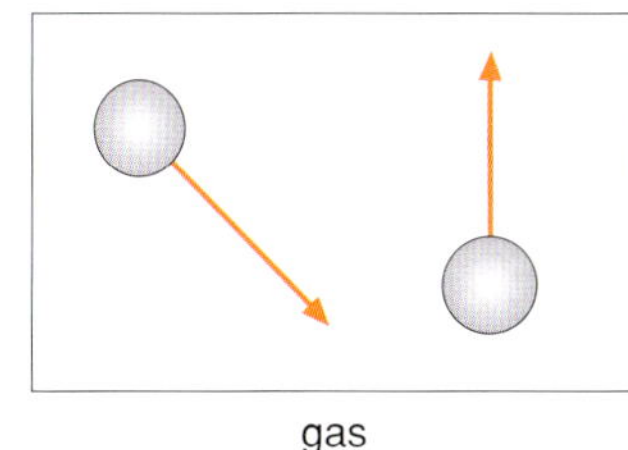

key: strength of attractive force between particles; → direction and speed of movement of particles

Figure 8.1 *The arrangement of particles in a solid, liquid and gas*

Tasks

1 Find the melting and boiling points of some of the compounds mentioned in this book.

2 Why don't calcium carbonate and copper sulphate crystals have a melting point?

Topic Questions

1 The table below gives the melting points and boiling points of six substances.

Substance	Melting point (°C)	Boiling point (°C)
A	10.5	338
B	−101	−34
C	801	1413
D	−78	−33
E	114	183
F	−7	58

a) For each substance, work out if they are a solid, liquid or gas at room temperature (20°C).
b) Sodium chloride cannot be boiled with a Bunsen burner. Which substance is sodium chloride?
c) Bromine is a volatile liquid. Which substance is bromine?
d) The inside of a deep freeze is −20°C. Which substances would be in a different state in a deep freeze from what they are at room temperature?

8.2 Atoms

Co-ordinated	Modular
SA 11.1	SA 16
DA 11.1	DA 08

All substances are made of atoms. Although there are billions of different substances there are only about 100 different sorts of atoms. A substance that contains only one sort of atom is called an **element**. A **compound** contains more than one type of atom.

An atom has a nucleus in the centre. The nucleus contains protons and neutrons. Electrons go round the nucleus. Figure 8.2 shows the structure of helium – one of the smallest atoms.

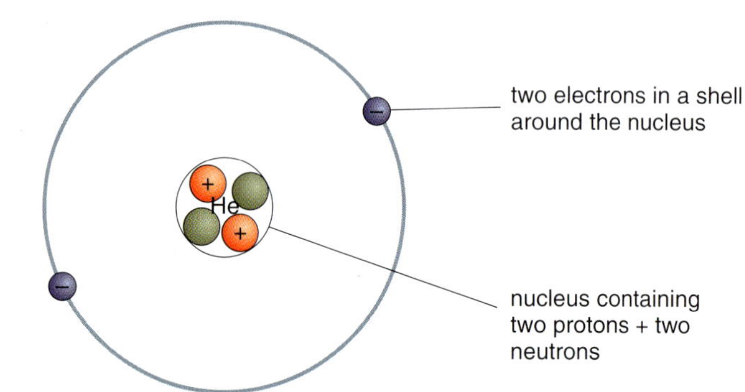

Figure 8.2
An atom of helium

The idea of atoms was first suggested in about 450 BC by a Greek called Democritus. The idea was not accepted because it conflicted with the religious beliefs of the time. In 1807, John Dalton re-introduced the idea of atoms. Using his 'atomic theory', Dalton was able to explain some of the scientific discoveries made by other scientists. Because of this his ideas were accepted. The theory was also able to predict what would happen in certain situations. This encouraged more scientists to accept it.

Did you know?

The idea of atoms had been used by Sir Isaac Newton in about 1650 but it was not widely accepted. John Dalton made the idea popular. But even then there were many important scientists who did not accept 'atomic theory'. These included Sir Humphry Davy.

Task

Read the Did you know? box on this page again. It mentions John Dalton and two other famous scientists. Find out what these two other scientists were famous for.

Topic Questions

1 How many different types of atoms are there?

2 Explain the difference between:
a) an element and a compound.
b) an atom and a molecule.

3 Draw a diagram of an atom with 1 proton, 1 neutron and 1 electron. Which part of this atom is the nucleus?

4 Explain why 'atomic theory' was accepted in the nineteenth century but had been rejected 2000 years earlier.

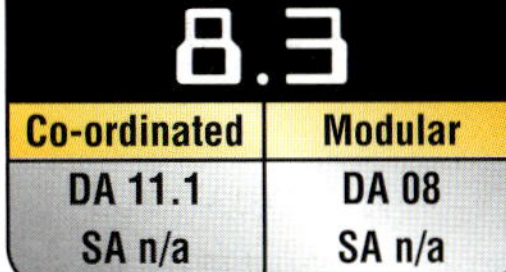

8.3	
Co-ordinated	Modular
DA 11.1	DA 08
SA n/a	SA n/a

Structure of the atom – protons, neutrons and electrons

The proton, neutron and electron are very small, but they do have some mass. In fact the proton and the neutron have about the same mass as a hydrogen atom. The proton and the electron also have a very small electrical charge. Figure 8.3 shows the mass and charge of these particles.

Figure 8.3
The mass and charge of the three particles in an atom

Particle	Mass	Charge
electron	almost 0	−1
neutron	1	0
proton	1	+1

The table in Figure 8.3 is easy to learn. If the particles are listed in alphabetical order, the mass and the charge are listed in numerical order. (Remember: -1 is less than 0.)

? Did you know?

No one has ever seen a proton, neutron or electron – they are far too small to be seen, even with the most powerful microscope. The existence of these particles has been deduced from observing the way materials behave. Making deductions from the way things behave is quite common. Most people will check to see if a matchbox has any matches in it by shaking the box to see if it rattles. It is quicker than opening the box. But it does not always give the right answer. The matchbox could contain used matches or even drawing pins! So sometimes an indirect test might not give the right answer. This means that our idea about the structure of the atom might not be quite right.

The proton and electron have an electrical charge. But atoms have no overall electrical charge. (If they did every time you touched something you'd get an electric shock.) This means that the number of protons in an atom equals the number of electrons in that atom.

All the atoms in an element are the same. This means that all the atoms of a particular element must have the same number of protons. But atoms of different elements are not the same; different elements have different numbers of protons in their atoms.

The number of protons in an atom is called the **atomic number** (sometimes called the proton number). So different elements have different atomic numbers. Because atoms have the same number of electrons as protons, the atomic number is also the number of electrons in an atom.

Every atom has mass, but the proton and the neutron are the only particles in an atom that have a significant mass. So the mass of an atom can be found by adding together the number of protons and neutrons. This is called the **mass number**.

In summary,

atomic number = **number of protons** = **number of electrons**

mass number = **number of protons** + **number of neutrons**

Atoms of the same element always have the same number of protons. But sometimes these atoms might not have the same number of neutrons. A difference in the number of neutrons does not affect its chemical properties, so the atom belongs to the same element. Atoms that have the same number of protons but different numbers of neutrons are called isotopes. So isotopes of the same element have the same atomic number but different mass numbers. Figure 8.4 shows the two main isotopes of lithium.

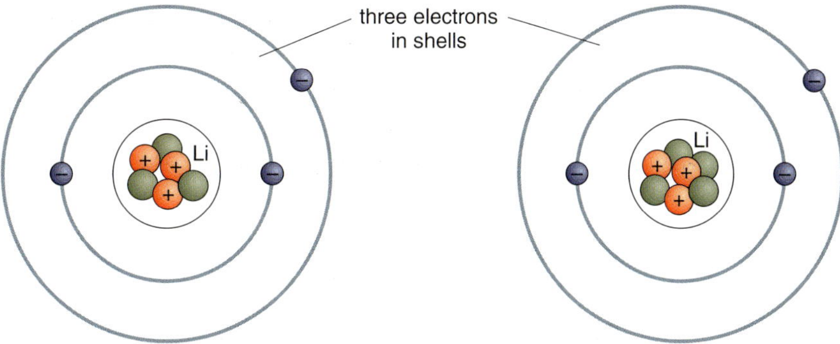

Figure 8.4
Isotopes of lithium – atomic number 3

> **Did you know?**
>
> Some elements only have one isotope. This means all the atoms have exactly the same number of neutrons. But many elements have more than one isotope. Tin has 10 isotopes.
>
> Chlorine has two isotopes. 75% of chlorine atoms have a mass number of 35 and 25% have a mass number of 37. The relative atomic mass is the average of these values. 75% of 35 = 26.25 and 25% of 37 = 9.25. This gives a relative atomic mass of 26.25 + 9.25 = 35.5.

Atoms can be represented by symbols. The mass number and atomic number can be included with the symbol.

$$^{\text{mass number}}_{\text{atomic number}}\text{Symbol}$$

Figure 8.5 shows how this system can be used to represent some atoms. The first two items in the table are different isotopes of hydrogen.

Figure 8.5
Representing atoms with their symbols, mass number and atomic number

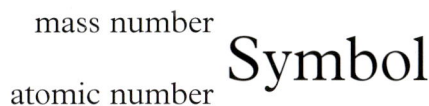

Symbol	Element	Mass Number	Atomic Number	Number of protons	Number of neutrons
${}^{1}_{1}H$	hydrogen	1	1	1	0
${}^{2}_{1}H$	hydrogen	2	1	1	1
${}^{23}_{11}Na$	sodium	23	11	11	12
${}^{35}_{17}Cl$	chlorine	35	17	17	18
${}^{235}_{92}U$	uranium	235	92	92	143

> **Did you know?**
>
> Although different isotopes have the same chemical properties they have slightly different physical properties. Water made with the ${}^{2}_{1}H$ isotope instead of the ${}^{1}_{1}H$ isotope is called 'heavy' water. The density of 'heavy' water is about 11% higher than that of normal water. Its melting and boiling points are also different.

Task

Pick three elements. Try to find the mass numbers of the isotopes of each of these elements. What is the number of protons and neutrons in an atom of each of these isotopes?

Topic Questions

1 Complete the following table.

Particle	Mass	Charge
electron		
		0

2 For each of the following statements, decide whether it is true or false.

a) In an atom the number of protons is equal to the number of neutrons.
b) In an atom the number of protons is equal to the number of electrons.
c) In an atom the number of electrons is equal to the number of neutrons.
d) The atomic number of an element equals the number of protons in each atom.
e) The atomic number of an element equals the number of neutrons in each atom.
f) The atomic number of an element equals the number of electrons in each atom.
g) The mass number of an element equals the number of protons plus the number of electrons in each atom.
h) The mass number of an element equals the number of protons plus the number of neutrons in each atom.
i) The mass number of an element equals the number of neutrons plus the number of electrons in each atom.

3 Explain what is meant by an isotope.

4 The diagram below is of the atom that can be represented as $^{4}_{2}He$.

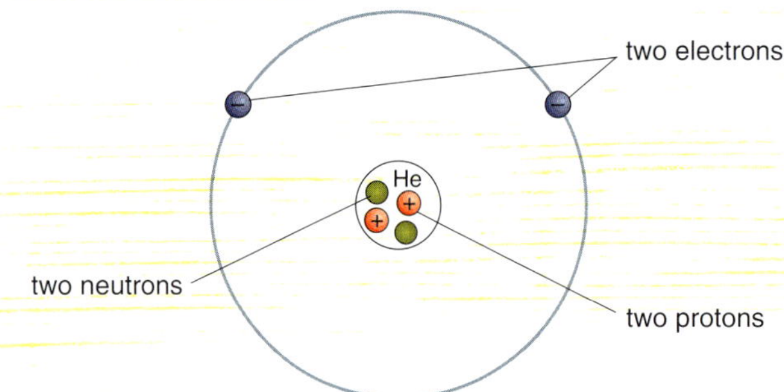

a) Draw a diagram of the atom that can be represented as $^{2}_{1}H$.
b) Write the full symbol showing atomic number and mass number for:
   i) carbon (symbol C; atomic number = 6; mass number = 12)
   ii) boron (symbol B; atomic number = 5; mass number = 11).

8.4	
Co-ordinated	Modular
SA 11.1	SA 16
DA 11.1	DA 08

Structure of the atom – electron energy levels

Electrons travel round the nucleus. They do not travel in circular orbits like planets round the Sun. Their path is much more complex. Electrons have different amounts of energy and their path round the nucleus is called an energy level. (This energy level is sometimes called a 'shell'.)

Did you know?

The picture represents the complex path of electrons around the nucleus. Electrons in low energy levels stay closer to the nucleus than those in higher energy levels.

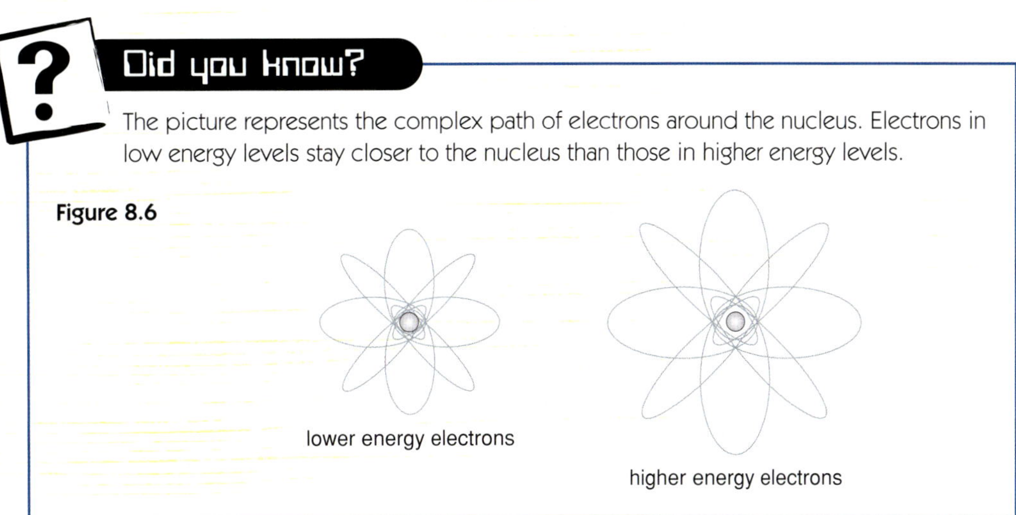

Figure 8.6

There are a number of different energy levels. For GCSE Science only four energy levels are considered.

The simplest atom is hydrogen. It has one proton and one electron. This electron goes in the first (lowest) energy level (see Figure 8.7). The next element is helium. Helium has two protons and two electrons. Both of these electrons go in the lowest energy level (see Figure 8.8). Once it has two electrons in it, the lowest energy level is full. Lithium is the next element. Lithium has three protons and three electrons. The first two electrons go in the first energy level. This energy level is now full. So the third electron goes in the second energy level (see Figure 8.9).

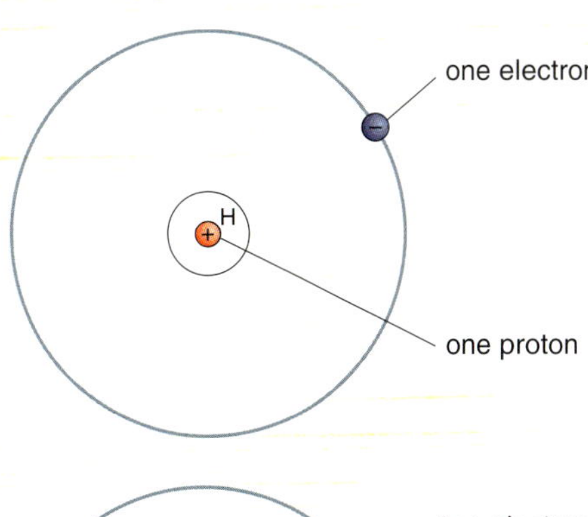

Figure 8.7 *The atomic structure of hydrogen*

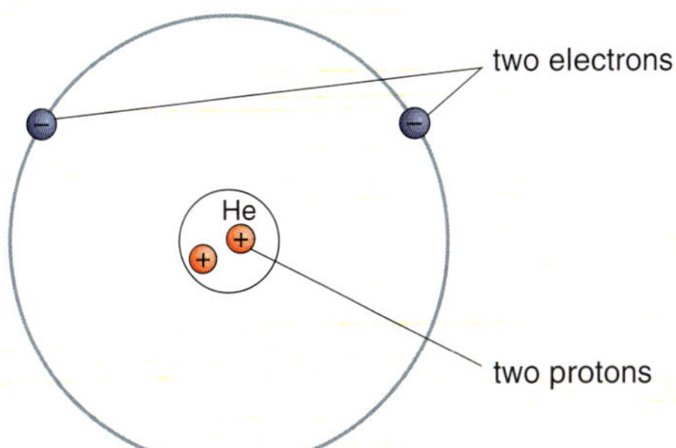

Figure 8.8 *The atomic structure of helium (no neutrons shown)*

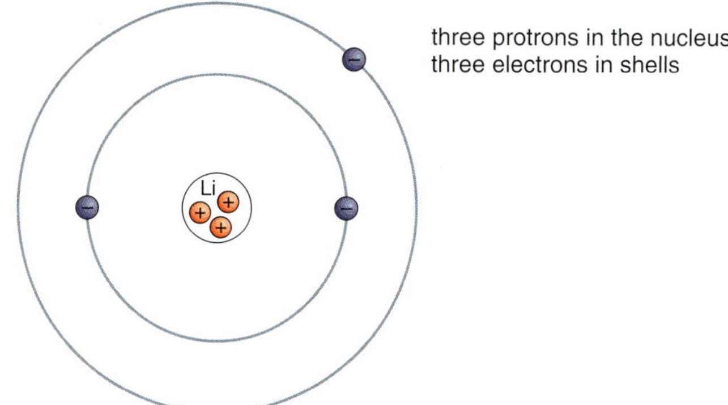

Figure 8.9 *The atomic structure of lithium (no neutrons shown)*

The next few atoms follow this pattern. They have two electrons in the first energy level and the extra electrons go into the second energy level (see Figure 8.10).

Figure 8.10 *The electronic structures of elements 3 to 10, without details of the contents of the nucleus*

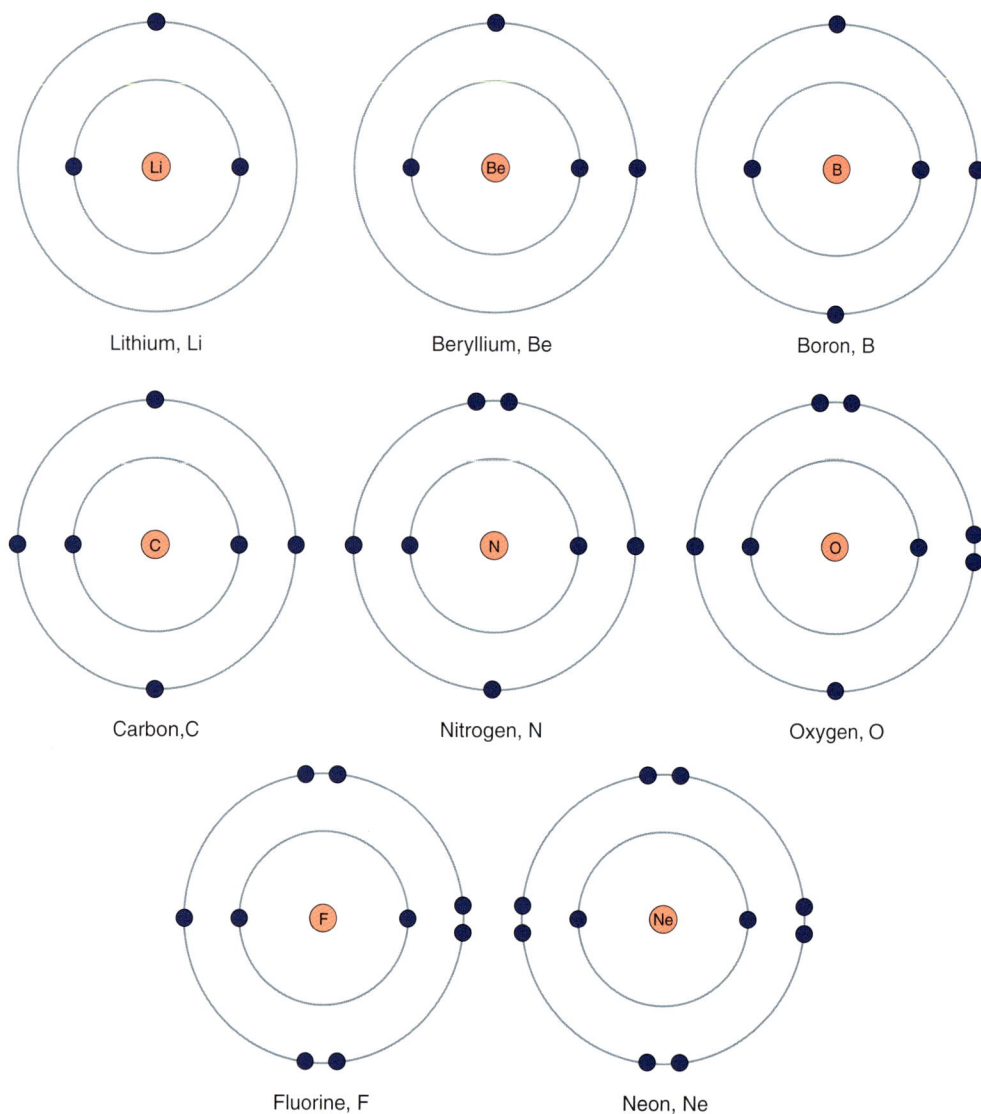

A neon atom has ten protons and ten electrons. There are two electrons in the first energy level and eight in the second energy level. Once it has eight electrons in it, the second energy level is full.

Figure 8.11 shows the electron structure of the next ten atoms. From sodium to argon the first two energy levels are full and the electrons begin to fill the third energy level. Like the second energy level this level is full when it contains eight electrons. So with potassium and calcium the electrons begin to fill the fourth energy level.

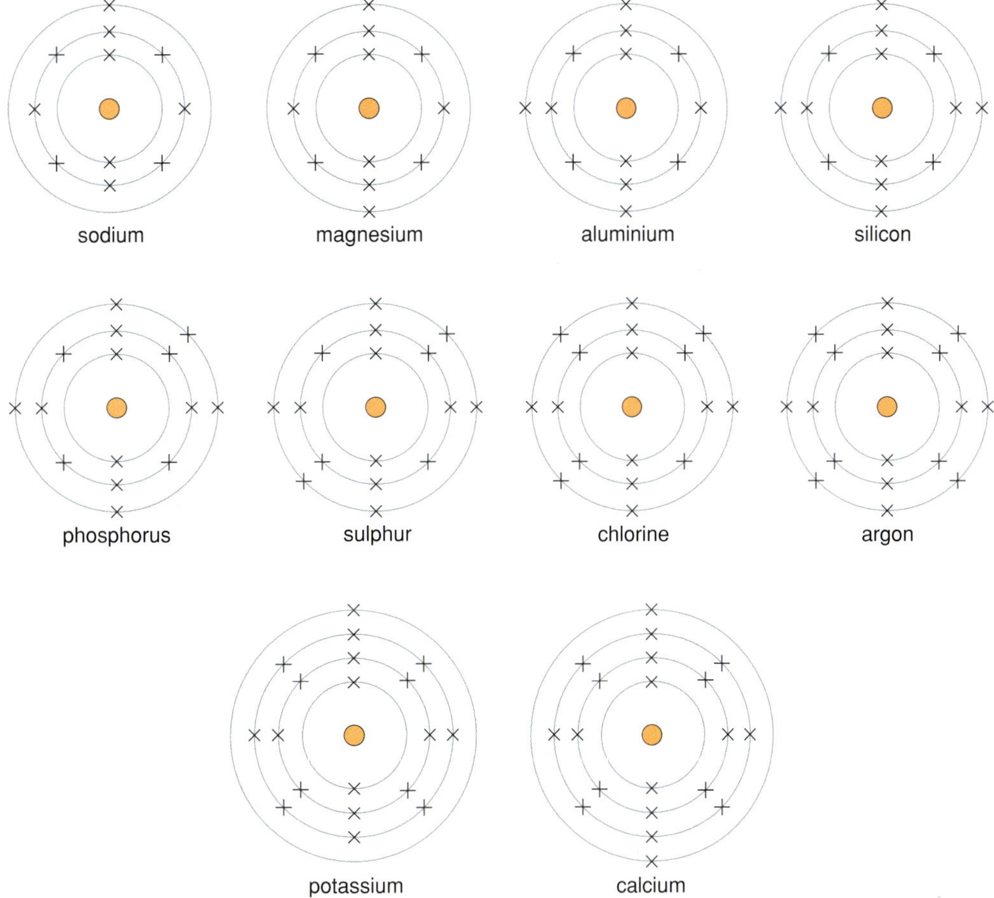

Figure 8.11
A 'dot and cross' diagram of atoms of the elements sodium to calcium

Atoms can also be represented in a simpler form. In this method the number of electrons in each energy level is written as a number. The different energy levels are separated by a comma. So lithium would be shown as Li: 2,1. Figure 8.12 shows this simpler form for four of the atoms in Figure 8.11.

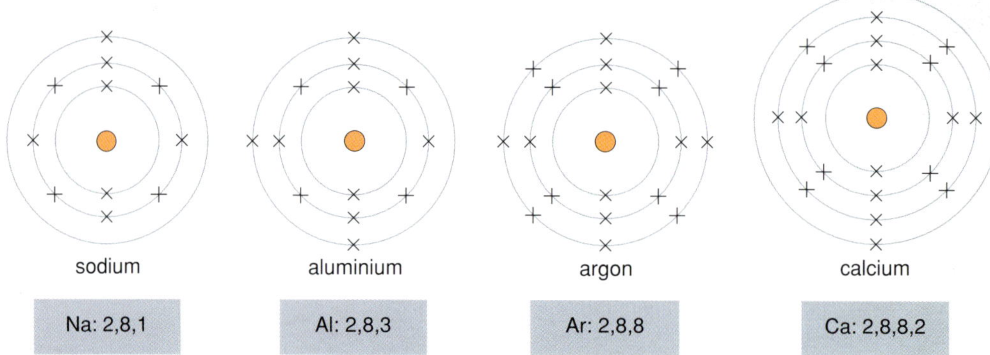

Figure 8.12
Using a simpler method for showing the electron structure of an atom

Tasks

1 You only need to be able to construct electron energy level diagrams for the first 20 elements. After that the way in which the electrons fill the energy levels becomes more complicated. Use the Periodic Table to help you decide how element 21 (scandium) differs from element 20 (calcium).

2 Below is a diagram of part of the Periodic Table. Some of the elements are included. Complete the table by filling in the blank boxes.

				H:			He:2
Na:2,8,1		Al:2,8,3					Ar:2,8,8
	Ca:2,8,8,2						

Topic Questions

1 The diagram is of a simple atom.

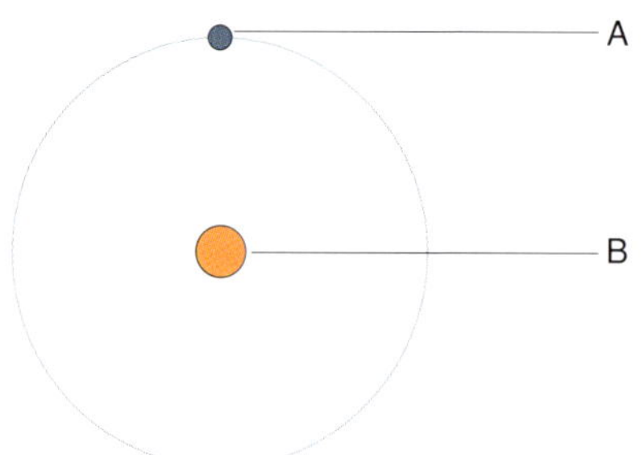

Name the parts of the atom labelled A and B.

2 In which energy level do electrons have the least energy?

3 What other name is given to energy levels?

4 Draw a 'dot and cross' diagram of the first ten elements.

5 How many electrons can go into each of the first three energy levels before they are full?

6 What do you notice about the elements whose atoms have all their energy levels full?

8.5 Ionic bonding

Co-ordinated	Modular
DA 11.2	DA 08
SA n/a	SA n/a

Most substances exist as compounds. They contain more than one type of element. In compounds the atoms of different elements are held together by chemical **bonds**.

One of the ways in which atoms bond together is called ionic bonding. In this process electrons move from one atom to another. Electrons have a negative charge. If an atom loses an electron it becomes positively charged. If an atom gains an electron it becomes negatively charged. Atoms that are charged are called **ions**. Ions with opposite charges attract each other. It is these strong attractive forces that bond the ions together. These bonds are called ionic bonds.

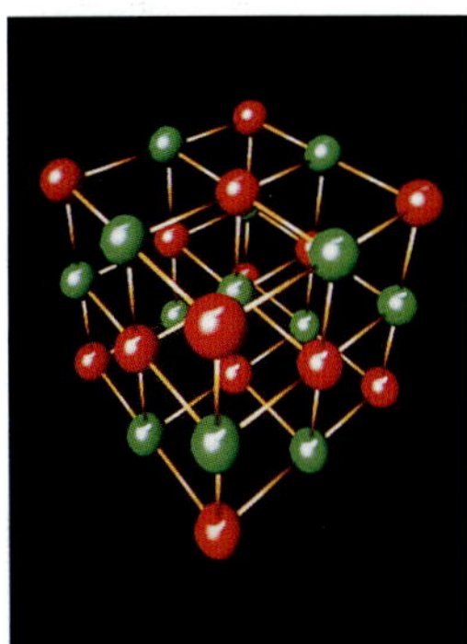

Figure 8.13
The arrangement of sodium (Na^+) ions and chloride (Cl^-) ions in sodium chloride

Compounds that are held together by ionic bonds form giant, three-dimensional structures (see Figure 8.13). Substances with giant structures have high melting and boiling points. So ionic compounds usually have high melting and boiling points.

Figure 8.14

Material	Formula	Melting point (°C)	Boiling point (°C)	Normal state at room temperature
sodium chloride	NaCl	801	1465	solid
copper oxide	CuO	1326	very high	solid
iron oxide	Fe_2O_3	1565	very high	solid
magnesium oxide	MgO	2800	3600	solid
potassium chloride	KCl	770	1407	solid
lead bromide	$PbBr_2$	373	916	solid
aluminium oxide	Al_2O_3	2045	2980	solid

In the process of **electrolysis**, ions are attracted to the **electrode** with the opposite charge. In an ionic solid the ions are held together and are unable to move out of their position. So trying to electrolyse an ionic solid doesn't work. But if the ionic compound is molten or dissolved in water then electrolysis does work. The positive ions are attracted to the negative electrode and move in that direction. The negative ions move towards the positive electrode. So ionic compounds conduct electricity when molten or in solution but not when they are solids.

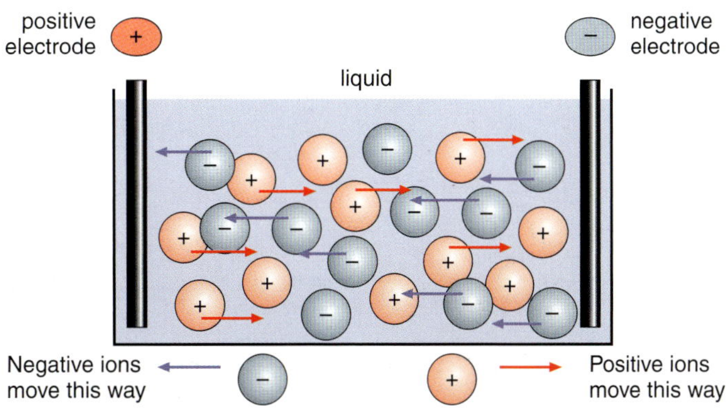

Figure 8.15
How the ions move in a liquid ionic compound

Did you know?

In electrolysis, the positive electrode is called the anode. Ions that move towards the anode are called anions. So an anion is negatively charged. The negative electrode is called the cathode. It attracts positive ions, which are called cations.

Elements whose atoms have full energy levels are chemically unreactive. This is because full energy levels are very stable. Helium, neon and argon have full electron energy levels. They are called the noble gases because they are so unreactive. When atoms form ions they gain or lose enough electrons to ensure that they only have full energy levels. The lower energy levels are always full so it is only the electrons in the highest energy level that are involved in making bonds.

For example, a sodium atom becomes a sodium ion by losing one electron:

$$\underset{\substack{\text{sodium atom}\\ \text{Na}}}{\text{Na: 2,8,1}} \longrightarrow \underset{\substack{\text{sodium ion}\\ \text{Na}^+}}{(\text{Na: 2,8})^+} + \text{one electron}$$

Figure 8.16 is a 'dot and cross' diagram of a sodium ion:

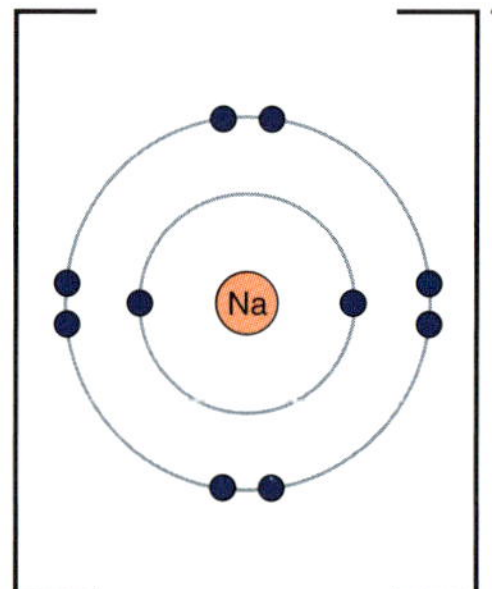

Figure 8.16
A 'dot and cross' diagram of a sodium ion

Notice that the sodium ion now has the electron structure of neon. Ions do not behave like atoms. So it is important to be able to tell an atom from an ion. For this reason the symbol for an ion includes the charge.

Chlorine has seven electrons in its outer energy level. It needs to gain one electron to get a full energy level. Notice that when chlorine becomes an ion its name changes. It is wrong to talk about a chlorine ion – the correct name is a chloride ion.

$$\underset{\substack{\text{chlorine atom}\\ \text{Cl}}}{\text{Cl: 2,8,7}} + \text{one electron} \longrightarrow \underset{\substack{\text{chloride ion}\\ \text{Cl}^-}}{(\text{Cl: 2,8,8})^-}$$

Figure 8.17 is a 'dot and cross' diagram of a chloride ion:

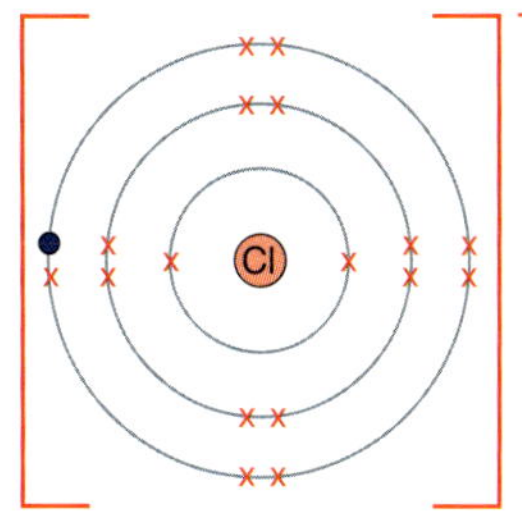

Figure 8.17
A 'dot and cross' diagram of a chloride ion

The chloride ion has the electron structure of argon.

Of course atoms cannot lose an electron unless there is somewhere for that electron to go. Nor can they gain an electron unless there is a source of supply. So ions are formed when elements react and electrons transfer from one atom to another. Figure 8.18 is a diagram showing this process for the reaction between sodium metal and chlorine gas to form sodium chloride (common salt).

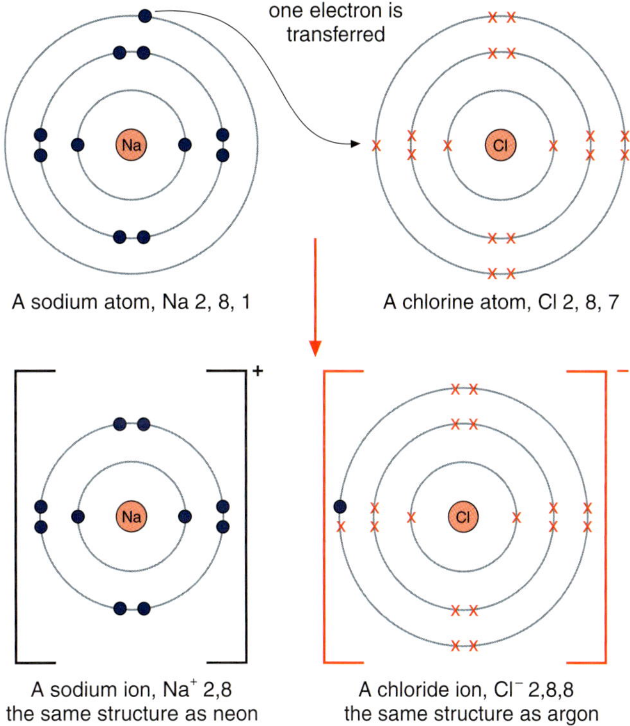

Figure 8.18
How sodium and chlorine atoms form an ionic bond

Some atoms need to lose or gain more than one electron. Magnesium has an electron structure of 2,8,2. It needs to lose two electrons to have all its energy levels full. Oxygen, with an electron structure of 2,6, needs to gain two electrons to get a stable, noble gas structure. Figure 8.19 shows the electron transfer between magnesium and oxygen to produce magnesium oxide.

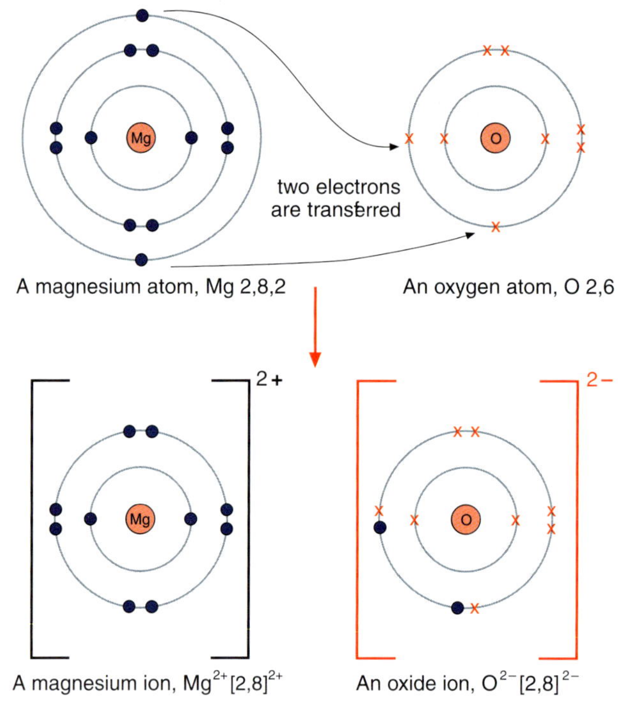

Figure 8.19
How the ionic bond is formed in magnesium oxide

Figure 8.20 shows the electron transfer between calcium and chlorine to produce calcium chloride. In this case each calcium atom can give away two electrons but each chlorine atom requires only one electron. So one calcium atom can supply electrons to two chlorine atoms. This is why the formula of calcium chloride is $CaCl_2$.

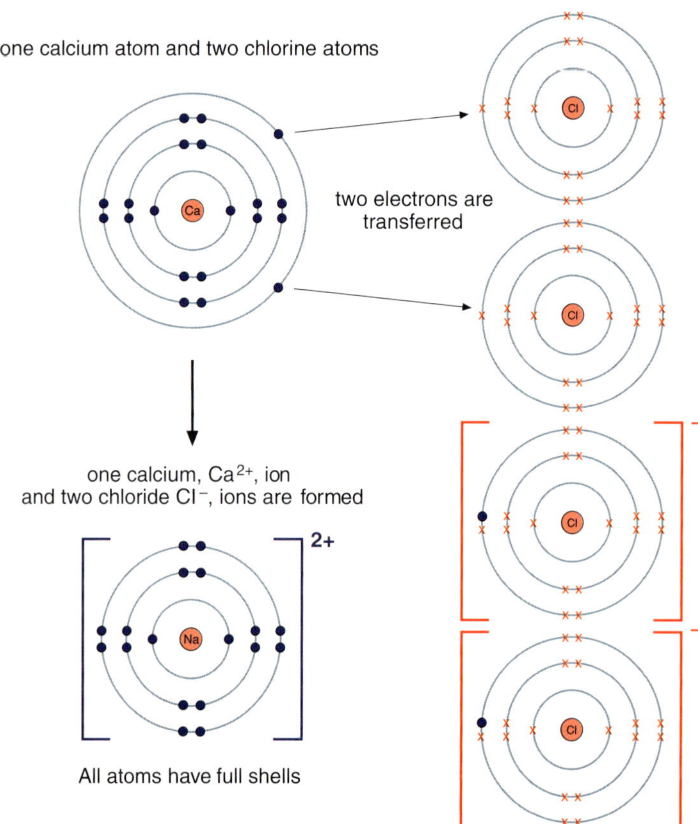

Figure 8.20
How the ionic bond is formed in calcium chloride

You may have noticed that all ionic compounds formed between two elements contain a metal and a non-metal.

Tasks

1. Find the names of the following ions – F^-, Br^-, I^-, O^{2-}, S^{2-}, N^{3-}, P^{3-}, Al^{3+}.
2. Try to find some compounds with giant structures that are not ionic.
3. Why do ionic compounds between two elements always contain a metal and a non-metal?

Topic Questions

1. What causes ions to have an electrical charge?
2. What happens to electrons when ionic compounds are formed?
3. For the following elements state whether they are electron donors or electron acceptors – potassium, fluorine, sulphur and beryllium.
4. Complete the following table for a particle with the electron structure 2,8,8.

Electron structure of particle	Charge on the particle	Name of the particle	Symbol of the particle
	−2		
	−1		
2,8,8	0		
	+1		
	+2		

5. Draw a diagram like those of Figures 8.19 and 8.20 to show electron transfers in:

   a) sodium oxide (Na_2O)
   b) aluminium chloride ($AlCl_3$)

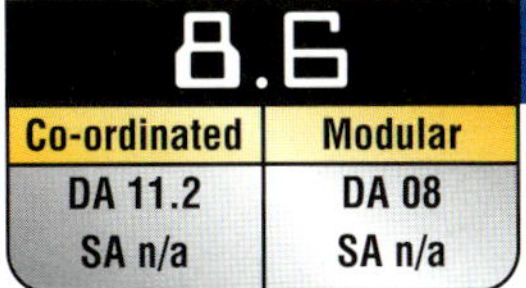

8.6
Co-ordinated
DA 11.2
SA n/a
Modular
DA 08
SA n/a

Metallic and covalent bonding

Metals have only a few electrons in their highest energy level. These electrons are not held firmly to the atom, they are free to move through the metal. Figure 8.21 is a diagram showing the structure of a metal.

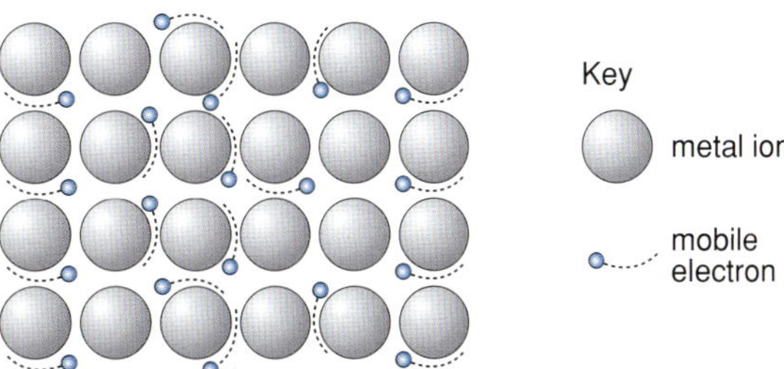

Figure 8.21
The structure of a metal showing the mobile electrons

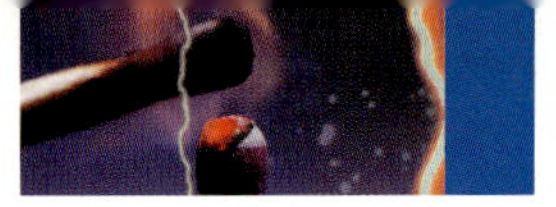

These free electrons hold the atoms together. But the atoms can move around because there are no fixed bonds holding them together. This is why metals can be pressed into shape or extruded (like toothpaste from a tube) to make wire. It is these mobile electrons that also allow metals to conduct electricity and heat. Because the metals are a giant structure they usually have quite high melting and boiling points.

Figure 8.22
Shaping metals

Did you know?

The word **malleable** is used to describe the ability metals have to be hammered into shape. The word mallet comes from the same stem. Metals are also **ductile**. This means they can be made to flow so they can be extruded. The words aqueduct, viaduct and conductor all come from the same stem.

Atoms of non-metals have outer energy levels that have a lot of electrons. These electrons are not free to move around like they are in metals. For these atoms the only way to get a full energy level is to share electrons. Figure 8.23 is a 'dot and cross' diagram showing how two hydrogen atoms bond. They do this by sharing electrons. Each atom contributes its electron to a common 'pool'. The two electrons belong to both atoms so each atom appears to have a full outer energy level (in this case a helium structure). This is called covalent bonding. With covalent bonding the atoms involved are strongly bonded together to form molecules.

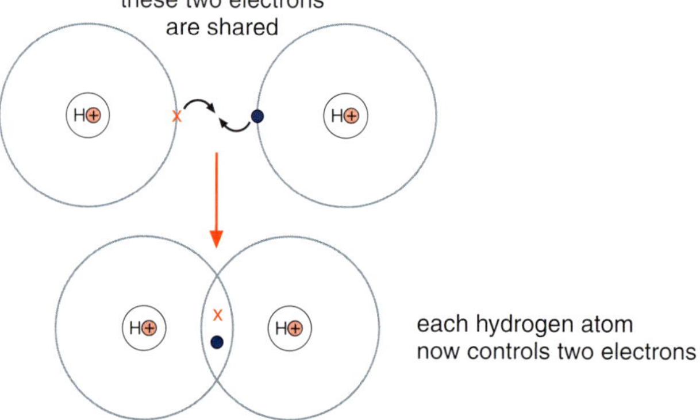

Figure 8.23
How two hydrogen atoms combine to make a hydrogen molecule

A similar thing happens with oxygen. In this case each atom donates two electrons. So there are four electrons involved. Each pair of electrons makes one bond so the oxygen atoms are held together by two bonds.

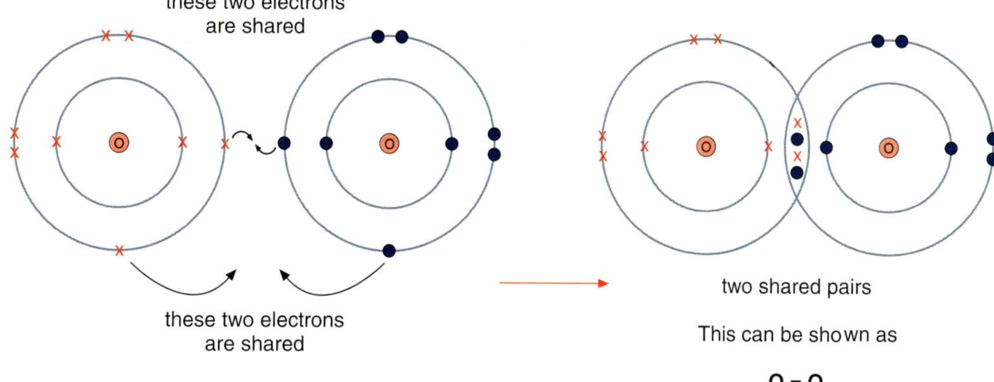

Figure 8.24 *How two oxygen atoms combine to form an oxygen molecule with a double covalent bond*

Covalent bonding also exists in compounds. Hydrogen chloride (HCl) has its atoms held together by covalent bonds (see Figure 8.25). Here too, each atom donates one electron into the common 'pool'. Hydrogen now has a share in two electrons – so its energy level is full (helium structure). The outer energy level of chlorine has seven electrons of its own and a share in one other electron making a total of 8 electrons (argon structure).

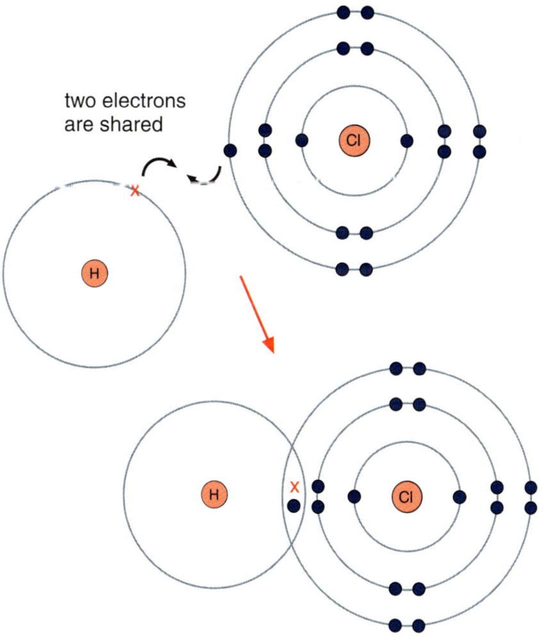

Figure 8.25 *The bonding in hydrogen chloride*

Water is a compound whose atoms are bonded covalently (see Figure 8.26). In this molecule each hydrogen has its electron paired with one of the oxygen electrons. So there are two hydrogen atoms bonded to each oxygen atom. This explains why the formula of water is H_2O.

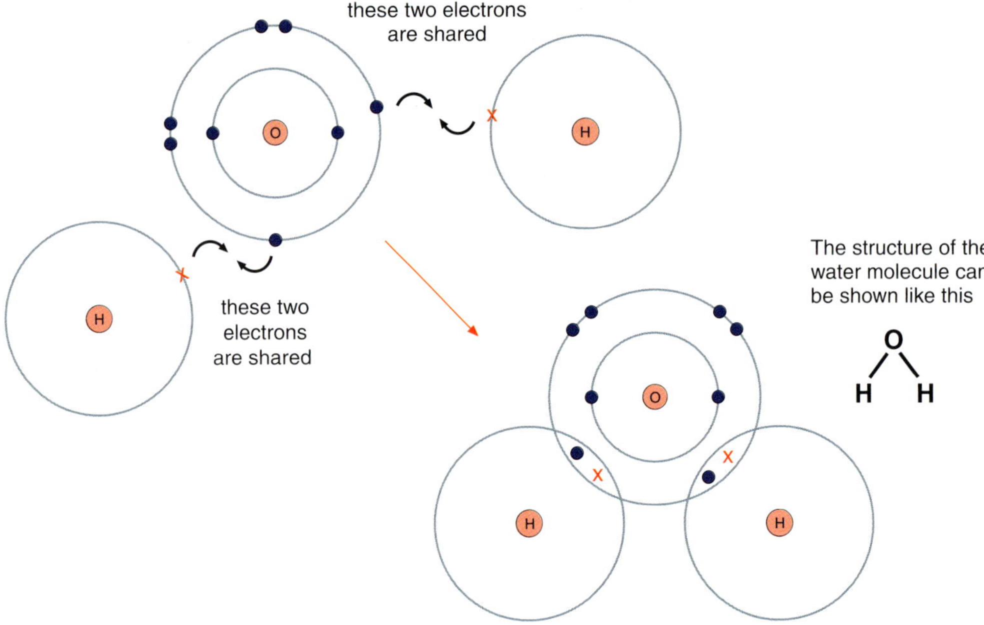

Figure 8.26
How two hydrogen atoms and one oxygen atom form a water molecule

Nitrogen needs three electrons to complete its outer energy level. To get a share in the three electrons it needs it has to 'lend' three electrons. Figure 8.27 is a 'dot and cross' diagram of a molecule of ammonia (NH_3). There are three bonds, each with one pair of electrons.

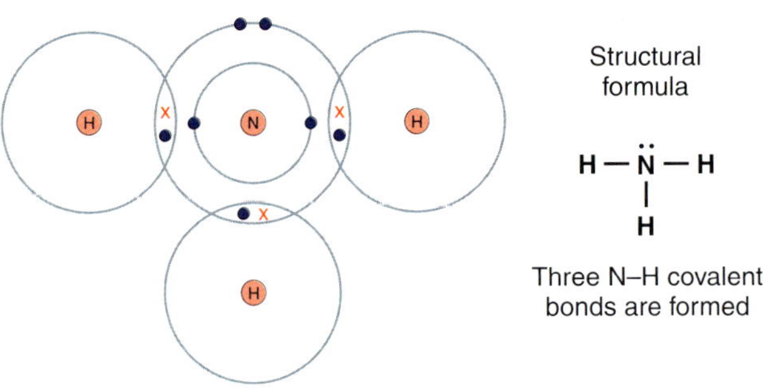

Figure 8.27
The bonding in ammonia

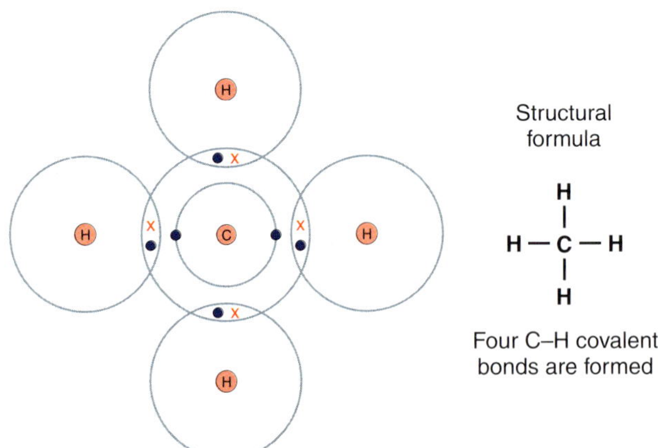

Figure 8.28
The bonding in methane

Each covalent bond requires a pair of electrons. Each atom in the bond contributes one electron. This means that an atom 'lends' as many electrons as it 'borrows'. Boron has three electrons in its outer energy level. To get a full energy level it would need to 'borrow' five electrons. Atoms can only 'borrow' as many as they can 'lend out'. So boron cannot form covalent bonds. For this reason none of the atoms in Groups 1, 2 and 3 can form covalent bonds.

Figure 8.29 gives the melting and boiling points of some compounds that have covalent bonds. As the table shows covalent compounds have low melting and boiling points. This is because covalent compounds form molecules. The bonds holding the atoms together within the molecule are very strong but the attractive forces between molecules are quite weak.

Figure 8.29 *The melting and boiling points of some covalent compounds*

Compound	Formula	Melting point (°C)	Boiling point (°C)
hydrogen chloride	HCl	−114	−85
water	H_2O	0	100
ammonia	NH_3	−78	−33
methane	CH_4	−182	−161

Tasks

1 Plotted on the chart below are the melting points of some of the compounds mentioned in this chapter.

Figure 8.30

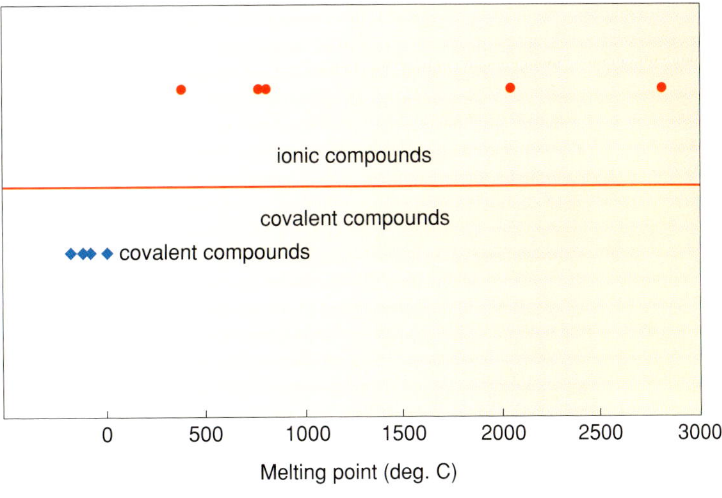

Make a list of some other substances and their melting points. Plot them on the chart. Other substances you could use are in the box below – but you can pick your own if you want to.

carbon dioxide	carbon monoxide	calcium oxide	copper chloride	ethane
glucose	iron sulphate	potassium carbonate	sodium hydroxide	

2 Although there are general rules about the way substances behave there are usually some anomalies (substances that don't seem to obey the rules).

a) Name a metal that does not have a high melting point.
b) Name a non-metal that conducts electricity.
c) Name a covalently bonded substance that has a melting point over 1000°C.

3 Wood's metal is an alloy of four other metals. It melts at 71°C. Find out the names of the other metals and their melting points.

Topic Questions

1 Name two elements that exist as covalent molecules.

2 Name four compounds that exist as covalent molecules.

3 Decide which substances in the list below have covalent bonds.

carbon dioxide	calcium chloride	copper oxide	gold	hydrogen sulphide
lead	phosphine (PH_3)	lithium chloride		

4 Use the idea of free, mobile electrons to explain why metals conduct electricity.

5 Why don't metals like sodium and magnesium form covalent bonds?

8.7 The Periodic Table

Co-ordinated	Modular
SA 11.3	SA 16
DA 11.11	DA 08

In Chapter 5 (Module 5), the Periodic Table was introduced. The elements in the table were arranged in order of **relative atomic mass**. This arrangement wasn't perfect; some elements (like argon and potassium) had to be reversed to make them 'fit' better. In the modern Periodic Table, the elements are arranged in order of their atomic number. When this is done elements do not need to be reversed. Figure 8.31 is a diagram of part of the Periodic Table. Only the first twenty elements are included. The elements are in order of their atomic number. For each element the electron structure is shown.

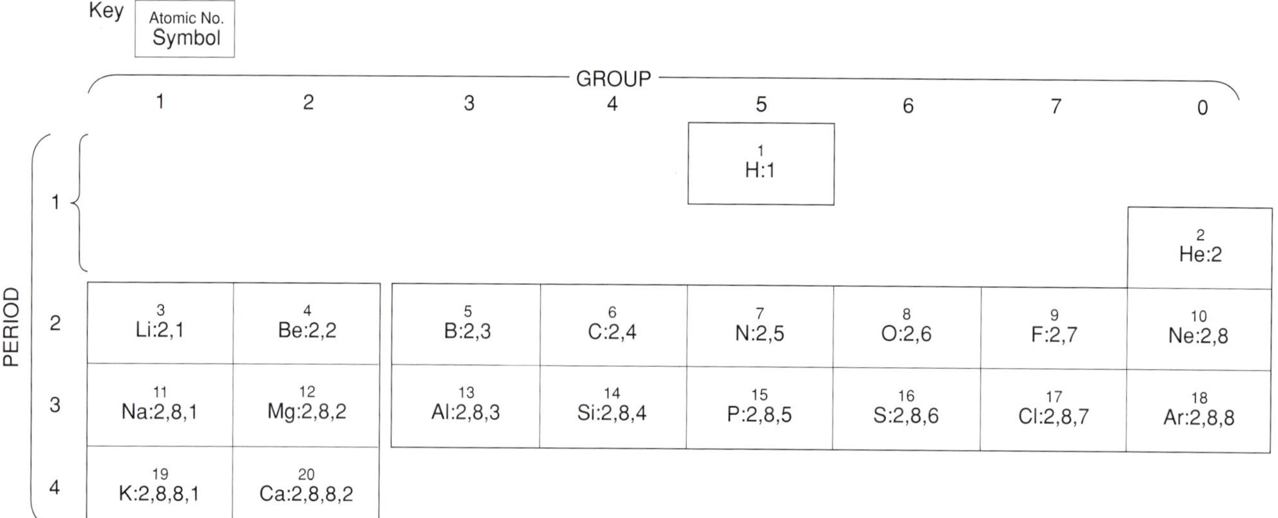

Figure 8.31
The electron structure of the first 20 elements of the Periodic Table

In the first **period** the electrons are filling the first energy level. In the second period it is the second energy level that is being filled and so on. In the first **group** there is one electron in the outer energy level, in the second group there are two electrons and so on. If all the energy levels are full then the element is in Group 0.

Elements in the same group have similar chemical properties. They also have the same number of electrons in their outer energy level. It is not surprising that the number of electrons in the outer energy level is related to the chemical properties. It is, after all, the outer electrons that are involved in making chemical bonds.

The non-metals are clustered on the right hand side of the Periodic Table. Less than a quarter of the elements are non-metals.

Scientists always try to classify things. It is much easier to understand what is happening if some sort of pattern can be observed. In 1817, a German chemist called Johann Wolfgang Döbereiner noticed that the atomic weight (which is what relative atomic mass used to be called) of strontium was almost exactly the average of the atomic weights of calcium and barium – elements which had similar chemical properties to strontium. He found other examples, for example chlorine, bromine and iodine. These groups of three elements were known as 'Döbereiner's triads'.

Figure 8.32 *Dimitri Mendeleev*

Nearly 50 years later in 1864, John Newlands, a British chemist, published what he called 'The Law of Octaves'. He observed that if the elements were arranged in order of atomic weight every eighth element had similar chemical properties. Scientists of the day laughed at him. Newlands' arrangement is shown in Figure 8.33.

Notice how similar Newlands' arrangement was to the present Periodic Table. The noble gases were not included because they were not discovered until 1895.

Figure 8.33 *Newlands' arrangement of some of the elements*

1	2	3	4	5	6	7
H	Li	Be	B	C	N	O
F	Na	Mg	Al	Si	P	S
Cl	K	Ca				

Five years later, Dimitri Mendeleev extended Newlands' idea and produced a version of the Periodic Table, shown in Figure 8.34.

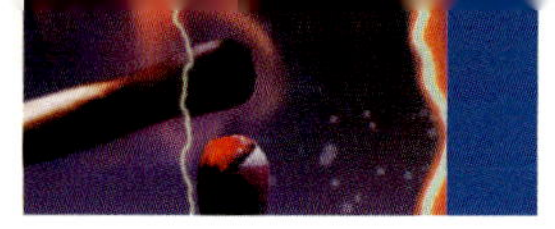

	Group 1		Group 2		Group 3		Group 4		Group 5		Group 6		Group 7	
Period 1	H (1)													
Period 2	Li (7)		Be (9.4)		B (11)		C (12)		N (14)		O (16)		F (19)	
Period 3	Na (23)		Mg (24)		Al (27.3)		Si (28)		P (31)		S (32)		Cl (35.5)	
Period 4	K (39)	Cu (63)	Ca (40)	Zn (65)	?	?	Ti (48)	?	V (51)	As (75)	Cr (52)	Se (78)	Mn (55)	Br (80)
Period 5	Rb (85)	Ag (108)	Sr (87)	Cd (112)	Y (88)	In (113)	Zr (90)	Sn (118)	Nb (94)	Sb (122)	Mo (96)	Te (125)	?	I (127)

Figure 8.34
Part of Mendeleev's 1869 Periodic Table. The numbers in brackets are the values for the atomic weights used by Mendeleev

At first Mendeleev's Periodic Table was treated as an interesting curiosity. But Mendeleev had done something different in his arrangement. In order to make the elements fit into his table he left gaps. He claimed that these gaps were for elements that had not yet been discovered. He predicted what these elements would be like and many of the properties of their compounds. When these elements were discovered and shown to agree closely with Mendeleev's predictions, the Periodic Table became accepted as an important addition to scientific knowledge and understanding.

? **Did you know?**

In 1894, two British scientists, John William Strutt (Lord Rayleigh) and William Ramsay, separated argon from the atmosphere. In the same year Rayleigh also discovered helium. He used Mendeleev's Periodic Table to predict the existence of neon between helium and argon and in 1898 finally discovered it.

Task

The atomic volume of an element can be calculated using the formula:

$$\text{atomic volume} = \frac{\text{relative atomic mass}}{\text{density of element (as a solid)}}$$

Find the data to complete the table below. (The densities of the solid for elements that are normally gases are provided in the table.)

Now calculate the atomic volume of each element.

Symbol	Na	Mg	Al	Si	P	S	Cl	Ar	K
Atomic number									
A_r									
Density (g/cm^3)							1.6	1.4	
Atomic volume									

Draw a graph of atomic volume (vertical axis) against atomic number (horizontal axis). Join the points with dots.

Topic Questions

1 In the Periodic Table what are the vertical columns called?

2 Look at the Periodic Table (Figure 5.1 on page 76).

a) How many of the elements are metals?
b) How many of the elements are non-metals?
c) What percentage of the elements are non-metals?

3 Complete the table below without referring to Figure 8.31.

Period	Group	Number of electrons in outer energy level	Number of electrons in total
2	2		
		3	13
3		6	
			10

4 Why is there a connection between the chemical properties of an element and the number of electrons in the outer energy level of its atom?

5 Döbereiner discovered that the relative atomic masses of some elements were the average of other elements with similar chemical properties. He called these groups of three elements 'triads'. Calculate the relative atomic masses of the elements left blank in the table below. Then compare the calculated value with the correct value. How good do you think Döbereiner's theory was?

Element	A_r	Element	A_r	Element	A_r
calcium	40.1	chlorine	35.5	sulphur	32.1
strontium		bromine		selenium	
barium	137.4	iodine	126.9	tellurium	127.6

6 Newlands' Law of Octaves was laughed at when he published it in 1864. Yet five years later when Mendeleev published his Periodic Table scientists eventually accepted it. Why was Mendeleev's work accepted when Newlands' was not?

8.8

Co-ordinated	Modular
DA 11.11	DA 08
SA n/a	SA n/a

Group 1 – The alkali metals

The Group 1 metals are called the **alkali metals**. This is because they react with water to form alkaline solutions (Figure 5.2). The solutions contain the hydroxide of the metal. Hydrogen gas is also given off. For lithium the reaction is:

$$\text{lithium} + \text{water} \longrightarrow \text{lithium hydroxide} + \text{hydrogen}$$
$$2Li + 2H_2O \longrightarrow 2LiOH + H_2$$

The other alkali metals react in exactly the same way.

For sodium the reaction is:

$$\text{sodium} + \text{water} \longrightarrow \text{sodium hydroxide} + \text{hydrogen}$$
$$2Na + 2H_2O \longrightarrow 2NaOH + H_2$$

For potassium the reaction is:

$$\text{potassium} + \text{water} \longrightarrow \text{potassium hydroxide} + \text{hydrogen}$$
$$2K + 2H_2O \longrightarrow 2KOH + H_2$$

The alkali metals also react with many non-metals. (In the chemical equations below the letter M is used to represent the symbol of the alkali metal. To get the equation for a particular alkali metal replace M with Li, Na or K.)

With oxygen:

$$\text{alkali metal} + \text{oxygen} \longrightarrow \text{alkali metal oxide}$$
$$4M + O_2 \longrightarrow 2M_2O$$

And with chlorine:

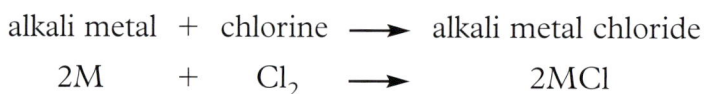

$$\text{alkali metal} + \text{chlorine} \longrightarrow \text{alkali metal chloride}$$
$$2M + Cl_2 \longrightarrow 2MCl$$

In each of these compounds the alkali metal exists as an ion. The ion has a single positive charge (Li^+, Na^+, K^+).

The melting and boiling points of the alkali metals decrease further down the group.

Figure 8.35 *Physical properties of lithium, sodium and potassium*

Element	Atomic number	Relative atomic mass	Melting point (°C)	Boiling point (°C)
lithium	3	6.9	180	1330
sodium	11	23.0	98	883
potassium	19	39.1	64	760

The metals also become more reactive further down the group. If a piece of lithium is dropped into water it fizzes (the correct word is 'effervesces') and gives off hydrogen. With sodium the effervescence is more vigorous. The reaction is exothermic and the sodium melts. If potassium is dropped

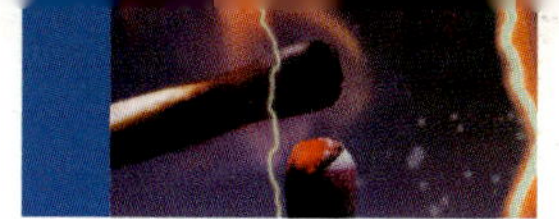

into water the reaction is so exothermic that the hydrogen gas ignites (see Figure 5.2 on page 78).

Testing for hydrogen

If a test tube containing hydrogen is placed by a flame, the hydrogen will burn with a loud 'pop'.

Task

Try to find out a use for one of the alkali metals. (It must be the metal itself, not one of its compounds.)

Topic Questions

1 What is the charge on the ion of a Group 1 metal?

2 Draw a 'dot and cross' diagram to show the electron movements in the following reactions:

a) lithium combining with chlorine
b) sodium combining with oxygen

3 Describe how you would test a gas to see if it was hydrogen. What would you notice if the gas were hydrogen?

4 Use the Periodic Table (Figure 5.1 on page 76) to answer the following questions.

a) What other alkali metals are there besides those mentioned in the text?
b) What, approximately, do you think the melting points of these metals will be?
c) Describe the reaction of each of these metals with water.
d) Write a chemical equation for the reaction of these metals with oxygen.

8.9	
Co-ordinated	Modular
DA 11.11 SA n/a	DA 08 SA n/a

Group 7 – The halogens

The halogens are non-metals. Their properties are typical of non-metals. They have low melting and boiling points (see Figure 8.36). They are brittle and crumbly when solid and they are poor conductors of electricity and heat.

? Did you know?

The word halogen comes from two Greek words. Halo-, from the word for sea-salt and -gen from the word for origin or creation. The same stem is used in the words Genesis and gene. So, a halogen is a 'salt-creator'.

Figure 8.36
The melting points, boiling points and appearance of the halogens

Element	symbol	Melting point (°C)	Boiling point (°C)	Appearance at room temperature	Appearance of vapour
fluorine	F	−220	−188	pale yellow gas	
chlorine	Cl	−110	−34	yellowish green gas	
bromine	Br	−7	58	dark reddish brown liquid	dark reddish brown
iodine	I	114	183	black solid	violet

The halogen elements exist as molecules. Each molecule has two atoms joined by covalent bonds. Figure 8.37 shows the bonding in fluorine.

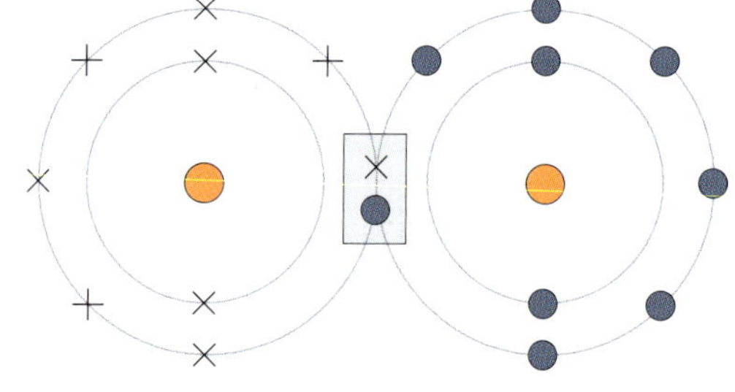

Figure 8.37
Covalent bonding in fluorine

Did you know?

Chlorine was used in the First World War as a poisonous gas.

The halogens form ionic salts with metals. Figure 8.18 shows the formation of the ionic bond between sodium and chlorine. All of these compounds contain the halogen as a negatively charged ion. For example, the chloride ion Cl^-.

Halogens also form compounds with many non-metals. These compounds exist as separate molecules because they have covalent bonds. Figure 8.25 shows the formation of a covalent bond in a molecule of hydrogen chloride.

As Figure 8.36 shows, as you go down Group 7 the higher the melting and boiling points of the halogens become. The halogens have their own reactivity series. The further down the group, the less reactive the halogen.

A more reactive halogen will displace a less reactive halogen from a solution of its salts. So chlorine will displace bromine from sodium bromide solution. It will also displace iodine from a solution containing iodide ions. Figure 8.38 shows chlorine displacing bromine and iodine.

Figure 8.38
Chlorine displacing bromine and iodine from a solution of their salts

The equations for these reactions are:

sodium bromide + chlorine ⟶ sodium chloride + bromine

$$2NaBr(aq) + Cl_2(g) \longrightarrow 2NaCl(aq) + Br_2(l)$$

and:

sodium iodide + chlorine ⟶ sodium chloride + iodine

$$2NaI(aq) + Cl_2(g) \longrightarrow 2NaCl(aq) + I_2(l)$$

Task

Astatine (At) is another halogen. It is below iodine in Group 7. It has been estimated that there is less than 1 gram of astatine on the Earth. Make some predictions about astatine. (For example, you could predict its appearance at room temperature, its melting point and boiling point and the formulae of its compounds with sodium and with hydrogen.)

Topic Questions

1 Draw a 'dot and cross' diagram of a chlorine molecule.

2 Draw a diagram of a fluoride ion.

3 Complete the following word equations. If the substances do not react, write 'No Reaction' in the space. Write balanced chemical equations for those cases where there is a reaction.

a) sodium fluoride solution + chlorine ⟶
b) potassium chloride solution + fluorine ⟶
c) potassium iodide solution + bromine ⟶
d) lithium bromide solution + iodine ⟶

4 Explain why the halogens are poor conductors of electricity.

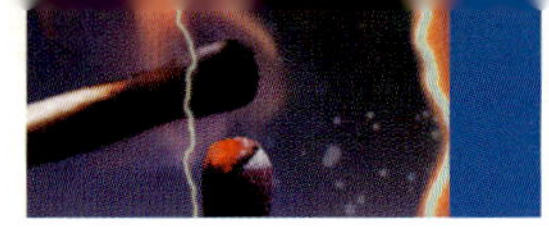

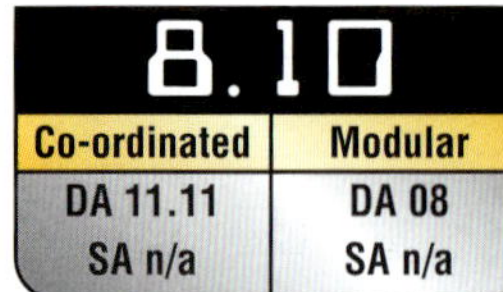

8.10

Co-ordinated	Modular
DA 11.11 SA n/a	DA 08 SA n/a

Group 0 – The noble gases

Like the halogens the noble gases are typical non-metals. They have very low melting and boiling points (see Figure 8.39). But unlike the halogens they are not reactive.

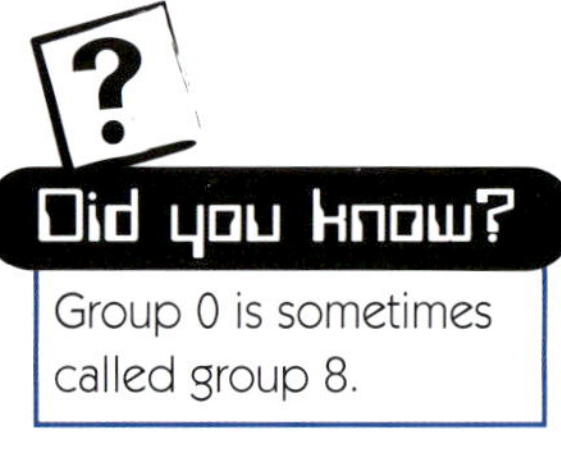

Did you know?

Group 0 is sometimes called group 8.

Figure 8.39
The melting and boiling points of the noble gases

Element	Symbol	Atomic number	Melting point (°C)	Boiling point (°C)
helium	He	2	−270	−269
neon	Ne	10	−249	−246
argon	Ar	18	−189	−186
krypton	Kr	36	−157	−153
xenon	Xe	54	−112	−108
radon	Rn	86	−71	−62

Did you know?

The noble gases used to be called the inert gases. Then it was discovered that xenon would react with fluorine (a very reactive element) if the conditions were correct. So the group name was changed.

The noble gases have all of their occupied energy levels full. This means they have a stable electron structure. This is why they exist as free atoms. All other gaseous elements exist as molecules containing two atoms.

Because they are so unreactive the noble gases are used in lamps. In fact electric discharge lamps are usually called 'neon' lights. Argon is used as the filling in normal electric light bulbs.

Helium, like hydrogen, is less dense than air. It is used in 'lighter-than-air' balloons instead of hydrogen because helium is not flammable.

Tasks

1 It is estimated that about 23% of the mass of the Universe is helium. Find out where most of the helium is found. Find out why helium was 'discovered' in 1868 but not found on the Earth for another 27 years.

2 An ordinary light bulb has a volume of about 150 cm^3. Estimate how many light bulbs could be filled from the argon in the room you are in now. (Clue: about 1% of the atmosphere is argon)

Topic Questions

1 Give **two** ways in which the noble gases are typical non-metals.

2 Draw a diagram of the electron structures of helium, neon and argon. Use this diagram to explain why the noble gases are so unreactive.

3 Helium is less dense than air. Why is it wrong to say helium is 'lighter' than air?

4 Draw a graph of the data in the table in Figure 8.39. Draw the melting point graph and the boiling point graph on the same axes. Compare this data with the melting and boiling points of some other gaseous elements, for example fluorine, chlorine. From this comparison decide what is particularly unusual about the noble gases.

8.11 Some halogen compounds

Co-ordinated	Modular
DA 11.12	DA 08
SA n/a	SA n/a

The properties of compounds are different from the properties of the elements in them. Sodium is a reactive metal. Chlorine is a toxic, green gas. But sodium chloride is a white, crystalline salt. It is 'common salt' – the salt used in cooking. Sodium chloride is found in the sea. It is also found as deposits underground.

Figure 8.40
Salt pans – the water evaporates and the salt remains

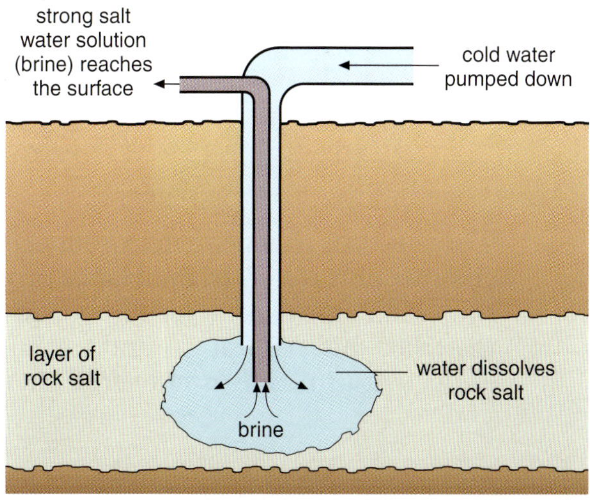

Figure 8.41
Solution mining

Did you know?

There are about 400 million million tonnes of salt in the sea.

Brine is a solution of sodium chloride. If brine is electrolysed chlorine gas comes off at the positive electrode and hydrogen at the negative electrode.

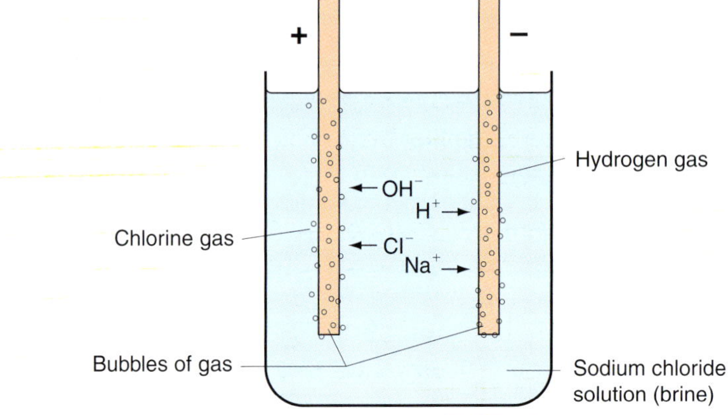

Figure 8.42
The electrolysis of sodium chloride solution (brine)

The positive ions (Na^+ and H^+) are attracted to the negative electrode. The hydrogen ions gain electrons from the electrode and form hydrogen gas. The sodium ions stay in solution.

The negative ions (Cl^- and OH^-) are attracted to the positive electrode. The chloride ions lose electrons to the electrode and form chlorine gas. The hydroxide ions stay in solution.

When electrolysis is finished the solution contains sodium ions and hydroxide ions. It is a solution of sodium hydroxide. The chlorine, hydrogen and sodium hydroxide produced are used to make other useful materials. Figure 8.43 summarises these uses.

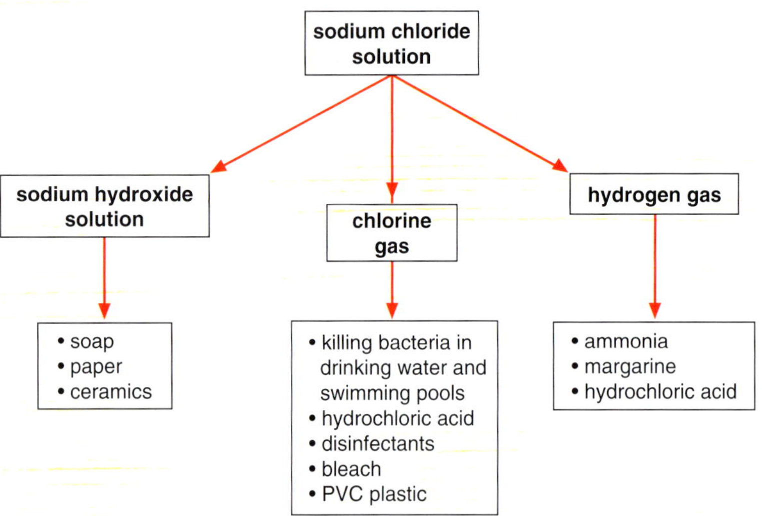

Figure 8.43
Useful products from the electrolysis of sodium chloride solution

These useful chemicals are produced from cheap materials – sodium chloride and water. In the Haber process (Section 7.8), useful chemicals (ammonia, nitric acid, fertilisers and explosives) are made from cheap materials (air, natural gas and water). Using cheap raw materials is an important principle in industrial chemistry.

Testing for chlorine

If chlorine gas is tested with a piece of damp litmus paper the paper is bleached. So if damp litmus paper goes white then the gas is chlorine.

Silver chloride, silver bromide and silver iodide are the silver halides. Light, X-rays and radiation from radioactive substances affect them; the silver ion is reduced to silver metal. This means the silver halides are light sensitive and can be used in photography. Photographic film and paper use silver halides.

The hydrogen halides are gases. Figure 8.25 shows how the covalent bond between hydrogen and chlorine is formed. But when hydrogen chloride dissolves in water it forms hydrochloric acid. This happens with all the hydrogen halides. So hydrogen bromide and hydrogen iodide are acidic gases. They dissolve in water to form hydrobromic and hydroiodic acids.

Tasks

1 The properties of sodium chloride are not like those of its elements. Compare the properties of water and copper sulphate with their elements. Try the same exercise with some other compounds (e.g. methane, carbon monoxide and silver bromide).

2 Find out how a) hydrogen is used in the manufacture of margarine, and b) sodium hydroxide is used to make soap.

3 PVC is a polymer. Find out the name and formula of the monomer.

Topic Questions

1 What is brine?

2 Describe the test for chlorine.

3 When sodium chloride solution is electrolysed one of the products is hydrogen. What method is used to get hydrogen for making ammonia?

4 The silver halides are ionic. Use the data sheet to work out the formulae of the silver halides.

5 How could you prove that hydrogen chloride was an acidic gas?

6 When sodium chloride solution is electrolysed the hydrogen ions and the sodium ions are attracted to the negative electrode. Why is hydrogen produced at the negative electrode not sodium?

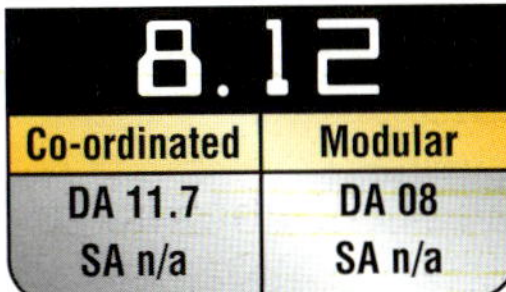

Symbols, formulae and equations

Symbols

All elements are represented by symbols. Some symbols are just one letter. It is always a capital letter. It is usually the first letter of the name of the element.

Figure 8.44
Symbols that are single letters

Symbol	Name	Symbol	Name	Symbol	Name
B	boron	H	hydrogen	O	oxygen
C	carbon	I	iodine	P	phosphorus
F	fluorine	N	nitrogen	S	sulphur

Most symbols contain two letters. The first is always a capital, the second always lower case. Often the letters are the first two letters of the element's name.

Figure 8.45
Symbols that are the first two letters of the element's name

Symbol	Name	Symbol	Name	Symbol	Name
Al	aluminium	Br	bromine	Li	lithium
Ar	argon	Ca	calcium	Ne	neon
Be	beryllium	He	helium	Si	silicon

Sometimes the second letter of the symbol is not the second letter of the name.

Figure 8.46
Symbols that are the first letter and one other letter of the element's name

Symbol	Name	Symbol	Name	Symbol	Name
Cl	chlorine	Mg	magnesium	Rn	radon
Cr	chromium	Pt	platinum	Zn	zinc

There are a number of elements with symbols based on the Latin names of the substance.

Figure 8.47
Symbols that are based on the Latin names of the element

Symbol	Name	Symbol	Name	Symbol	Name
Ag	silver	Fe	iron	Na	sodium
Au	gold	Hg	mercury	Pb	lead
Cu	copper	K	potassium	Sn	tin

Formulae

Formulae are the 'symbols' of chemical compounds. They are made up of the symbols of the elements in the compound. The formula also shows how many atoms of each element are present. The formulae of some compounds are shown in Figure 8.48.

Figure 8.48
The formulae of some compounds

Formula	Elements (Quantity)	Formula	Elements (Quantity)
CO	carbon (1), oxygen (1)	NH_3	nitrogen (1), hydrogen (3)
$NaCl$	sodium (1), chlorine (1)	CH_4	carbon (1), hydrogen (4)
H_2O	hydrogen (2), oxygen (1)	Fe_2O_3	iron (2), oxygen (3)
CO_2	carbon (1), oxygen (2)	P_2O_5	phosphorus (2), oxygen (5)

Most compounds have names that indicate what the compound contains, for example sodium chloride. Others have names that tell you how much of a particular element is present, for example carbon monoxide, carbon dioxide, phosphorus pentoxide. Some compounds have 'common' names

that tell you nothing about them, for example water, ammonia and methane.

In many cases it is possible to work out the formula of a compound. Using the Data Sheet makes it possible to work out the formulae of ionic compounds.

Example 1 What is the formula of silver chloride?

Step 1 Write down the symbols of the ions.

$$Ag^{+} \quad Cl^{-}$$

Step 2 Vary the numbers of each symbol so there are as many positive charges as negative charges.

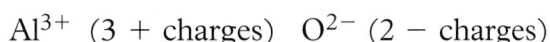

Ag^{+} (1 + charge) Cl^{-} (1 − charge)

The charges are the same so the formula is:

$$AgCl$$

Example 2 What is the formula of copper carbonate?

Step 1 Write down the symbols of the ions.

$$Cu^{2+} \quad CO_3^{2-}$$

Step 2 Vary the numbers of each symbol so there are as many positive charges as negative charges.

Cu^{2+} (2 + charges) CO_3^{2-} (2 − charges)

The charges are the same so the formula is:

$$CuCO_3$$

Example 3 What is the formula of potassium sulphide?

Step 1 Write down the symbols of the ions.

$$K^{+} \quad S^{2-}$$

Step 2 Vary the numbers of each symbol so there are as many positive charges as negative charges.

K^{+} (1 + charge) S^{2-} (2 − charges)

Two potassium ions make a total of 2 + charges. This matches the 2 – charges. So the formula is:

$$K_2S$$

Example 4 What is the formula of aluminium oxide?

Step 1 Write down the symbols of the ions.

$$Al^{3+} \quad O^{2-}$$

Step 2 Vary the numbers of each symbol so there are as many positive charges as negative charges.

Al^{3+} (3 + charges) O^{2-} (2 − charges)

Two aluminium ions make a total of 6 + charges. Three oxide ions make a total of 6 − charges. So the formula is:

$$Al_2O_3$$

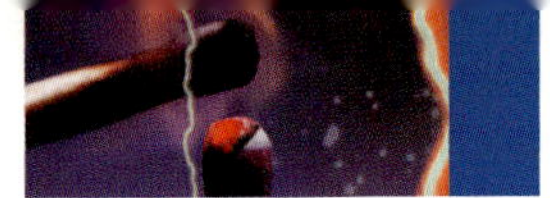

Chemical equations

Chemical equations are a simplified way of describing a chemical reaction.

In a description of a reaction phrases like *reacts with, burns in* and *produces* are used. In a word equation the names of the **reactants** and **products** are the same but the other phrases are replaced with symbols.

In a chemical equation the names of the reactants and products are replaced with their chemical symbols or formulae. In a chemical reaction there is no change in total mass. This is because every atom present in the reactants is still there in the products.

Chemical equations MUST balance. **That means that the total number of each atom in the products must be the same as the total number of each atom in the reactants.** To do this certain rules must be kept.

1 The formulae of the reactants and products must not be changed.
2 The reactants used and the products formed must not be changed.
3 The only thing that can be changed is the number of particles of reactant and product in the equation.

Example 1 Carbon burns in oxygen to produce carbon dioxide

(In the table below the reactants and products are in bold.)

Sentence	**Carbon** burns in **oxygen** to produce **carbon dioxide**
Word equation	carbon + oxygen ⟶ carbon dioxide
Chemical equation	$C + O_2 \longrightarrow CO_2$

In this case they already balance. There is one carbon and two oxygen atoms in both the reactants and products.

Example 2 Calcium carbonate reacting with hydrochloric acid

(In the table below the reactants and products are in bold.)

Sentence	**Calcium carbonate** fizzes in **hydrochloric acid**. **Carbon dioxide** is given off and a solution of **calcium chloride** is left. **Water** is also formed.
Word equation	calcium carbonate + hydrochloric acid ⟶ carbon dioxide + calcium chloride + water
Chemical equation	$CaCO_3 + HCl \longrightarrow CO_2 + CaCl_2 + H_2O$

Reactants		Products	
Element	No. of atoms	Element	No. of atoms
Ca	1	Ca	1
C	1	C	1
O	3	O	3
H	**1**	H	**2**
Cl	**1**	Cl	**2**

In this case the equation does not balance. This can be checked by adding up the number of atoms of each type in the reactants and products.

To make the equation balance one more hydrogen and one more chlorine atom are needed in the reactants. This can be done by having '2HCl' in the equation. So we can change the previous chemical equation to:

Chemical equation	$CaCO_3 + \mathbf{2HCl} \longrightarrow CO_2 + CaCl_2 + H_2O$

Now the equation balances. There are the same number of atoms of each type in the reactants and products.

So the steps are:

1 Write a word equation.
2 Write a chemical equation.
3 Balance the chemical equation.

Example 3 Methane burns in oxygen to produce carbon dioxide and water vapour

(In the table below the reactants and products are in bold.)

Steps 1 and 2

Sentence	**Methane** burns in **oxygen** to produce **carbon dioxide** and **water vapour**
Word equation	methane + oxygen ⟶ carbon dioxide + water vapour
Chemical equation	$CH_4 + O_2 \longrightarrow CO_2 + H_2O$

Step 3

Reactants		Products	
Element	No. of atoms	Element	No. of atoms
C	1	C	1
H	4	H	2
O	2	O	3

To make the hydrogen atoms balance put '$2H_2O$' to the products.

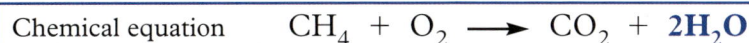

Chemical equation	$CH_4 + O_2 \longrightarrow CO_2 + \mathbf{2H_2O}$

Repeat Step 3

Reactants		Products	
Element	No. of atoms	Element	No. of atoms
C	1	C	1
H	4	H	4
O	2	O	4

But the oxygen atoms still do not balance. They can be made to balance by putting '$2O_2$' to the reactants.

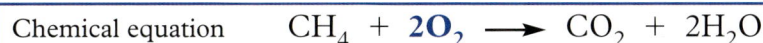

Chemical equation	$CH_4 + \mathbf{2O_2} \longrightarrow CO_2 + 2H_2O$

Repeat Step 3 to check

Reactants		Products	
Element	No. of atoms	Element	No. of atoms
C	1	C	1
H	4	H	4
O	4	O	4

The equation now balances.

State symbols

It is sometimes important to know if a substance is a solid, liquid or gas or if it is in solution. To do this state symbols are used.

Figure 8.49 *State symbols*

State	State symbol
Solid	(s)
Liquid	(l)
Gas	(g)
In solution	(aq)

The state symbols are written in the equation after the formulae. The two equations below have state symbols included.

$$C(s) + O_2(g) \longrightarrow CO_2(g)$$

$$CaCO_3(s) + 2HCl(aq) \longrightarrow CO_2(g) + CaCl_2(aq) + H_2O(l)$$

Tasks

1 The formula of water is written as H_2O, not HHO, and calcium carbonate is $CaCO_3$, not CaCOOO. Why is it an advantage to use subscript numbers rather than write all the symbols out?

2 Find out the Latin names for some of the elements in Figure 8.47. Find out the French words for the elements gold, iron, lead and silver.

3 What were some of the original symbols for elements used by John Dalton? One of Dalton's elements was called 'azote'. What is it called now? Why was it called 'azote'?

Topic Questions

1 In the table below are some elements with possible symbols. Circle the correct symbol for the element. (You should be able to work them out from the information in this section. **You do not need to learn these symbols**.)

barium	B	Ba	BA	Br	b
caesium	Ca	CA	Cs	C	c
cerium	Ce	C	c	CR	Cr
nickel	ni	Ni	N	Ne	NE
niobium	Ni	NB	ni	Nb	N
strontium	Sr	S	s	SR	Si

2 Name the compounds with the following formulae.

a) HI　b) MgO　c) CaS
d) H_2S　e) NaF　f) $Cu(NO_3)_2$
g) $Ca(OH)_2$　h) Li_2CO_3

3 The formula of nitrogen monoxide is NO. What are the formulae of:

a) nitrogen dioxide　b) dinitrogen monoxide
c) dinitrogen trioxide　d) dinitrogen tetroxide

4 Complete the following table. The first one has been done for you.

Element or 'group' Symbol	Quantity	Element or 'group' Symbol	Quantity	Formula	Name
Si	1	Cl	4	$SiCl_4$	silicon tetrachloride
P	2	O	3		
P	2	O	5		
Mg	1	OH	2		

5 Use the Data Sheet to find the formulae of the following:

a) zinc carbonate　b) calcium iodide　c) silver sulphate
d) aluminium oxide　e) ammonium sulphate

6 One of the compounds in Figure 8.48 is Fe_2O_3. What is the charge on the iron ion in this compound?

7 Ammonia will react with oxygen to produce nitrogen monoxide and water vapour. Which of the following is a correctly balanced chemical equation for this reaction?

a) $NH_3 + O_2 \longrightarrow NO + H_3O$
b) $NH_2 + O_2 \longrightarrow NO + H_2O + H$
c) $NH_3 + O_2 \longrightarrow NO + H_2O$
d) $2NH_3 + 2O_2 \longrightarrow 2NO + 3H_2O$
e) $4NH_3 + 5O_2 \longrightarrow 4NO + 6H_2O$

Give a reason why each of the other equations is wrong.

8 Write balanced chemical equations for the following reactions.

a) $C_2H_4 + O_2 \longrightarrow CO_2 + H_2O$
b) $C_3H_8 + O_2 \longrightarrow CO_2 + H_2O$
c) $C_5H_{12} + O_2 \longrightarrow CO_2 + H_2O$
d) $C_4H_{10} + O_2 \longrightarrow CO_2 + H_2O$

9 Write balanced chemical equations for the following reactions. Include state symbols.

a) When calcium carbonate solid is heated, carbon dioxide gas is given off and calcium oxide powder is left behind.
b) Sodium hydroxide solution is neutralised by dilute sulphuric acid solution. One of the products is sodium sulphate solution.
c) Aluminium powder burns in oxygen to produce solid aluminium oxide.
d) If copper nitrate crystals are heated it gives off dinitrogen tetroxide gas and oxygen. Black copper oxide powder is left behind.

Summary

- The Kinetic Theory explains the behaviour of solids, liquids and gases in terms of the particles in them.
- Atoms are the smallest part of an element that can exist.
- Compounds contain more than one sort of element.
- The nucleus of an atom contains protons and neutrons.
- Protons and neutrons have a mass of 1.
- Protons have a positive electrical charge.
- Neutrons have no electrical charge.
- Electrons orbit the nucleus of an atom within energy levels.
- Electrons have a negative electrical charge and a very small mass.
- In an electrically neutral atom the number of electrons equals the number of protons.
- The atomic number of an atom is the number of protons it contains.
- The mass number of an atom is the total number of protons and neutrons it contains.
- Each electron energy level has a maximum number of electrons it can contain. For the first energy level it is two electrons. For the second and third energy levels it is eight electrons.
- The noble gases (Group 0) are unreactive because they have full energy levels. This makes the atom very stable.
- Other elements try to get full energy levels by gaining, losing or sharing electrons.
- An ion is formed when an atom gains or loses electrons to form a charged particle with only full energy levels.
- An ionic bond is formed when electrons move from one atom to another forming two oppositely charged ions. These ions are held together by the attraction between their opposite charges.
- A covalent bond is formed when atoms share electrons to form full energy levels in each atom.
- In metals the atoms are held together by the freely moving outer electrons from each atom.
- The Periodic Table is an arrangement of atoms in order of their atomic number.
- Elements with the same number of electrons in their outer energy level are in the same group of the Periodic Table and have similar chemical properties.
- The alkali metals are in Group 1 of the Periodic Table. They are very reactive and form ions with a charge of +1.
- Fluorine, chlorine, bromine and iodine are called the halogens. They are in Group 7 of the Periodic Table.
- The halogens are non-metals with similar chemical properties. They react with many metals and form ions with a charge of −1. Halogens can form covalent bonds with other non-metals, for example hydrogen.
- The noble gases are in Group 0 of the Periodic Table. They are very unreactive.
- Sodium chloride (common salt) is an abundant chemical. The useful substances hydrogen, chlorine and sodium hydroxide can be obtained by the electrolysis of sodium chloride solution.
- Silver halides are used to make photographic film and paper.
- All chemical elements can be represented by symbols.
- All chemical compounds can be represented by formulae.
- All chemical reactions can be represented by equations.
- Chemical equations must be 'balanced'. (They must have the same number of each type of atom in the reactants and in the products.)
- State symbols after the formula of a substance indicate whether that substance is a solid, liquid, gas or in solution in a reaction.

Examination Questions

1 a) The diagram below shows an atom. Draw lines joining the labels to the correct part of the diagram. *(3 marks)*

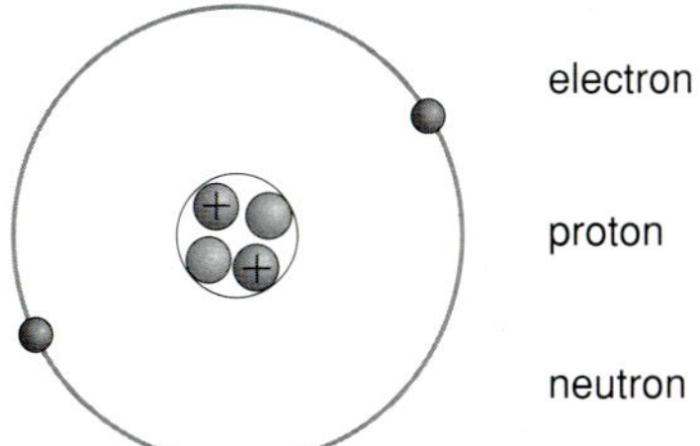

b) Complete the following table showing the properties of the parts of an atom. *(6 marks)*

Name of part of atom	Mass	Charge
	very small	
	1	
		+1

2 Some elements exist as atoms and some elements exist as molecules. But compounds can never exist as atoms.

a) Name a non-metal that exists as atoms. *(1 mark)*

b) Name a non-metal that exists as molecules. *(1 mark)*

c) Why can compounds never exist as atoms? *(3 marks)*

3 Using the method shown for calcium, complete the structures of the elements for which the symbols are given in the diagram of the Periodic Table below.

4 Elements in the Periodic Table are arranged in order of their atomic number.

a) What is meant by atomic number? *(2 marks)*

The mass number of an element is usually larger than the atomic number.

b) What is meant by mass number? *(2 marks)*

c) Hydrogen has an atomic number of 1 and a mass number of 1. What particles are present in a hydrogen atom? *(4 marks)*

5 The electron structure of an element can be represented in two ways. The diagram below shows those ways.

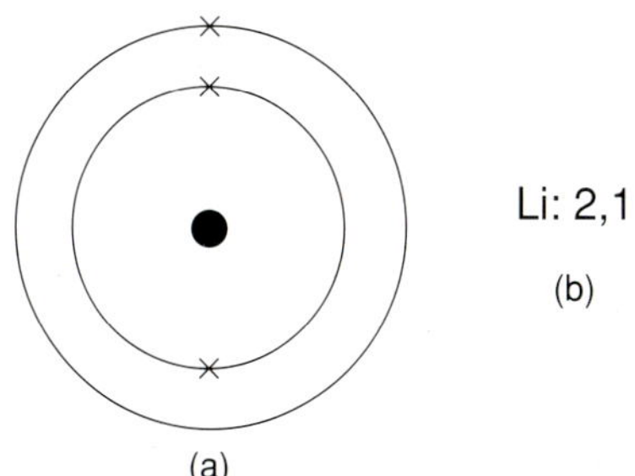

Diagram a) is called a 'dot and cross' diagram.

a) Draw 'dot and cross' diagrams for the following atoms.
- (i) He: 2,8
- (ii) Al: 2,8,3
- (iii) K: 2,8,8,1 *(3 marks)*

b) For each of the elements in part a):
- (i) What is the name of the element?
- (ii) What is the atomic number of the element?
- (iii) In which group of the Periodic Table is the element found? *(6 marks)*

H:

			C:				Ne:
	Mg:				S:		Ar:
K:	Ca:2,8,8,2						

(7 marks)

6 The diagram shows atoms of lithium and fluorine. Lithium and fluorine react to form lithium fluoride. Use the diagram to show how the bond between lithium and fluorine is formed. (*3 marks*)

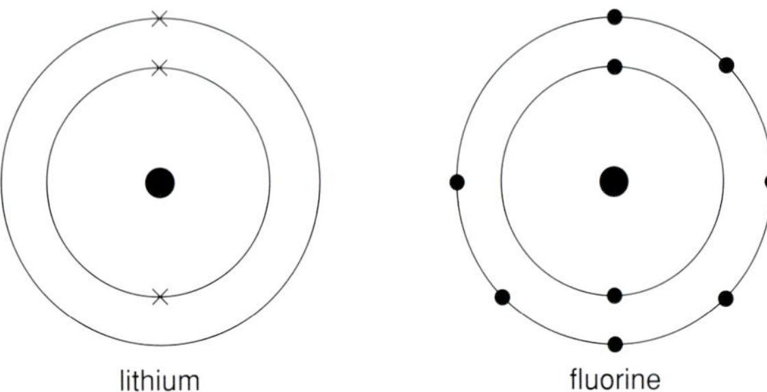

7 The 'dot and cross' diagram shows the covalent bond between two hydrogen atoms and an oxygen atom in water. The structure can also be represented as H–O–H.

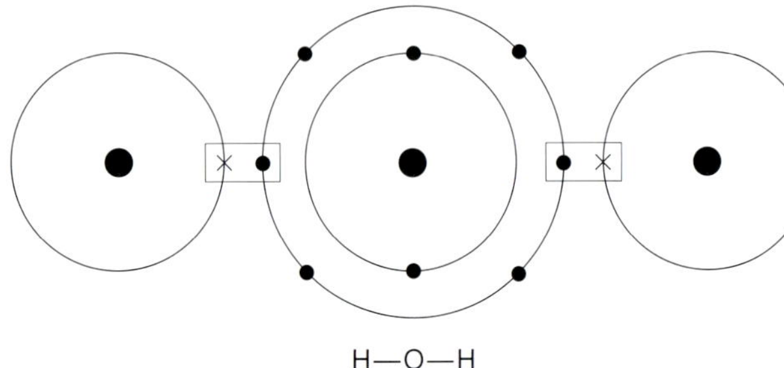

Draw a similar 'dot and cross' diagram to show the covalent bonds in the following molecules.

a) hydrogen fluoride H—F (*2 marks*)
b) carbon dioxide O=C=O (*4 marks*)

8 In 1894, Rayleigh and Ramsay discovered argon. Later in the year Rayleigh discovered helium. He then predicted the existence of neon. The relative atomic masses of the gases are: He = 4, Ar = 40.

a) Predict the relative atomic mass of neon. (*2 marks*)
b) Why was Rayleigh able to predict the existence of neon? (*4 marks*)

9 Potassium metal reacts with water. A solution of potassium hydroxide is formed and hydrogen gas is given off. In the reaction the hydrogen catches fire.

a) Write a word equation for the reaction. (*2 marks*)
b) Write a balanced chemical equation for the reaction. Include state symbols. (*4 marks*)
c) Is the reaction exothermic or endothermic? Give a reason for your answer. (*2 marks*)
d) Universal indicator is green in water. As the potassium reacts the Universal indicator changes colour.
   (i) What colour does the Universal indicator become?
   (ii) Why does it change colour?
   (iii) Name a substance that could be added to the container to turn the colour back to green. (*3 marks*)

10 Chlorine is a halogen.

a) Name two other halogens. (*2 marks*)
b) The halogens form compounds with many other elements. Complete the table below.

Name of element	Name of chloride of the element	Is the bond ionic or covalent?
Hydrogen		
Silver	Silver chloride	
Sodium		

(*5 marks*)

c) Give a use for silver chloride. (*1 mark*)
d) What are the three important chemicals obtained by the electrolysis of brine (a solution of common salt)? (*3 marks*)

Chapter 9

Energy

Key terms efficiency • fossil fuel • geothermal energy • global warming • hydroelectric • kilowatt hour • non-renewable energy • power • radiation • renewable energy • turbine

9.1 Heat energy transfer

Co-ordinated	Modular
DA 12.16	DA 9
SA 12.10	SA 17

What you know

- Heat energy always moves from places at a high temperature to places at a low temperature. So when different parts of a substance are at different temperatures, heat energy is transferred by the substance from the higher temperature places to the lower temperature places.
- Conduction is the process by which heat energy is transferred by a substance without the substance moving.
- If one end of a metal bar is heated in a Bunsen flame, heat energy from the flame moves quickly along the bar from the hot end to the cold end. The metal is a good conductor. Non-metals, like wood, glass and plastic, are poor conductors. They are good insulators. Gases, such as air, are very poor conductors.
- Convection is a form of heat transfer that happens in liquids and gases. Liquids and gases can flow. Convection happens when liquids and gases flow from higher to lower temperature places.
- All objects give out and take in heat energy by radiation. Heat energy is transferred through empty space (a vacuum) by radiation.

Energy transfer by radiation

All objects transfer energy by **radiation**. The hotter the object the more radiation it gives out.

Figure 9.1
An electric fire emits mainly infra red radiation

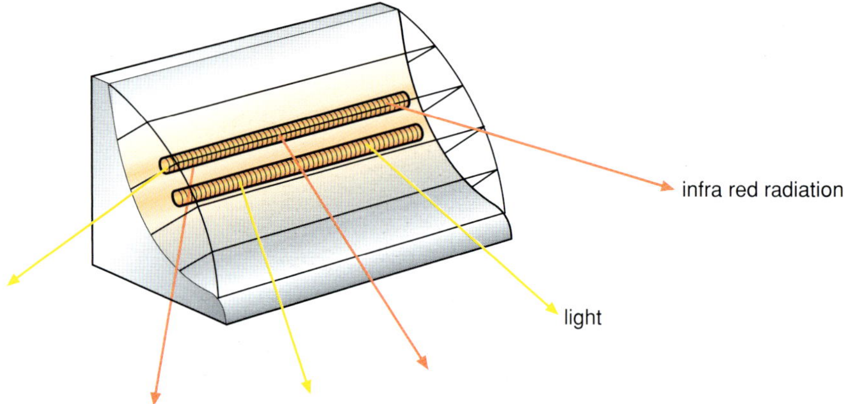

The energy given out by a hot body is mainly infra red radiation. Infra red radiation is one of the family of waves that makes up the electromagnetic spectrum (see Section 12.7). So particles of matter are not involved in the transfer of energy by radiation.

Taking in radiation (absorbing radiation)

Objects that absorb infra red radiation heat up.

The outside metal of a dark coloured car warms up quicker in the Sun than the outside metal of a light coloured car. A polished shiny car will not get as hot as an unpolished car.

Figure 9.2

This happens because:

- dark coloured surfaces absorb infra red radiation faster than light coloured surfaces
- shiny surfaces reflect more infra red radiation than matt surfaces.

Giving out radiation (emitting radiation)

The radiation given out by a hot object can be detected using a thermometer that has its bulb painted black.

- Hot tea in a light coloured teapot will give out radiation slower than hot tea in a dark coloured teapot.
- Hot tea in a shiny teapot will give out radiation slower than hot tea in a dull, non-shiny teapot of the same colour. (A dull, non-shiny surface is called a matt surface.)

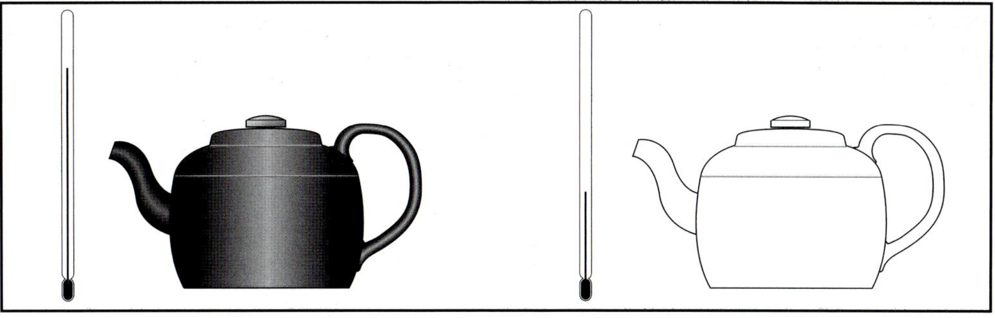

Figure 9.3
Tea in the shiny white teapot keeps hotter than tea in the matt black tea pot

So, dark coloured, matt surfaces are good emitters of infra red radiation. Light coloured, shiny surfaces are poor emitters of infra red radiation.

Figure 9.4
A summary of radiation

Good emitters of infra red radiation		Poor emitters of infra red radiation	
If cold these warm up quickly	**If hot these cool down quickly**	**If cold these warm up slowly**	**If hot these cool down slowly**
dark coloured surfaces	dark coloured surfaces	light coloured surfaces	light coloured surfaces
non-shiny (matt) surfaces	non-shiny (matt) surfaces	shiny surfaces	shiny surfaces

Topic Questions

1 Copy and complete the following questions.

a) Heat energy is transferred through a vacuum by ______________ .
b) A shiny surface is a poor ______________ of radiation but a good ______________ of radiation.
c) A white matt surface is a ____________ absorber of radiation than a black matt surface.

2 What happens to the temperature of an object that emits more infra red radiation than it absorbs?

3 The window blinds in a school laboratory are matt black. Explain why, on sunny days with the blinds closed the temperature in the laboratory goes up.

4 Explain each of the following statements.

a) Houses in hot countries are often painted white.
b) Petrol storage tanks are painted a shiny silver colour.
c) Hot food from a take away restaurant is packed in shiny aluminium containers.
d) Car radiators are usually painted matt black.

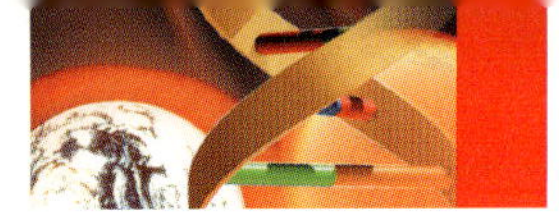

9.2

Co-ordinated	Modular
DA 12.16	DA 9
SA 12.10	SA 17

Keeping your home warm

Keeping your home warm is not just about turning on the central heating or lighting a fire. It's also about reducing the amount of heat transferred from inside your home to the air outside.

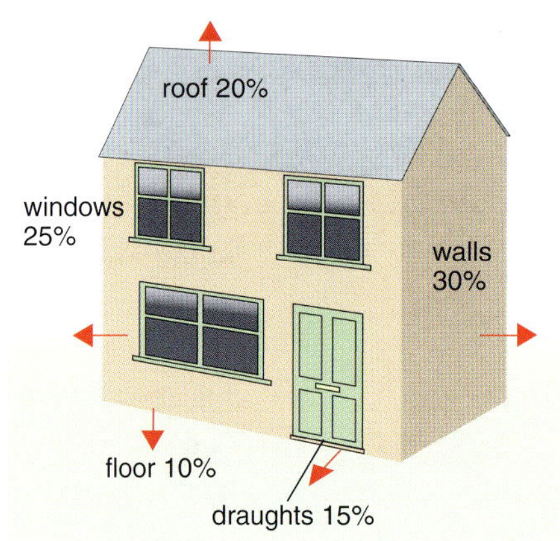

Figure 9.5 *Heat loss from an average home*

Different amounts of heat energy are transferred through the roof, the walls, the windows, the floor and the doors of a home. These are shown in Figure 9.5.

Most of the heat energy is transferred by conduction and convection. So reducing heat loss is about trying to stop conduction and convection. Figure 9.6 shows some of the methods used to reduce the heat loss from our homes.

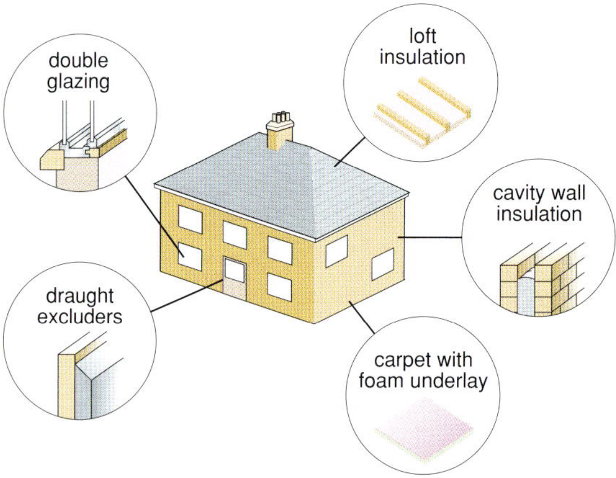

Figure 9.6 *Reducing heat loss from a home*

Most methods of reducing heat loss work by trapping air. If the air is trapped in small pockets it cannot move far, so heat loss by convection is reduced. Also the still air is a good insulator.

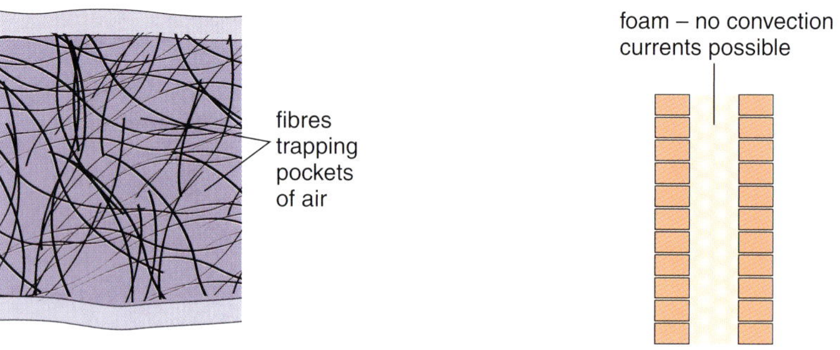

Figure 9.7 *Air trapped by fibre wool and by foam*

With double glazing both the air and the glass are good insulators. To keep heat loss by convection low the layer of air between the glass sheets must be thin.

Draught excluders trap warm air inside the home and stop cold air from coming in.

Saving energy, saving money

Installing double glazing or insulation or draught proofing costs money. However, once installed less energy will be needed to keep the home warm. This means that the bills for heating will be less, so you save money.

How long it takes to get back the cost of the installation from the money saved on the heating bills can be called the 'pay-back time'.

Example: Gurpal has just installed loft insulation. It cost £180. The insulation will save Gurpal £45 each year on his heating bill. What is the 'pay-back time' for the insulation?

$$\text{pay-back time} = \frac{180}{45} = 4 \text{ years}$$

So Gurpal will get back the cost of installing the loft insulation in 4 years.

Did you know?

Turning the central heating down by just 1°C can reduce heating bills by up to 10%.

More energy saving ideas

- **Close the curtains**

Thick curtains stop cold air blowing into a room. They also trap air between the window and the curtain so there is less heat loss by conduction.

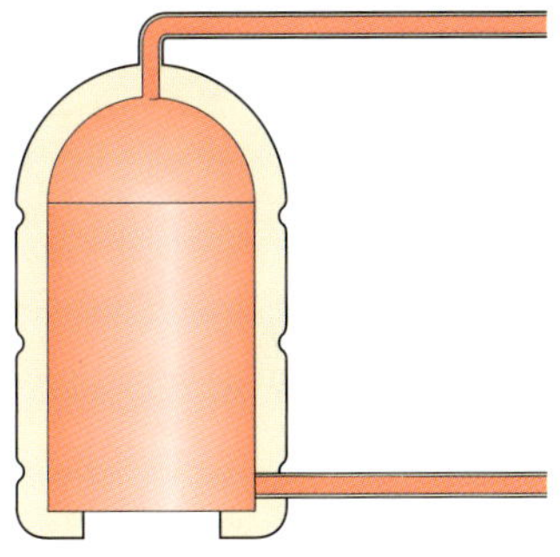

- **Insulate the hot water tank**

Fit a thick jacket around the hot water tank. The fibres in the jacket trap small pockets of air.

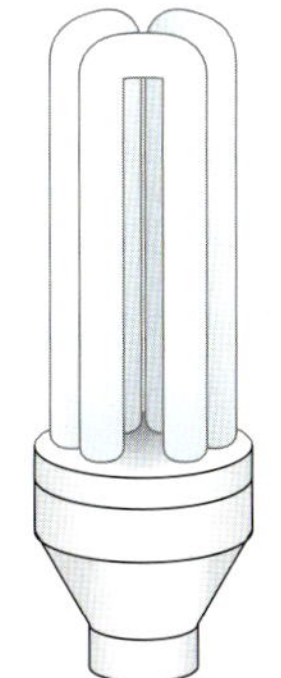

- **Use low-energy light bulbs**

Low-energy light bulbs transfer very little electrical energy into unwanted heat. (Also switch lights off when you leave a room.)

Topic Questions

1 Copy and complete the following sentences.

a) Most heat energy is transferred from a home by ____________________ and ____________________ .

b) Foam is a good ________________ because it traps small pockets of air.

c) Draught excluders stop ___________ air from getting into a home and _________ air from leaving.

2 The picture shows two houses. They are the same size and have the same types of insulation. Which of the two houses will be the most expensive to keep warm? Give a reason for you choice.

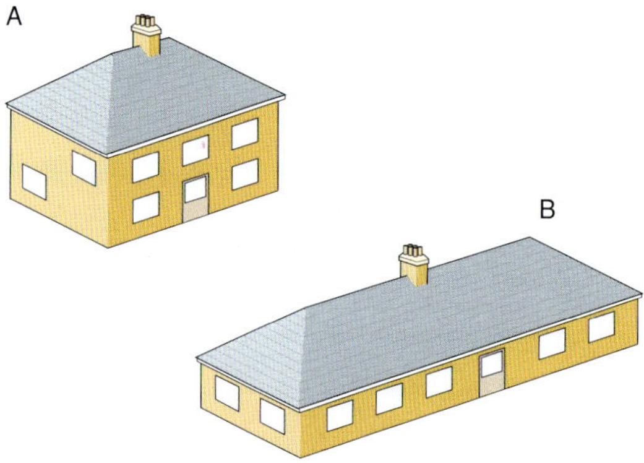

3 The table gives the costs and savings for different methods of reducing heat loss from a home.

Method of reducing heat loss	Cost to install	Savings made to the yearly heating bill
Draught proofing	£50	£25
Cavity wall insulation	£480	£120
Hot water jacket	£16	£8
Double glazing	£4500	£150

Work out the 'pay-back time' for each method of reducing heat loss.

4 Write a list of ten ways to reduce the amount of energy used in a home. Each energy saving tip must be cost free.

Co-ordinated	Modular
DA 12.4	DA 9
SA 12.4	SA 17

9.3 Using electrical energy

What you know

- There are lots of different forms of energy, for example:
  - movement (kinetic) energy
  - electrical energy
  - sound energy
  - light energy
  - heat (thermal) energy.
- Energy can be transferred from one place to another.
- Energy can be transformed (changed) from one form into another. For example, a firework will transfer chemical energy as light, heat and sound energy. The energy has been transferred and transformed.

  This energy transfer can be shown as:

  chemical energy $\xrightarrow{\text{firework}}$ light + heat + sound

In most homes there are many different devices, also called appliances, that work by transferring electrical energy as other forms of energy.

In industry electrical energy is a widely used source of energy.

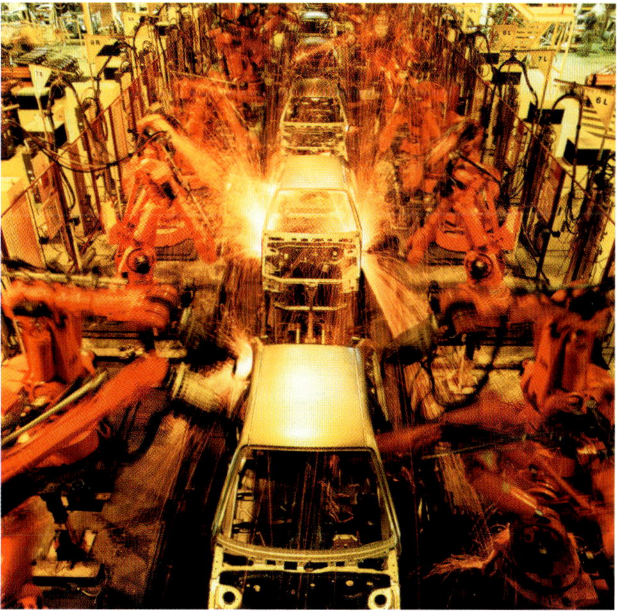

Figure 9.8 *Many machines work by transferring electrical energy*

Electrical energy is such a useful form of energy because it can so easily be transferred as other forms of energy, such as:

- heat (thermal energy)
- light
- sound
- movement (kinetic energy).

This is why so many appliances are designed to work from the electricity mains supply.

Figure 9.9 shows six everyday electrical appliances and the energy transfers they are designed to make.

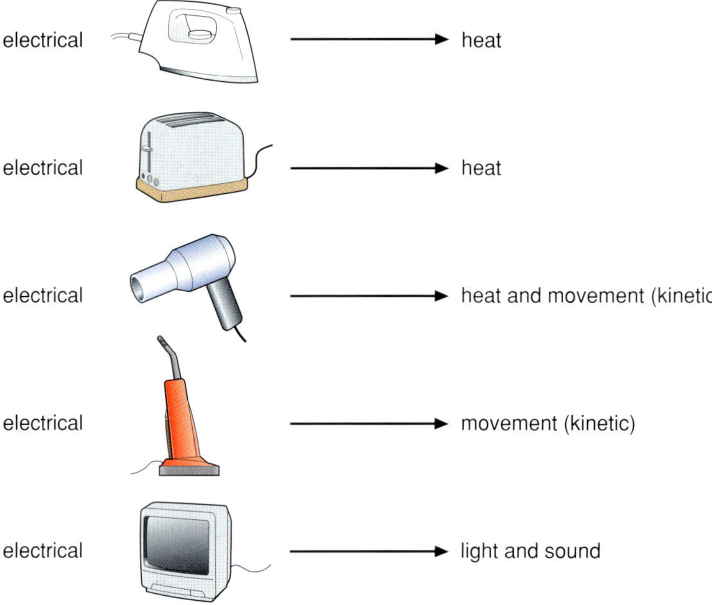

Figure 9.9 *Some appliances designed to transfer energy*

How much electrical energy an appliance transfers depends on two things:

1 How much time the appliance is switched on for.
2 How quickly the appliance transfers energy.

Electrical energy can also be transferred as gravitational potential energy. Objects that have been lifted store energy as gravitational potential energy. So using an electric pump to lift water from a lower lake to a higher lake is a way of changing electrical energy to gravitational potential energy. This idea is used in a pumped storage power station (see Section 9.8).

Power

How quickly an appliance transfers energy is called its **power**.

The power of an appliance can be worked out using the equation:

$$\text{power (W)} = \frac{\text{energy transferred (J)}}{\text{time taken (s)}}$$

Power is measured in watts (W). An appliance that transfers 1 joule of energy in 1 second has a power of 1 watt.

So a 350 watt electric jig-saw transfers 350 joules of electrical energy as other forms of energy every second it is switched on.

Electrical appliances will have an information label that gives its power.

Figure 9.10
Information label

Figure 9.11
Jig-saw

Power can also be measured in kilowatts (kW).

1 kilowatt (kW) = 1000 watts (W)

Figure 9.12 gives the power of the appliances seen earlier in this section.

Figure 9.12
The power ratings of some common household appliances

Appliance	Power rating
Lamp	60 W
Television	150 W
Toaster	250 W
Vacuum cleaner	770 W
Hairdryer	1200 W
Iron	1800 W

Topic Questions

1 Copy and complete the following sentences.

a) Electrical energy can be easily transferred as ____________ , ____________ and ____________ (thermal) energy.
b) A 600 watt electric drill transfers ________ joules of ____________ every ____________ it is switched on.

2 Each of the following appliances is used for 10 minutes. Which one transfers the most energy? Give a reason for your choice.

- 100 W light bulb
- 20 W radio
- 1200 W hairdryer
- 900 W vacuum cleaner

3 Energy is supplied to a laptop computer by electricity from batteries.

Copy and complete the following sentences using words in the box to help you.

heat	kinetic	light	sound

a) The screen transfers energy to the surroundings mainly as ____________ .
b) The loudspeaker transfers energy to the surroundings mainly as ____________.
c) An electric motor transfers energy to the disc drive mainly as ____________ .
d) When the computer is switched on it becomes warm. This is because some energy is transferred as ____________ energy.

4 Write the names of five appliances used in most homes that are designed to transfer electrical energy as heat (thermal) energy.

5 Write the names of five appliances used in most homes or workshops that are designed to transfer electrical energy as movement (kinetic) energy.

9.4	
Co-ordinated	Modular
DA 12.4	DA 9
SA 12.4	SA 17

Paying for electricity

Energy is transferred from the mains electricity supply every time an appliance is plugged in and switched on. This energy must be paid for.

The cost of using an appliance depends on the energy transfer the appliance is designed to make. Appliances designed to transfer electrical energy as heat usually have a high power (see Section 9.3). The greater the power, the more energy is taken from the mains each second. So this type of appliance costs a lot to use.

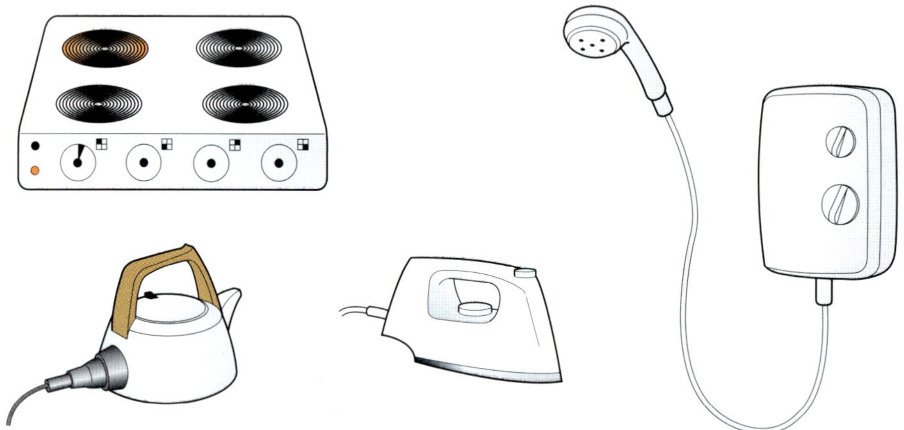

Figure 9.13 *These appliances cost a lot to use*

The cost also depends on how long the appliance is switched on for. The longer the appliance is on, the more it costs.

The energy transferred by an appliance can be worked out using this equation:

$$\text{energy transferred (J)} = \text{power (W)} \times \text{time (s)}$$

Example: A 2500 W kettle is switched on for 5 minutes. How much electrical energy is transferred from the mains supply to the kettle?

$$\begin{aligned}\text{energy transferred (J)} &= \text{power (W)} \times \text{time (s)}\\ &= 2500 \times (5 \times 60)\\ &= 750\,000 \text{ joules (J)}\end{aligned}$$

This is a big number but not a lot of energy. One joule is a very small amount of energy.

If electricity companies measured the total amount of energy transferred to a home in joules, the numbers would be enormous. So they measure the energy transferred using a much larger unit, the kilowatt hour (kWh).

The equation is the same.

energy transferred = power × time

BUT, energy transferred is in kilowatt hours (kWh)

power is in kilowatts (kW)

time is in hours (h)

1 kilowatt hour is the same as 3 600 000 joules of energy.

Example: A 4000 watt cooker is switched on for 3 hours. How much electrical energy is transferred from the mains supply to the cooker?

$$\text{energy transferred (kWh)} = \text{power (kW)} \times \text{time (h)}$$
$$= 4 \times 3$$
$$= 12 \text{ kilowatt hours (kWh)}$$

Working out the bill

In most homes there is an electricity meter. This records the total electrical energy being supplied by the electricity company. The meter records the energy supplied in kilowatt hours.

One kilowatt hour is called one Unit of electrical energy.

Figure 9.14 *Two electricity meter readings, three months apart*

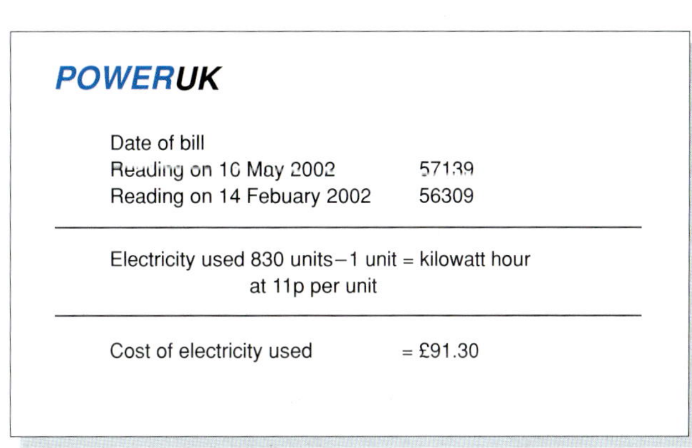

POWERUK

Date of bill

Reading on 10 May 2002	57139
Reading on 14 Febuary 2002	56309

Electricity used 830 units–1 unit = kilowatt hour
at 11p per unit

Cost of electricity used = £91.30

Figure 9.15 *The bill for the electricity used*

Number of Units of energy transferred between February and May

= 57 139 – 56 309

= 830 Units

The cost of this energy can be worked out using the equation:

total cost = number of Units × cost per Unit

So if the cost for each unit of energy is 11p, then:

between February and May the total cost = 830 × 11p

= 9130 p

= £91. 30

Topic Questions

1 Copy and complete the following sentences.

a) One Unit of electrical energy is the same as one ______________________.
b) The cost of using an appliance depends on the ___________ of the appliance and how long it is _______________ on.

2 How many kilowatt hours of electrical energy are transferred to:

a) a 2 kW heater switched on for 4 hours?
b) a 3 kW water heater switched on for 2 hours?
c) a 2 kW kettle switched on for 6 minutes?
d) a 100 W light bulb switched on for 5 hours?

3 What is the cost of using each of the appliances in Question 2? Take the cost of 1 Unit of electrical energy to be 11p.

4 At the beginning of the month, the reading on an electricity meter was 23 330. At the end of the month the same meter read 23 500. Each Unit of electrical energy costs 11p.

a) How many Units of electrical energy were used during the month?
b) How much was the electrical energy bill for the month?

5 Which two appliances do you think use the most electrical energy in your home? Explain the reason for your choice.

9.5 Energy efficiency

Co-ordinated	Modular
DA 12.17	DA 9
SA 12.11	SA 17

Any device that transfers energy also wastes energy. The device transfers only part of the energy to where it is wanted and in the form that is wanted. This is the useful energy transfer. The rest of the energy is changed into forms that are not wanted. This transfer is not useful and the energy is wasted.

Figure 9.16

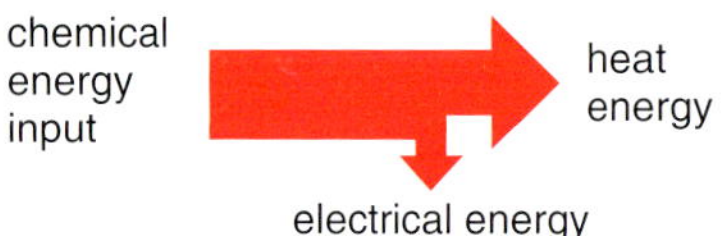

The wider the arrow the bigger the energy transfer

A coal-burning power station is designed to transfer chemical energy from the coal as electrical energy. But it also transfers a lot of energy as heat. This energy is wasted.

Figure 9.17

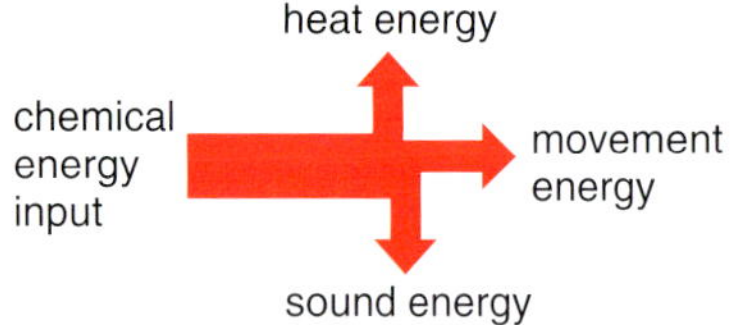

A motorbike engine is designed to transfer chemical energy from the petrol as movement (kinetic) energy. But it also transfers a lot of energy as heat and sound. This energy is wasted.

Figure 9.18

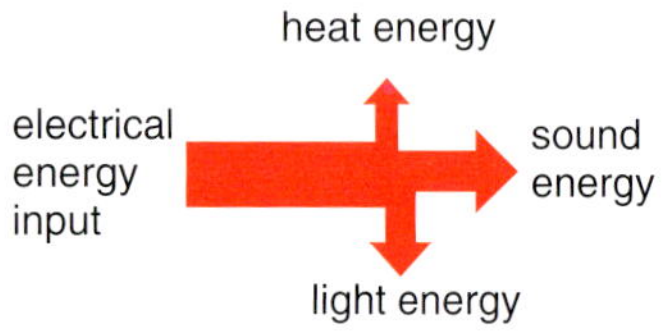

A television usefully transfers electrical energy as light and sound. But it also transfers some energy as heat. This energy is wasted.

Eventually in any energy transfer, all energy both useful and wasted is transferred to the surroundings. This makes the surroundings become warmer. Because the energy spreads out, any increase in temperature is usually too small to notice.

The more spread out the energy becomes, the more difficult it is to do anything useful with it.

Efficiency

Efficiency is a way of saying how good a device is at transferring energy as a useful form or forms. A device is 100% efficient if the total energy going in is the same as the useful energy transferred. No device can be more than 100% efficient.

The efficiency of a device can be worked out using the equation:

$$\text{efficiency} = \frac{\text{useful energy transferred by device}}{\text{total energy supplied to device}}$$

This will give efficiency as a decimal number. To change to a percentage (%), multiply the decimal number by 100.

Example: A vacuum cleaner is designed to transfer electrical energy as movement (kinetic) energy. But it also transfers energy as heat and sound.

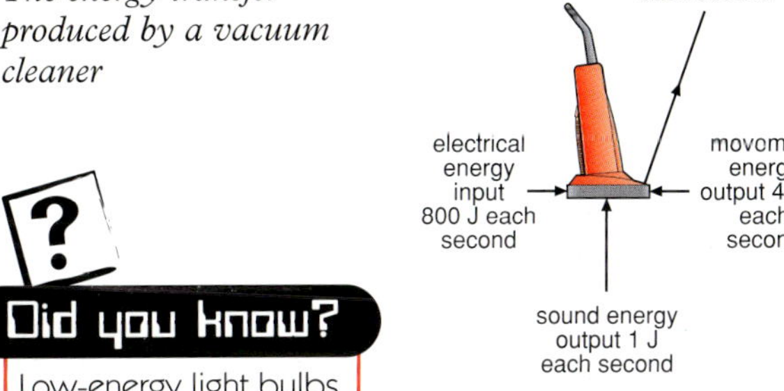

Figure 9.19 *The energy transfer produced by a vacuum cleaner*

$$\text{efficiency} = \frac{\text{useful energy transferred by device}}{\text{total energy supplied to device}}$$

$$\text{efficiency} = \frac{480}{800} = 0.6 = 60\%$$

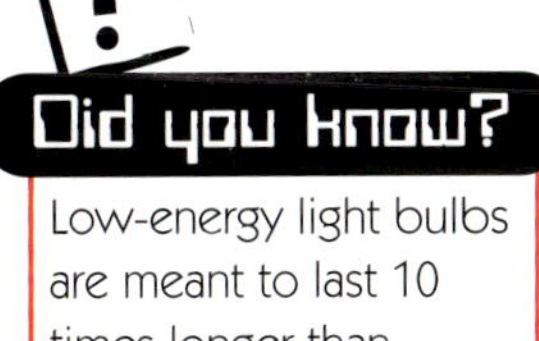

Low-energy light bulbs are meant to last 10 times longer than filament light bulbs. Filament light bulbs give about 1000 hours of use. So a low-energy light bulb should last about 10 000 hours. That's the same as leaving it switched on for 417 days

Low-energy light bulbs

A 100 W filament lamp is only about 5% efficient. Most of the electrical energy is transferred as heat. This energy is wasted.

Low-energy light bulbs do not have a filament that gets hot. So very little energy is transferred into unwanted heat. These light bulbs are much more efficient at transferring electrical energy as light energy. This means that a 20 W low-energy bulb can give out as much light as a 100 W filament lamp.

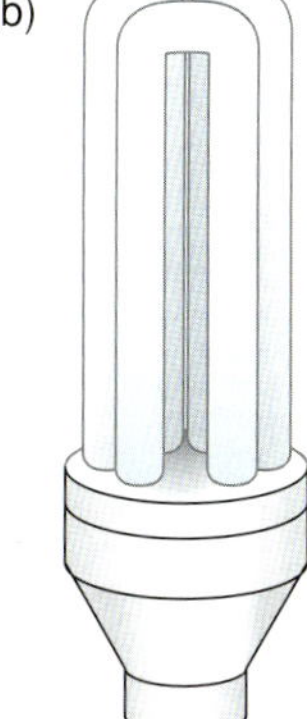

Figure 9.20 *a) Filament light bulb and b) a low-energy light bulb*

Topic Questions

1 Copy and complete the following sentences using words from the box.

chemical	electrical	heat	less	more	wasted

An electric fire is ____________ efficient than an electric motor. The fire is designed to transfer ____________ energy as heat energy, but the ____________ energy produced in a motor is ____________ .

2 Explain why a low-energy light bulb is more efficient than a filament lamp.

3 An electric oven is described as being 70% efficient. What does this mean?

4 In a traditional power station, 70% of the energy input is transferred and wasted as heat. The diagram shows the energy transfers in a combined heat and power (CHP) station.

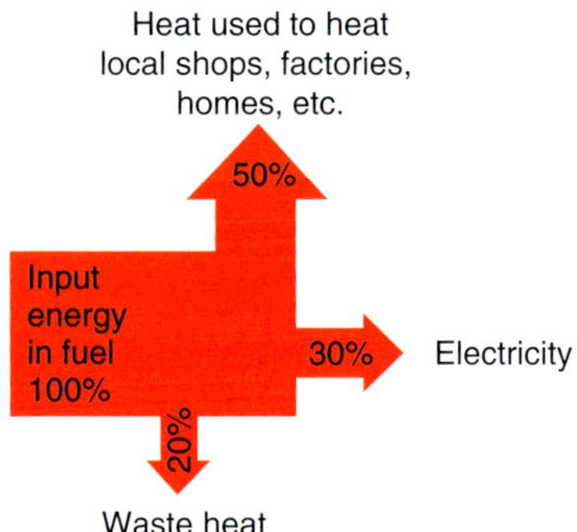

Explain why the CHP station is more efficient than a traditional power station.

9.6 Energy resources and electricity

Co-ordinated	Modular
DA 12.18	DA 9
SA 12.12	SA 17

What you know

Coal, oil and gas are **fossil fuels**. Like all fuels, they store energy. But to release their energy, fossil fuels must be burned.

Figure 9.21 *Burning coal transfers the stored energy to heat and light energy*

To replace the fossil fuels that have already been used will take millions of years. This means that the fuels are not being replaced as fast as they are being used. So the fuels are **non-renewable energy resources**. Once gone they are gone forever. Although nuclear fuels are not burnt, they are also a non-renewable energy resource. Fuel reserves are limited. Although some companies are always trying to find new reserves, fossil fuels (and nuclear fuels) will eventually run out. Figure 9.22 shows how long we can expect the different fuels to last if we carry on using them at the current rate.

Figure 9.22
How long fuel reserves may last

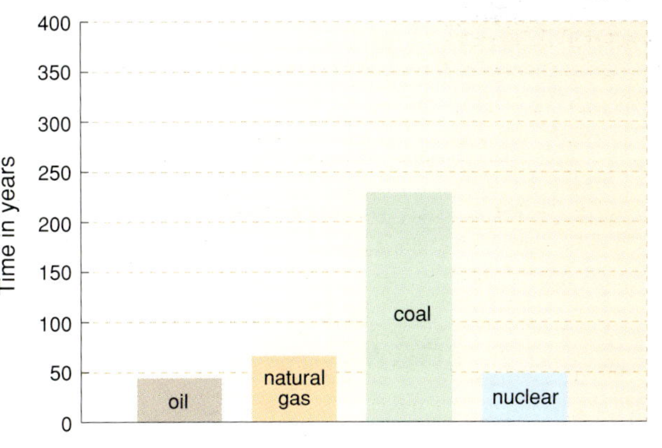

To make the Earth's fuel supplies last longer they must be used more efficiently and not wasted.

Figure 9.23
Non-renewable energy sources

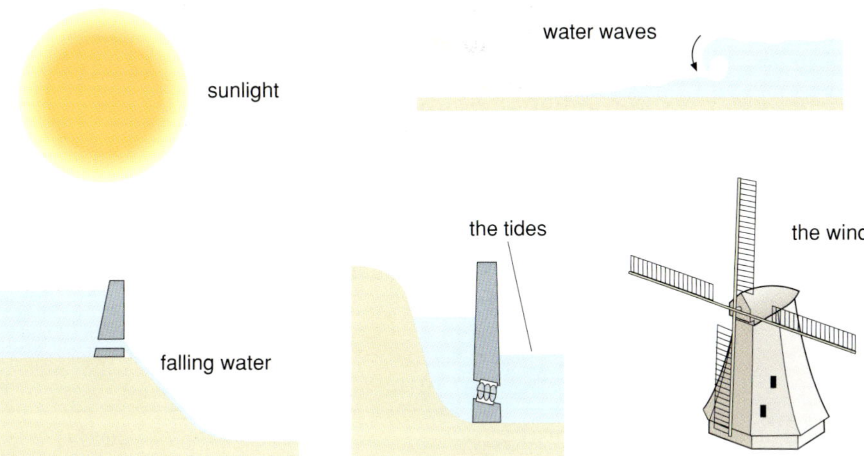

Wood is also a fuel. But when a tree is chopped down another can be grown in its place.

So wood is a **renewable energy** resource.

Renewable energy resources will not run out. They are energy resources that can be replaced. Figure 9.23 gives some examples of renewable energy resources.

Figure 9.24
Generating electricity from coal or oil

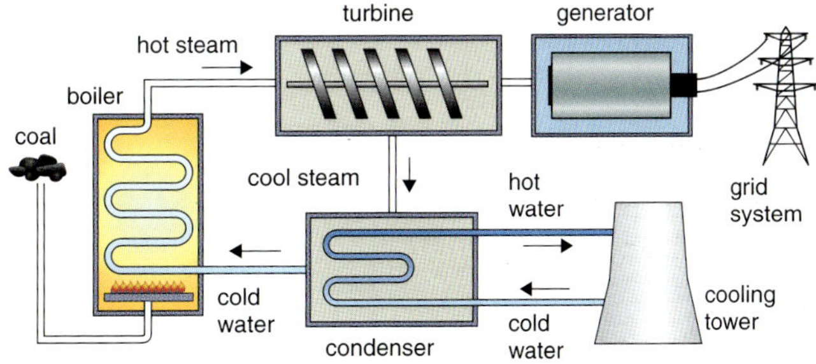

Figure 9.25
The blades of a power station turbine

Both renewable and non-renewable energy resources can be used to produce electricity. But in Britain most electricity is generated in power stations using non-renewable fossil fuels.

Coal or oil is burned and the heat used to turn water into steam. The steam is made to turn turbines. The turbines turn generators. The generators produce electricity.

In a gas-burning power station there is no need to produce steam. The heat from the burning gas turns the turbine directly.

A nuclear power station generates electricity in the same way as a coal-burning power station. However, the fuel (mainly uranium or plutonium) is not burned. The heat needed to turn water into steam is given out when the uranium or plutonium atoms are made to split inside the nuclear reactor.

1 kilogram of nuclear fuel can release the same amount of energy as 20 000 kilograms of coal.

Why are non-renewable fuels used to generate electricity?

Power stations that use non-renewable fuels can generate electricity at any time. It doesn't matter if it is night or day, summer or winter. Provided the fuel keeps arriving the power station can keep generating. This makes them reliable energy sources.

Start-up times

After a power station has been closed down, perhaps for maintenance, it will need to be started up again. The time it takes to start up a power station so that it begins to generate electricity depends on the type of power station.

shortest start-up time → longest start-up time

natural gas — oil — coal — nuclear

Topic Questions

1 Copy and complete the following sentences using the words in the box.

electricity	movement (kinetic)	steam	uranium

a) The energy source for a nuclear power station is ________________ .
b) The turbine in a coal-burning power station is turned by ________________ .
c) A turbine transfers ________________ energy to the generator.
d) The generator transfers energy as ________________ .

2 What is a renewable energy resource?

3 Which type of power station has the longest start-up time?

4 Give a reason why a gas-burning power station has a shorter start-up time than an oil-burning power station.

5 In what way is the production of electricity different in a fossil fuel power station to a nuclear power station?

6 Find out when the first nuclear power station started to generate electricity in Britain.

9.7

Co-ordinated	Modular
DA 12.18	DA 9
SA 12.12	SA 17

Non-renewable fuels and the environment

Fossil fuels and global warming

Some of the heat (infra red) energy from the Sun that hits the Earth's surface is absorbed. Some of it is reflected back into space.

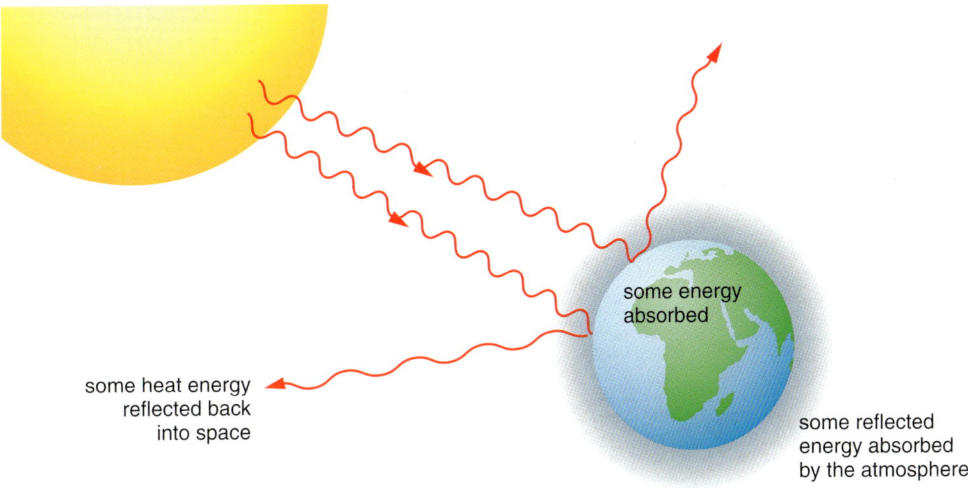

Figure 9.26
Some of the Sun's heat energy is reflected back into space

The carbon dioxide in the atmosphere acts like a blanket around the Earth. It lets through the heat energy (infra red) from the very hot Sun but it traps some of the heat energy radiated by the much cooler Earth, stopping it from travelling into space. This is called the greenhouse effect.

Fuel-burning power stations produce a lot of waste gases. One of these gases is carbon dioxide. So as more and more fossil fuels are burned, more carbon dioxide is put into the atmosphere. This may lead to more heat energy being trapped and the Earth's temperature increasing slowly. This warming of the Earth's atmosphere is called **global warming**. An increase in global temperatures of only 1°C would be enough to melt part of the polar ice caps.

Each fossil fuel burns to produce different amounts of carbon dioxide. Figure 9.27 shows how much carbon dioxide is produced for each kilowatt of power generated using the different fuels.

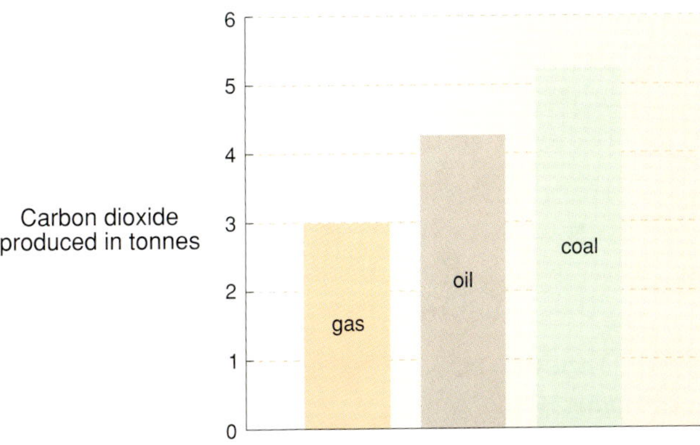

Figure 9.27

Acid rain

When coal and oil are burned sulphur dioxide is produced. This gas, which dissolves in water, goes into the atmosphere. If the sulphur dioxide dissolves in rain, the rain becomes a weak acid.

Acid rain can damage buildings, kill trees and pollute rivers and lakes.

Figure 9.28 *A building damaged by acid rain*

Figure 9.29 *Trees killed by acid rain*

Did you know?

Generating electricity using nuclear fuels reduces the amount of carbon dioxide that would be produced in Britain by over 100 million tonnes each year.

The problem of acid rain can be reduced, but it's not cheap. Either the sulphur can be taken from the fuels before they are burnt. Or the sulphur dioxide can be taken from the waste gases before they go into the atmosphere. Both ways add to the cost of the electricity generated.

Burning natural gas does not produce sulphur dioxide.

Nuclear fuels and the environment

Nuclear fuels do not produce carbon dioxide or sulphur dioxide. So nuclear power stations do not add to global warming or acid rain.

Some people do worry that nuclear power stations may leak radiation. But when the power stations are working normally, little or no radiation or radioactive materials are put into the environment.

Happily, serious accidents at nuclear power stations do not happen very often. But when they do, radiation can be carried by the wind over a very wide area. Radiation released after the accident at the Chernobyl nuclear power station (see Section 12.14) in the Ukraine was carried as far as Wales.

Nuclear power stations do produce radioactive waste. This must be stored safely for a long time, sometimes for thousands of years.

Did you know?

The amount of solid radioactive waste produced by a nuclear power station in one year would only fill one large classroom.

Topic Questions

1 Copy and complete the following sentences.

a) An oil-burning power station adds ___________ ___________ and ___________ ___________ to the atmosphere.

b) A ___________ power station adds no waste gases to the atmosphere.

c) Radiation can be carried over a wide area by the ___________.

2 What harmful effect can sulphur dioxide have on the environment?

3 Why does the waste from a nuclear power station need to be stored for a long time?

4 Which type of fuel-burning power station adds the smallest amount of waste gases to the atmosphere? Explain the reason for your choice.

5 What is the effect of putting large amounts of carbon dioxide into the atmosphere?

9.8

Co-ordinated	Modular
DA 12.18	DA 9
SA 12.12	SA 17

Using renewable energy resources to generate electricity (1)

Renewable energy resources are used to generate electricity. There are no fuels to burn. The turbines that turn the generators use energy straight from the renewable resource.

Using the tides

Every day tides rise and fall. The tides cause water to move in and out of river estuaries. (The estuary is where the river meets the sea.) The energy of the moving water can be used to generate electricity.

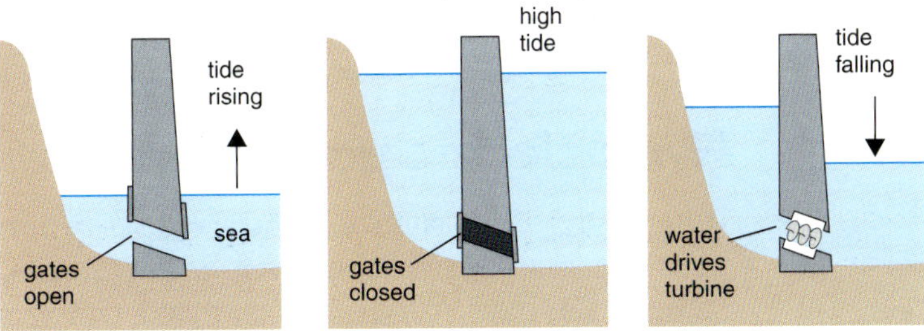

Figure 9.30
A tidal barrage

A barrage, which is like a dam, is built across a river estuary. The barrage contains some underwater gates. The gates are opened as the tide comes in and then closed to trap the water behind the barrage. When the tide goes out the gates are opened. Sea water rushes through the gates where it turns turbines. The turbines turn generators. The generators produce electricity. This system can be made more efficient by also using the incoming tides to turn turbines.

The amount of electricity produced depends on the tides. The tides not only vary during each day but the height of the tides change on a monthly cycle. So, to some extent using the tides to generate electricity is unreliable.

Did you know?

At the moment there are no tidal barrages in Britain. But one possible site for a barrage is across the river Severn. It is estimated that this could produce 6% of Britain's electricity.

Figure 9.31 *River Awe hydro-electric power station, Strathclyde, Scotland*

Using waves

The power stations are built on cliffs at narrow sea inlets. The up and down movement of the waves forces air to turn a turbine, which then turns a generator.

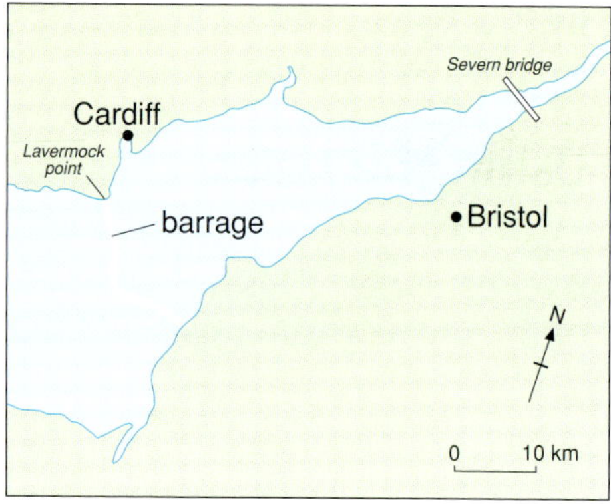

Figure 9.32

Hydroelectric power stations

The energy of a river can be used to generate electricity. To do this a dam must be built across the river. The water is then trapped and forms a lake behind the dam. When the trapped water is released it rushes downhill. The energy of the falling water is used to turn turbines, which then turn generators.

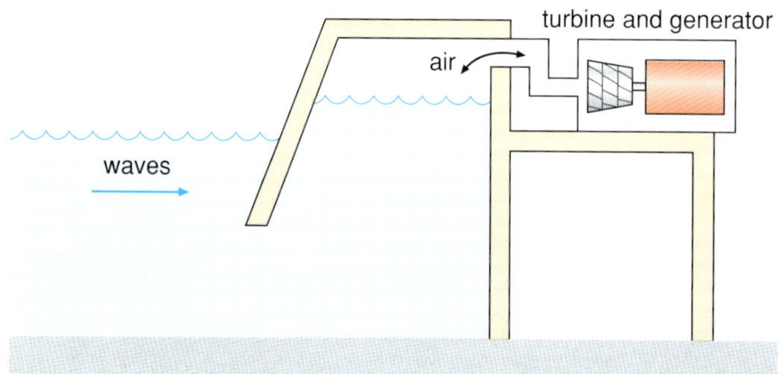

Figure 9.33 *A hydroelectric power station*

Hydroelectric power stations generate about 2% of Britain's electricity. But they produce nearly 10% of the world's electricity.

The demand for electricity changes during the day. When demand is high a hydroelectric pumped storage system can provide the extra electricity.

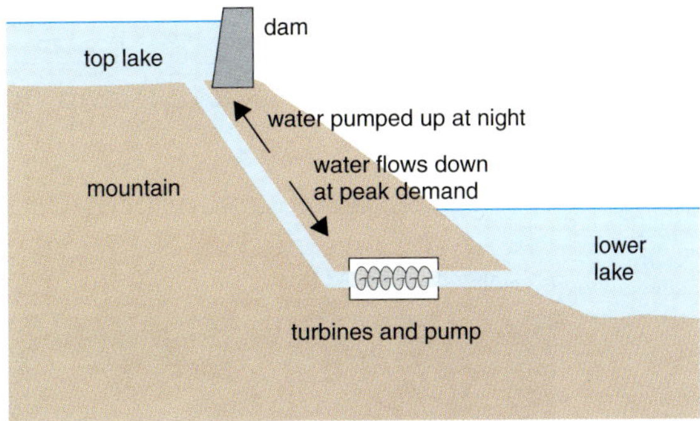

Figure 9.34
A pumped storage power station

In just a few seconds water can be released to flow from the top lake to the bottom lake. As the water falls it turns a turbine, which then turns a generator.

At night when there is more electricity being generated than is needed, electricity is used to pump the water back to the top lake. This means that energy from the spare electricity is transferred to the water rather than being wasted. The power station is also ready to generate again when it is next needed.

Topic Questions

1 Copy and complete the following sentences.

a) Renewable energy resources can drive a ____________ directly so there is no need to _________ any fuels.
b) Turbines are used to turn ______________, which then produce ___________.
c) In a ________________ power station falling water transfers gravitational _____________ energy as ______________ energy.

2 Suggest why the demand for electricity is likely to go up at half time in a televised football match.

3 How is generating electricity using fossil fuels different to generating electricity using a renewable energy resource?

4 Why does a tidal power station not cause acid rain?

5 Write out the following sentences in the right order.

- The turbine turns a generator.
- Wave-powered electricity generators are built on cliffs.
- The generator transfers movement (kinetic) energy to electrical energy.
- The movement of the waves pushes air through a turbine.

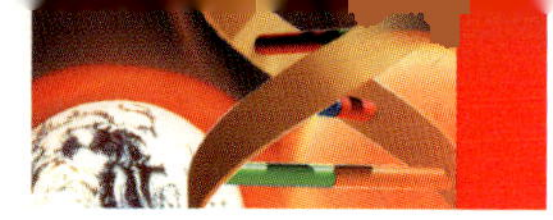

9.9	
Co-ordinated	Modular
DA 12.18	DA 9
SA 12.12	SA 17

Using renewable energy resources to generate electricity (2)

Using the wind

Using energy from the wind is not a new idea. Over a thousand years ago windmills were used for grinding grain. In some countries the power of the wind is still used to pump up water from natural underground reservoirs.

A modern wind generator is designed to transfer the movement (kinetic) energy of the wind as electrical energy.

Figure 9.35

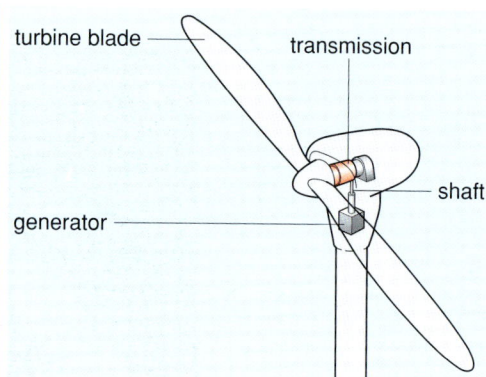

turbine blade
transmission
shaft
generator

Figure 9.36

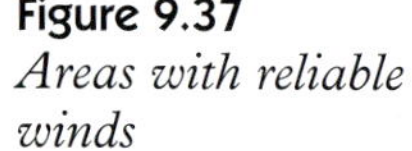

Figure 9.37
Areas with reliable winds

As the wind blows, the turbine blades rotate. The turbine turns the generator. The generator produces electricity.

One problem with wind generators is that the amount of electricity generated changes with the strength of the wind. If the wind is not very strong little or no electricity may be produced. This makes the wind an unreliable energy resource.

Generating electricity from the wind can be made more reliable by putting the wind generators in places where there is usually a strong wind. This is often around the coast or on high ground.

Direct from the Sun

Solar cells transfer energy from the Sun directly into electricity.

The amount of electricity produced by a solar cell depends on the brightness of the sunlight. If it's dark or cloudy little or no electricity may be produced. So solar cells are in some ways an unreliable way to produce electricity (see Section 9.11).

The cost of producing a Unit of electricity using solar cells is high. (Only a Unit of electricity from a non-renewable battery costs more.) But sometimes they are the best way to produce electricity. Solar cells are often used in remote areas, on satellites or in devices where only a small amount of electricity is needed.

Figure 9.38
Solar cells produce the electricity to run the satellite

Figure 9.39
A solar cell can provide the small amount of electricity a calculator needs

From the Earth

Uranium and other radioactive elements are found inside the Earth. When atoms of these elements decay (see Section 12.13) they transfer heat to the surrounding rocks. This is called **geothermal energy**.

In areas where there are volcanoes, geysers and hot springs, water heated by geothermal energy can reach the surface as steam. This steam can be used to turn turbines that are connected to electricity generators.

? Did you know?

The world's first geothermal power station was in Italy. It started producing power in 1904. The power station is still working and produces 400 megawatts (400 000 000 watts) of electricity.

Geothermal power stations are also found in New Zealand, Iceland and the USA.

Topic Questions

1 Copy and complete the following sentences.

a) Wind generators need to be at places where there is nearly always a __________ wind.

b) A solar cell is designed to transfer __________ energy as ______________ energy.

c) Geothermal energy is produced by the decay of _________________ isotopes.

2 Why are solar cells a useful energy resource for remote areas?

3 Why would solar cells be a more useful way to produce electricity for a country on the equator than for England?

4 Write down the energy transfer a wind generator is designed to make.

5 Find out which countries in Europe produce more than 50% of their electricity from renewable energy resources.

9.10 Renewable energy resources and the environment

Co-ordinated	Modular
DA 12.18	DA 9
SA 12.12	SA 17

Did you know?

At the moment Britain produces less than 3% of its electricity using renewable energy resources. The target, set by the government, is to increase this to 25% by the year 2025.

Most wind generators are designed to produce about 500 kilowatts of power. But some will produce up to 3000 kilowatts.

The average coal-burning power station generates over 1000 megawatts (1000 000 kilowatts) of power. So hundreds of wind generators would be needed in order to generate the same amount of electricity as just one coal-burning power station.

A large number of wind generators grouped together is called a wind farm.

Example: A group of people has suggested that a wind farm should replace a small coal-burning power station. The power station generates 420 megawatts of power.

How many 3000 kilowatt wind generators would the wind farm need?

$$420 \text{ megawatts} = 420\,000 \text{ kilowatts}$$

$$\text{number of wind turbines needed} = \frac{420\,000}{3000} = 140$$

Wind farms need to be on hills or along the coast (see Section 9.9). Some people think that wind farms are ugly and spoil the view. So to some people wind farms cause visual pollution.

People living near a wind generator may find them noisy. There is the noise from the rotating turbine blades and noise from the machinery inside the generator. So for some people wind farms cause noise pollution.

Energy from the tides

Because tidal barrages need to be built across river estuaries they can disturb the flow of the river. This may destroy the habitats of wild life such as wading birds and the mud-living organisms on which they feed.

Hydroelectric power stations

Many of these schemes involve flooding large areas of land. This may mean forests are cut down and farm land lost. The habitat of different birds and animals will be destroyed.

Figure 9.40
Between 1964 and 1968, the ancient temple of Abu Simbel was moved, stone by stone, to higher ground. This was to stop it being flooded by the lake created by the Aswan High Dam

Did you know?

Figure 9.41

The New Aswan High Dam built in Egypt between 1960 and 1970 is 111 metres high and at its base almost 1000 metres thick. The lake created by the dam, Lake Nasser, is 480 kilometres long and up to 16 kilometres wide. At the time of building it was the largest artificial lake in the world. The hydroelectric power station inside the dam is designed to produce up to 2100 megawatts of power, almost enough for the whole of the country.

Topic Questions

1 Copy the energy resource statements. Match each statement to one of the methods of producing electricity.

Energy resource statements	Methods of producing electricity
Produces a lot of noise	Tidal barrage
Land may be flooded	Wind generator
Built across a river estuary	Hydroelectric

2 A tidal power station in France produces 240 megawatts of power. How many 3 megawatt wind generators would be needed to produce the same amount of power?

3 What is a disadvantage of building a tidal barrage across a river estuary?

4 Why might people object to a wind farm being built on the edge of a National Park?

5 Write a letter to the government suggesting ways in which it could increase the amount of electricity produced using renewable energy resources.

9.11	
Co-ordinated	Modular
DA 12.18	DA 9
SA 12.12	SA 17

Using energy resources – advantages and disadvantages

Each energy resource used to generate electricity has its good and bad points. These have been mentioned in the previous sections. This double page gives a summary of the reasons FOR and AGAINST using the different energy resources to generate electricity.

Fossil fuels – coal, oil and natural gas

FOR

- Produce electricity when it is needed, they do not depend on the weather.
- There are large reserves of coal.
- Gas-burning power stations can be switched on and off quickly.

AGAINST

- Non-renewable energy resource.
- Burning coal, oil and natural gas adds carbon dioxide, a greenhouse gas, to the atmosphere
- Burning coal and oil adds sulphur dioxide to the atmosphere causing acid rain.

Nuclear fuels

FOR

- No polluting gases produced.
- Advances in technology will make the fuel reserves last much longer.
- A small amount of fuel gives a large amount of energy.

AGAINST

- Non-renewable energy resource.
- Some radioactive waste must be stored for thousands of years.
- Serious accidents may release radiation over a large area.
- Concerns over the safety of transporting the fuel.

Wind

FOR

- Renewable energy resource.
- Low running costs.
- No polluting gases produced.

AGAINST

- Wind generators can spoil the landscape and cause unwanted noise.
- Unreliable – electricity is only generated when there is a strong enough wind.
- A wind generator has a small power output compared to a coal-burning power station.

Tide

FOR

- Renewable energy resource.
- Low running costs.
- No polluting gases produced.

AGAINST

- Very expensive to build.
- Destroys the habitat of wading birds.
- Electricity is only generated at certain times of the day.

Hydroelectric schemes

FOR

- Renewable energy resource.
- Can be switched on and off quickly.
- Can be used to store energy.
- No polluting gases produced.

AGAINST

- Large areas of land may be flooded.
- Can affect plant and animal life.

Waves

FOR

- Renewable energy resource.
- Low running costs.
- No polluting gases produced.

AGAINST

- Difficult to built a power station strong enough to stand up to the power of rough seas.

Solar

FOR

- Renewable energy resource.
- Ideal for remote places.
- Ideal when only a small amount of electricity is needed.
- No polluting gases produced.

AGAINST

- Expensive to set up in large numbers.
- The amount of electricity produced depends on the brightness of the light.

Geothermal

FOR

- Massive amounts of energy are available.

AGAINST

- Unless there are natural geysers it can be expensive and inefficient to generate electricity.
- The steam may be contaminated with unpleasant gases.

Summary

- Radiation is the transfer of heat energy by electromagnetic waves, mainly infra red.
- Radiation can travel through empty space (a vacuum).
- Dark-coloured surfaces are better absorbers of radiation than light-coloured surfaces.
- Shiny surfaces reflect more radiation than matt surfaces.
- Materials that trap small pockets of air are good heat insulators.
- Many appliances work by transferring electrical energy into other forms of energy.
- Power is the rate at which energy is transferred.
- The power of an appliance is measured in watts (W) or kilowatts (kW).
- The amount of energy transferred from the mains is measured in kilowatt hours (kWh) or Units.
- Efficiency is the proportion of the energy input that is usefully transferred by a device or appliance.
- Renewable energy resources will not run out.
- Non-renewable energy resources, once used, cannot be replaced.
- Most electricity is generated using the energy from coal, oil, natural gas or nuclear fuels.
- Burning fossil fuels produces waste gases that pollute the atmosphere.
- Nuclear fuels and renewable energy resources produce no polluting waste gases.
- Electricity is produced when a turbine turns a generator.
- Most electricity is generated using the heat from a fuel to produce steam or hot gases which then turn a turbine.
- Renewable energy resources turn turbines directly.

Examination Questions

1 A firefighter wears a shiny, white, fire-retarding suit. Which two statements explain the reasons for choosing shiny, white suits rather than matt, black ones?

- Matt, black suits are good absorbers of radiation.
- Matt, black suits are good reflectors of radiation.
- Shiny, white suits are good absorbers of radiation.
- Shiny, white suits are good conductors of thermal energy.
- Shiny, white suits are good reflectors of radiation.

2 Electricity can be generated from different energy resources.
Match each resource from the list with the numbers 1–4 in the table.

coal	geothermal	hydroelectric	uranium

Energy resource	How the resource is used in power stations
1	Used as the fuel in a nuclear power station
2	Burned in a conventional power station
3	Uses energy released in volcanic areas
4	Uses liquid water to drive turbines

3 Match words from the list below with the spaces 1–4 in the sentences.

heat (thermal energy)	light	movement (kinetic energy)	sound

A fan is designed to transfer electrical energy as __1__
An iron is designed to transfer electrical energy as __2__
A lamp is designed to transfer electrical energy as __3__
A loudspeaker is designed to transfer electrical energy as __4__

4 Electrical appliances usefully transfer part of the energy that is supplied to them. The rest of the energy is wasted.
A vacuum cleaner blows air out of one end so that it can suck air and dust into the other end.

a) A vacuum cleaner usefully transfers energy as
A heat (thermal energy).
B light.
C movement (kinetic) energy.
D sound.

b) The energy that is not usefully transferred by the vacuum cleaner is wasted as
A heat (thermal energy) only.
B sound only.
C heat (thermal energy) and sound.
D sound and movement (kinetic) energy.

c) Which of the following statements is false?
A The energy transferred by the vacuum cleaner becomes difficult to use for other energy transfers.
B The energy transferred by the vacuum cleaner ends up making the surroundings a little warmer.
C The energy transferred by the vacuum cleaner ends up very spread out.
D The energy transferred by the vacuum cleaner no longer exists.

d) A newer vacuum cleaner transfers useful energy at the same rate as the old one, but it wastes less energy.
This means that the newer vacuum cleaner
A costs more per minute to run.
B is more efficient.
C is less efficient.
D transfers energy at a faster rate.

5 If we use renewable energy sources we will not need to burn so much fossil fuel. But capturing renewable energy sources can also cause problems.
Match the words in the list with the numbers 1–4 in the table.

dams (for hydroelectricity)	solar cells	tidal barrages	wind farms

What is used to capture energy?	Problem caused
1	Can often be seen from a long way away and look unsightly to some people.
2	Destroy muddy areas in river estuaries where wading birds feed.
3	Land that could be used for farming or forests is flooded.
4	Very high cost for each Unit of electricity generated during lifetime.

6 Heat energy is lost from a house.

a) Complete the sentences by choosing the correct words from the box. Each word may be used once or not at all.

conduction	conductor	convection	electric evaporation	insulator

The amount of heat energy lost through the windows by ______ can be reduced by using thick curtains. The curtains trap a layer of air and air is a good ______ The curtains will also stop ______ currents pulling cold air into the room through small gaps in the window.

b) Write down one other way of reducing heat loss from a house.

____

____

7 Using the words from the list, copy and complete the sentences about generating electricity.

energy	gas	generator	smoke	steam	transformer	turbine	water

In a coal-fired power station, coal is burnt to release ______. This is used to change ________ into __________ which drives a ________. Electricity is produced by a ________.

Chapter 10

Electricity

Key terms

alternating current (a.c.) • circuit breaker • component • direct current (d.c.) • earthed • electric current • electrolysis • electromagnet • electron • friction • hertz • ion • induced current • light dependent resistor (LDR) • magnetic field • thermistor

10.1 Circuits – starting out

Co-ordinated	Modular
DA 12.1	DA 10
SA 12.1	SA 17

What you know

- For an electric current to flow through a lamp or any other **component** in a circuit, there must be a voltage across the component. This means the circuit must include a battery (or other energy supply) and the circuit must be complete. If the circuit breaks the current will stop flowing.
- Voltage is also called potential difference.

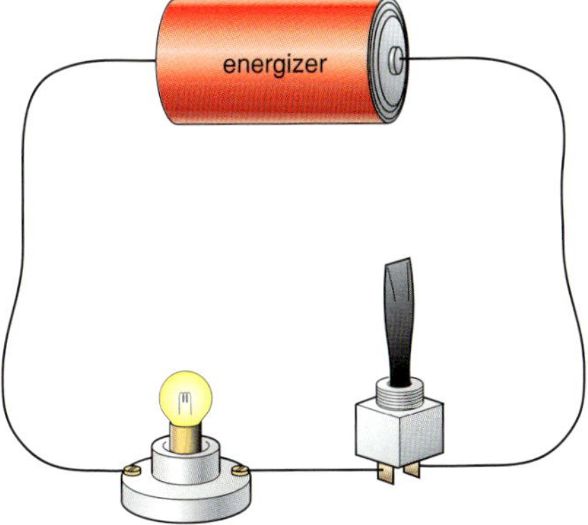

Figure 10.1

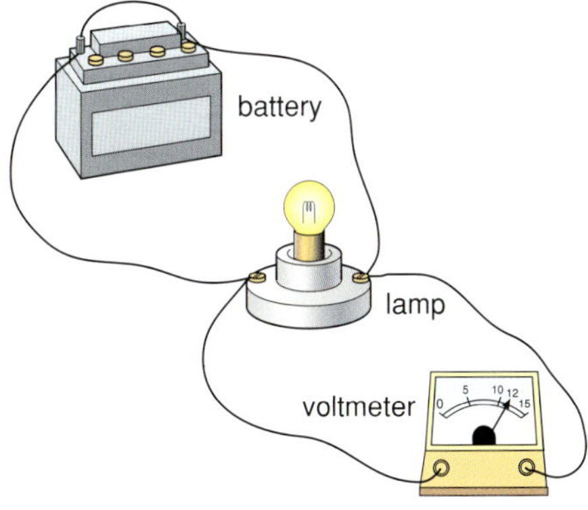

Figure 10.2

- Closing the switch completes the circuit. There is a voltage across the lamp. So an electric current flows and the lamp lights up.

Opening the switch breaks the circuit. There is no voltage across the lamp. An electric current can not flow so the lamp goes out.

- The smaller the voltage across a component, the smaller the current flowing through the component.
- An electric current is measured with an ammeter.
- Ammeters are always joined in series in a circuit. This is so that the current flowing around the circuit also flows through the ammeter.

- Electric current (I) is measured in amperes (A), often called amps for short.
- The voltage (potential difference) is measured with a voltmeter.
- Voltmeters are always joined across a component. So voltmeters are always joined in parallel with a component.
- Potential difference (p.d. for short) is measured in volts (V).
- All components resist the flow of an electric current. The easier a current flows through a component the less resistance the component has.
- For the same potential difference, a component with a high resistance will let less current flow through it than a component with a lower resistance.

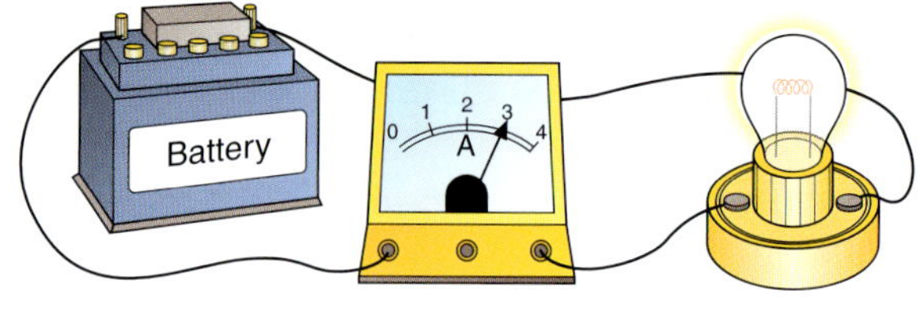

Figure 10.3

Each component in a circuit has a standard symbol. Rather than drawing a complicated picture the symbols are used to draw a circuit diagram. This makes life much easier for us and for electricians.

Figure 10.4 shows the symbols we use with the name of the components they represent.

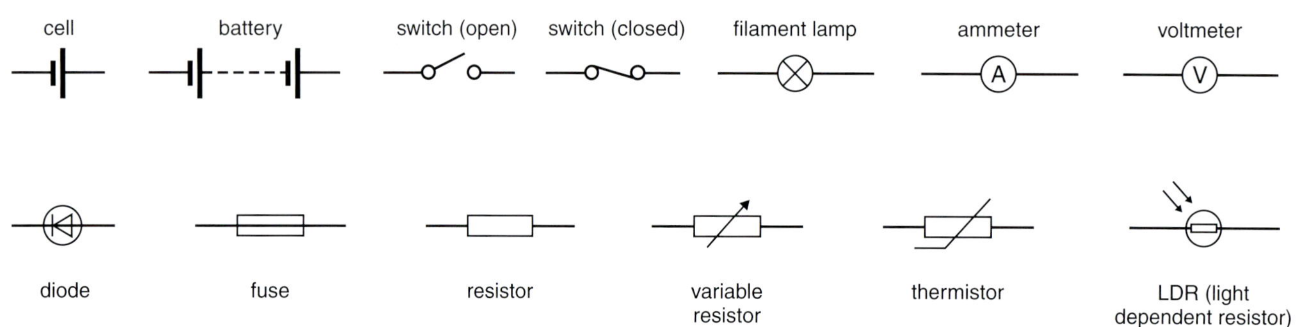

Figure 10.4
Electrical components and their symbols

It is common for people to call one cell on its own a battery. Really this is wrong; a battery is a set of cells joined together.

Figure 10.5
A 12 V car battery has six 2 V cells joined together

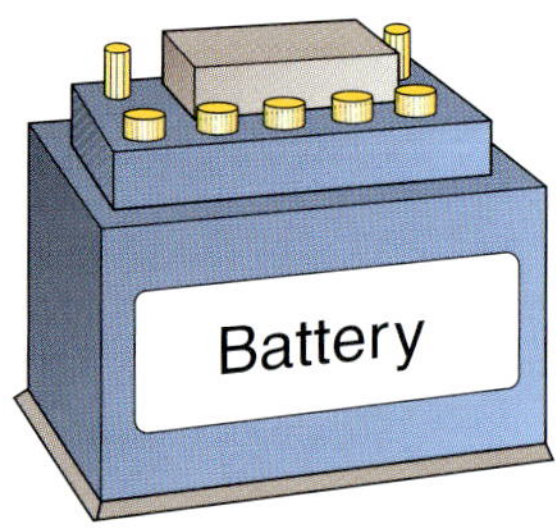

Topic Questions

1 Copy the following circuit. Label components A–D with their correct names.

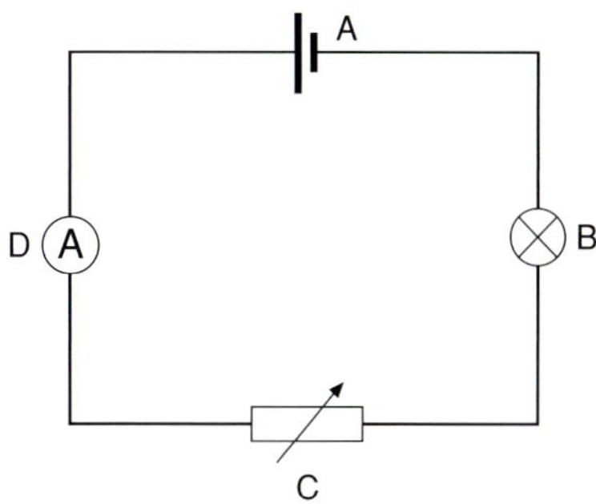

2 Which two of the following circuits are connected correctly? Give reasons for your choices.

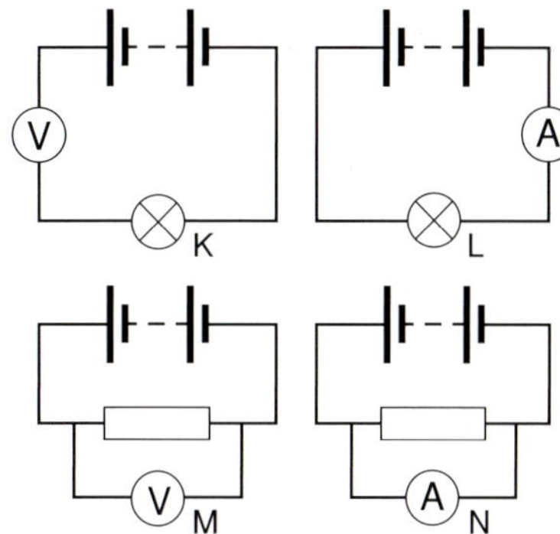

3 Which switches have to be closed to make these lamps light?

a) P and Q only
b) R and T only
c) R only

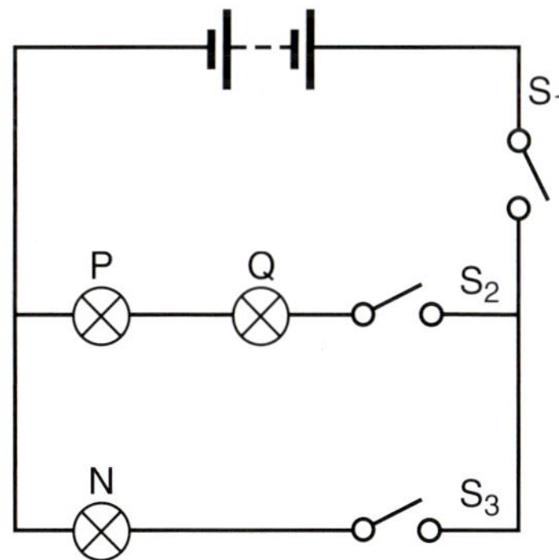

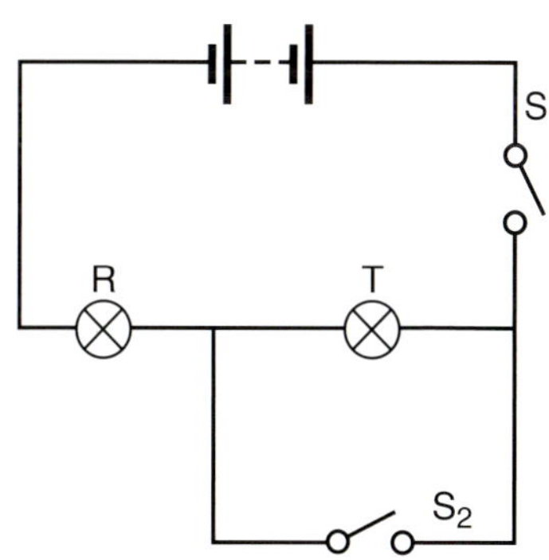

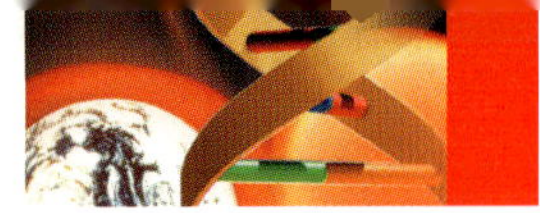

10.2 Series circuits

Co-ordinated	Modular
DA 12.1	DA 10
SA 12.1	SA 17

Components joined in series follow each other in one complete loop. The current has only one path it can take. So the same current must flow through each component and wire in the circuit.

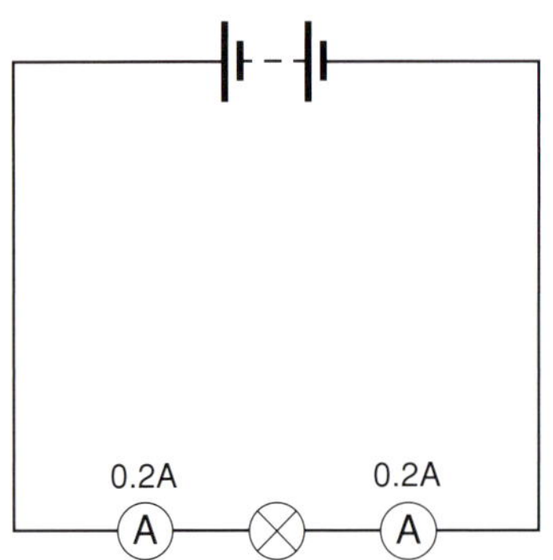

Figure 10.6

The two ammeters in Figure 10.6 both show the same reading. This is because the current is the same at all points in a series circuit. So only one ammeter is needed in a series circuit.

Adding components in series

Adding a second lamp to the circuit in Figure 10.7 makes the current go down. Which means the resistance of the circuit has gone up.

Since all components have resistance, adding extra components to a series circuit increases the total resistance of the circuit.

The total resistance of a series circuit is worked out by adding together all the separate resistances. The total resistance is the sum of the individual resistances.

Resistance is measured in ohms (Ω)

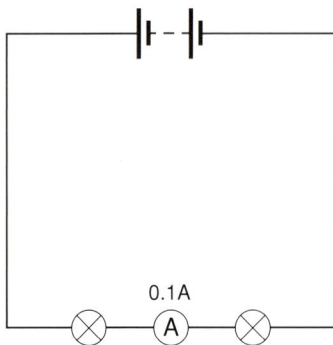

Figure 10.7

Example:

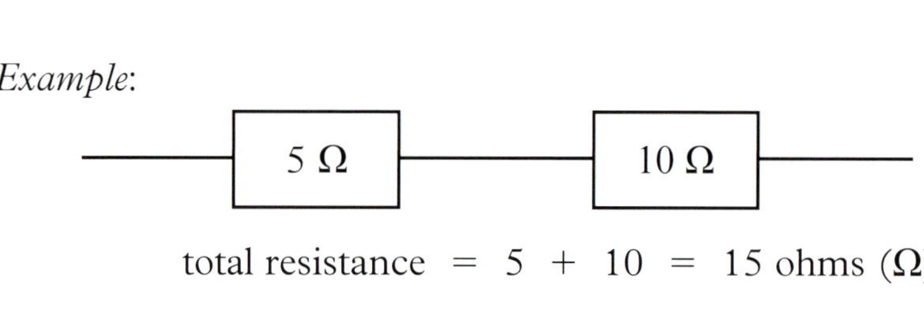

total resistance = 5 + 10 = 15 ohms (Ω)

Potential difference – components in series

Remember potential difference and voltage are the same thing.

The current through components joined in series is the same. But the potential difference across the components may be different. The potential differences are the same only if the components have the same resistance.

In Figure 10.8, the voltmeters are measuring the potential difference across the lamp and the resistor.

Figure 10.8

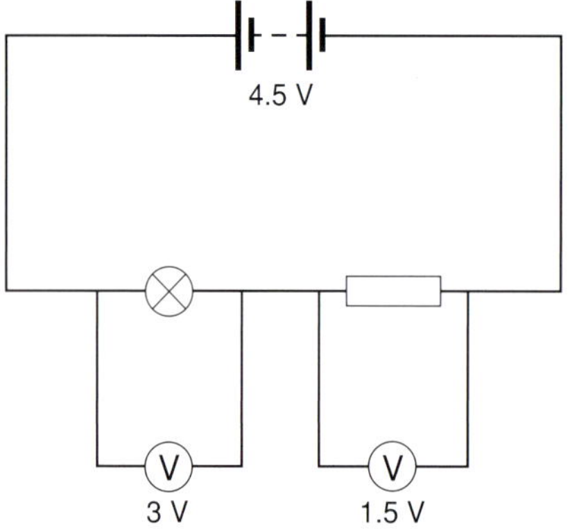

The lamp and the resistor cannot have the same resistance. The voltmeters are giving different readings. The lamp gives the biggest reading, so the lamp has the biggest resistance.

Adding the two voltmeter readings gives the total potential difference across the circuit components.

total potential difference = 3 + 1.5
= 4.5 V

This is the same as the potential difference (voltage) of the battery. So the potential difference of the battery has been shared between the two components.

For components joined in series, the potential differences across the components add up to equal the potential difference of the electricity supply.

Example:

Figure 10.9

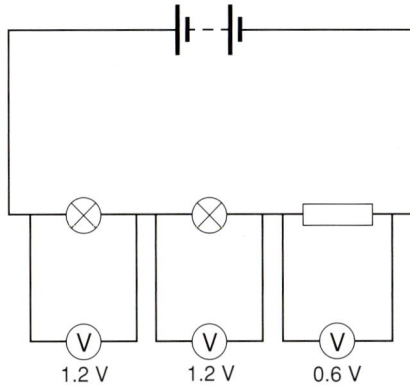

What is the potential difference of the battery in this circuit?

p.d. of the battery = p.d. across each component added together

p.d. of the battery = 1.2 + 1.2 + 0.6
= 3 V

Potential difference – cells in series

For cells joined in series, the total potential difference is worked out by adding the separate potential differences together. This only works if the cells are joined together correctly, positive (+) to negative (−).

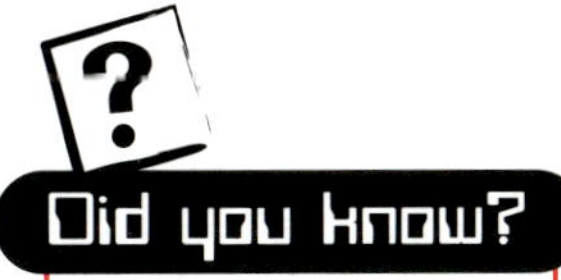

Messages are carried between different parts of your body by nerve impulses. Each nerve impulse is a tiny separate pulse of electricity.

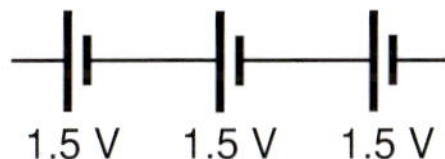

Figure 10.10
The cells are joined correctly. The separate potential differences add to give 4.5 V

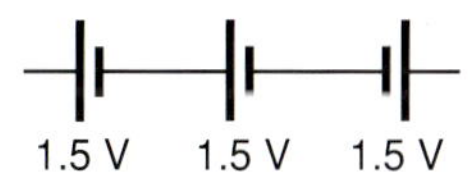

Figure 10.11
One cell is round the wrong way. The p.d. of this cell cancels out the p.d. of one of the other cells. The total p.d. is only 1.5 V

Topic Questions

1 Copy and complete the following sentences.

a) When components are joined in ____________ the potential difference of the ______________ is shared between them.

b) Adding extra components to a series circuit makes the total __________ go up.

c) The potential difference of a battery is worked out by ____________ the potential differences of the separate __________ .

2 Work out the total resistance of the following arrangements.

a) b)

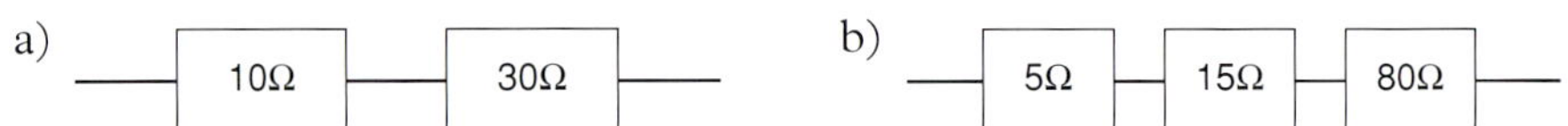

c)

3 The diagram shows a resistor joined in series to a lamp. The current through the lamp is 0.3 A. The potential difference across the lamp is 4 V.

a) What is the current through the resistor?

b) The potential difference across the battery is 6 V. What is the potential difference across the resistor?

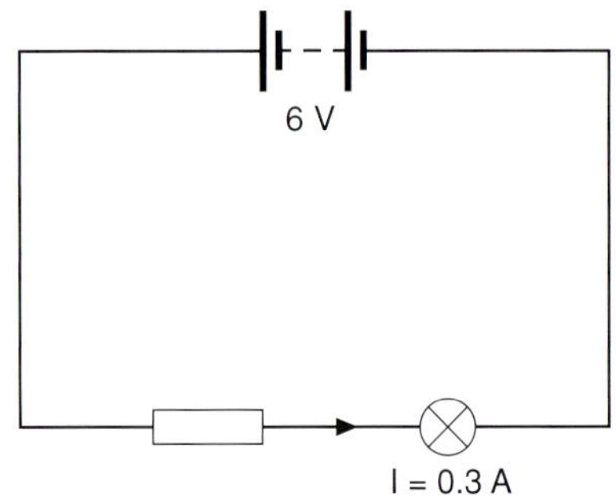

4 Work out the total potential difference of each cell combination.

a) b) c)

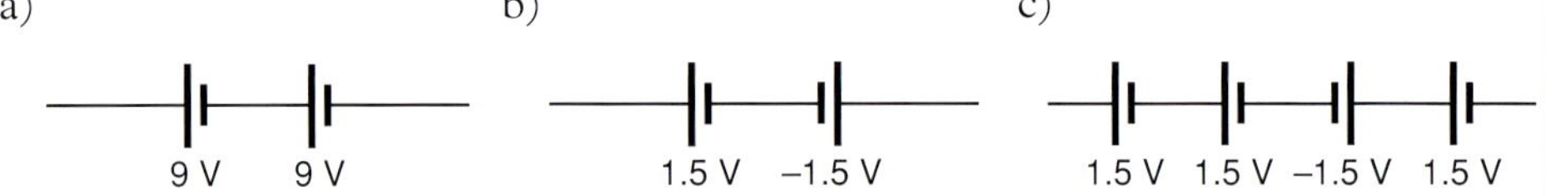

10.3

Co-ordinated	Modular
DA 12.1	DA 10
SA 12.1	SA 17

Parallel circuits

When components are joined in parallel the potential difference across each component is the same.

Joining components in parallel gives the electric current a number of different paths. Some current will go down each path.

How much current goes through each component depends on the resistance of the component. Since each potential difference is the same, the component with the biggest resistance will have the smallest current flowing through it.

Figure 10.12

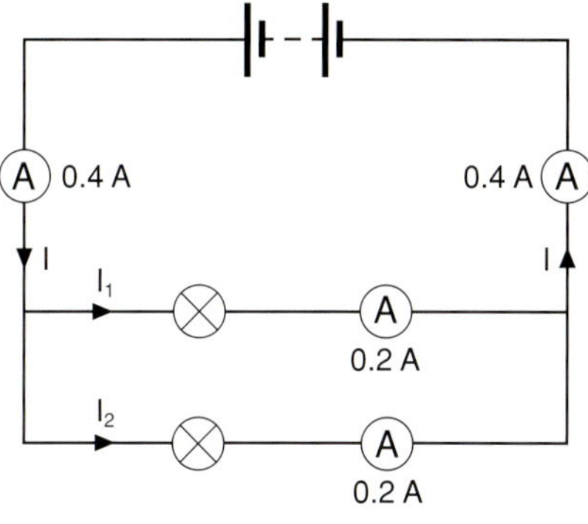

In Figure 10.12, half the total current flows through each lamp. So the resistance of the lamps must be the same because the current has split into two equal parts. The current joins back together where the parallel paths meet.

In a parallel circuit, the total current flowing through the battery can be worked out by adding up the currents flowing through the separate components.

Figure 10.13

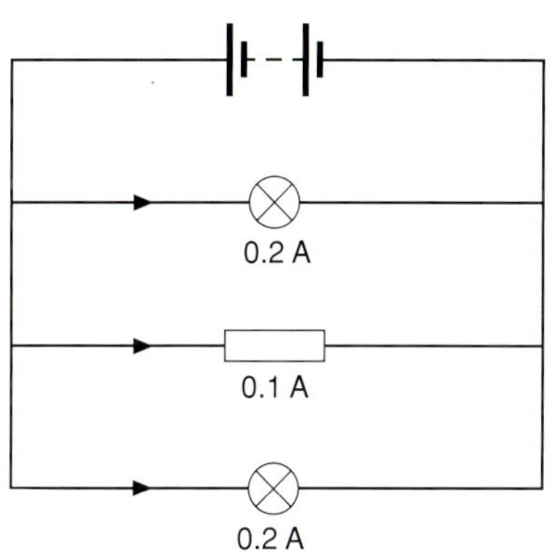

Example:

total current = 0.2 + 0.1 + 0.2
= 0.5 A

Overloaded circuits

An adaptor plug lets you plug more than one appliance into a mains socket.

Figure 10.14
Overloaded sockets

Appliance	Working current
computer	1 A
hairdryer	4 A
television	2 A
stereo	1 A
heater	10 A

The appliances are in parallel. So the total current taken from the socket is the sum of the currents flowing through each appliance.

If the computer, stereo and television are plugged into the adaptor:

total current = 1 + 1 + 2 = 4 A

This will not be a problem. The wire connecting the socket to the mains supply can safely have up to 13 amps flowing through it.

But if the hairdryer, television and heater are plugged into the adaptor:

$$\text{total current} = 4 + 2 + 10 = 16\text{ A}$$

This is a problem! The total current is too large for the connecting wire. It will get very hot and could cause a fire.

Lighting circuits

The lights in a home are joined in a parallel circuit.

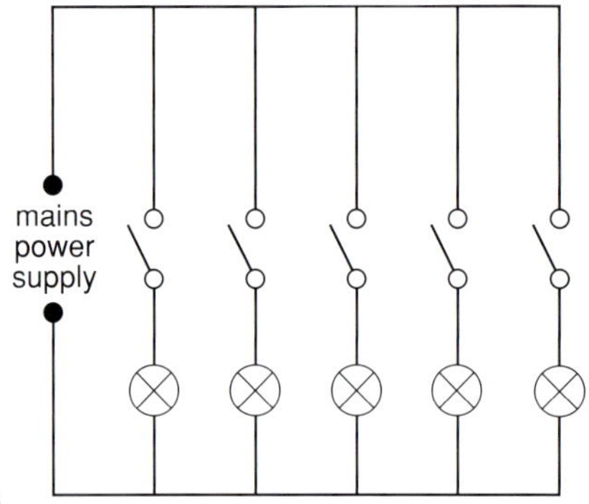

Figure 10.15
Lamps joined in parallel

Because they are in parallel:

- If one of the lamps 'blows' (burns out) it doesn't stop the others from working. The current has other paths it can take.
- If each lamp has its own switch they can be turned on and off separately.
- Each lamp that is switched on has the same potential difference across it as the mains power supply: 230 volts.

Did you know?

Some low voltage lighting circuits only use a 12 V power supply.

Topic Questions

1 Copy and complete the following sentences.

a) In a parallel circuit the current ______________ so that some current flows through each ________________ in the circuit.

b) The current through two components joined in parallel will only be the same if the two components have the same ___________________ .

2 The diagram shows two lamps, K and L connected in parallel. The potential difference across K is 6 volts. The current from the battery is 2 amps.

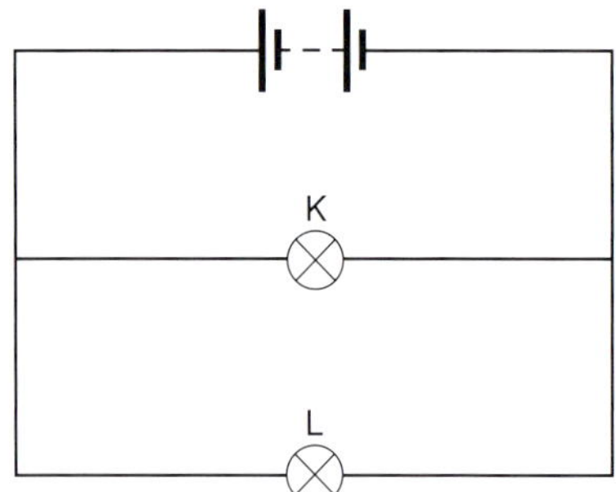

a) What is the potential difference across lamp L?

b) The current through lamp K is 1.5 amps. What is the current through lamp L?

c) How can you tell that the resistance of the two lamps is not the same?

3 An adaptor lets up to four appliances be used at the same time from one socket. Use the information in Figure 10.15 to work out the current that will flow through the socket when:

a) the computer, hairdryer, television and stereo are all plugged in.

b) only the heater and television are plugged in.

4 The diagram shows a wire and lamp connected in parallel with a battery. The wire is making a 'short circuit'.

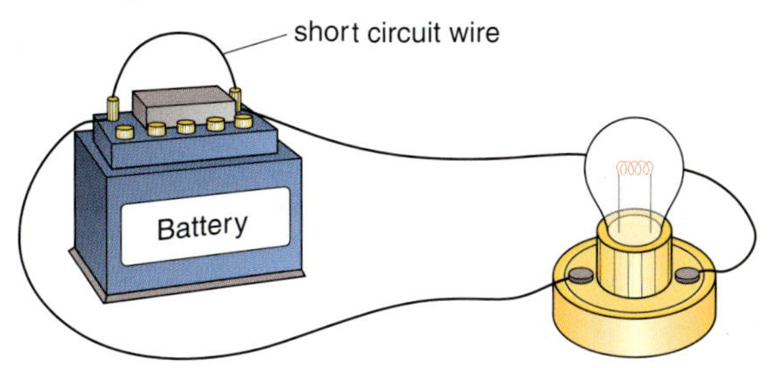

Explain why:

a) the lamp is so dim that it looks like it's not working.

b) the short circuit wire becomes dangerously hot.

10.4 Resistance

Co-ordinated	Modular
DA 12.1	DA 10
SA 12.1	SA 17

Current passes more easily through some components than through others. Some components let a large current flow through. These components have a low resistance to the flow of current. For the same potential difference, a high resistance component lets only a small current flow through.

The potential difference across a component, the current through a component and the resistance of a component are linked by the equation:

$$\text{potential difference (V)} = \text{current (A)} \times \text{resistance } (\Omega)$$

Figure 10.16

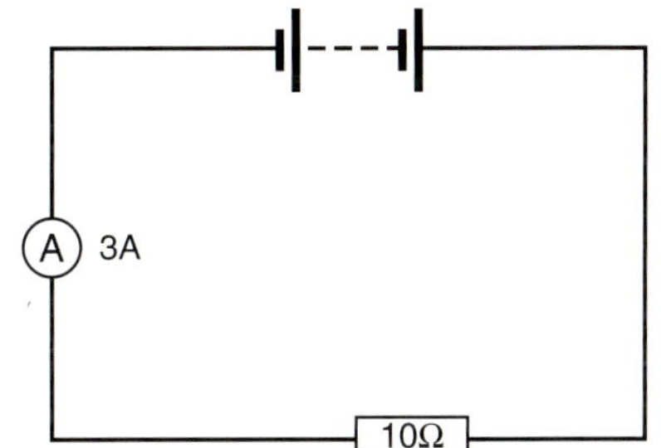

Example: What is the potential difference across the 10 ohm resistor in Figure 10.17?

$$\text{p.d.} = \text{current} \times \text{resistance}$$
$$= 3 \times 10$$
$$= 30\,\text{V}$$

To measure resistance, the current flowing through a component and the potential difference across the component must be known.

Figure 10.17

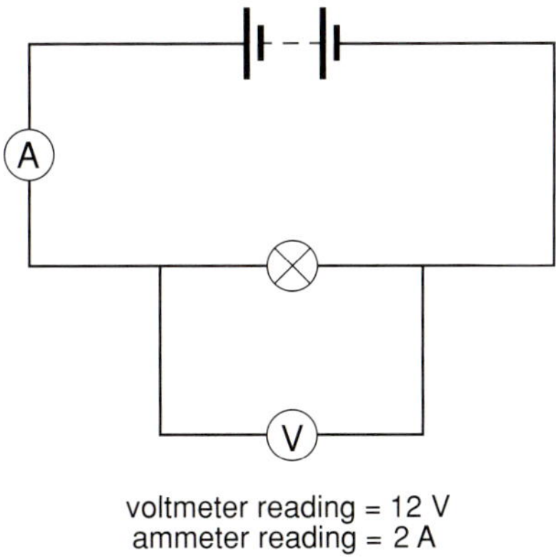

The circuit drawn in Figure 10.17 shows how the resistance of a component can be measured. In this case the component is a lamp. (*Remember*: The ammeter measures the current flowing through the lamp. The voltmeter measures the potential difference across the lamp.)

The resistance of the lamp can then be found by using:

$$\text{potential difference} = \text{current} \times \text{resistance}$$
$$12 = 2 \times R$$

So the resistance of the lamp, R, must equal 6 ohms.

The resistance of most components is not fixed, it can change. For example, the resistance of a wire goes up as the wire gets hot.

Variable resistor

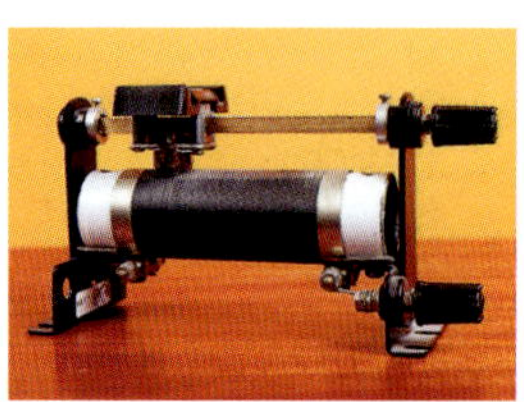

Figure 10.18
A variable resistor

By moving the sliding contact, the resistance of a variable resistor can be changed from 0 ohms up to its maximum value. This means that a variable resistor can be used to change the current in a circuit.

Light dependent resistor (LDR)

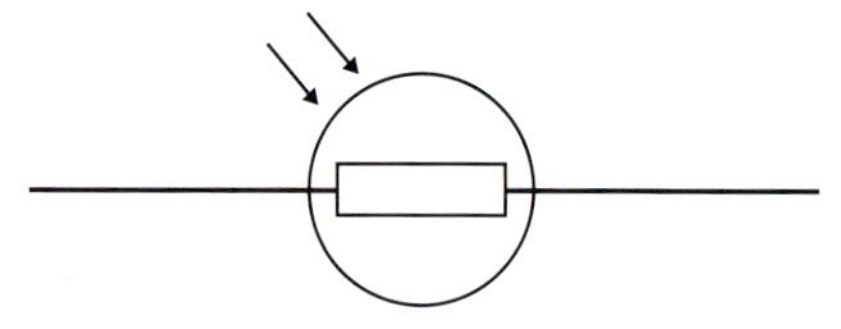

Figure 10.19
Symbol for an LDR

The resistance of a **light dependent resistor (LDR)** changes as the intensity (brightness) of the light shining on it changes. The resistance goes down as the intensity of the light goes up.

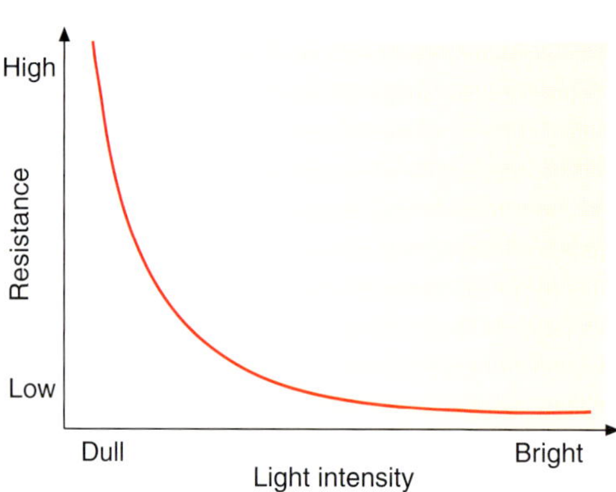

Figure 10.20
The LDR resistance changes with light intensity

Light level	Resistance, in ohms
dark	1 000 000
bright	500

The LDR is used as a sensor in light operated circuits, such as security lighting.

Thermistor

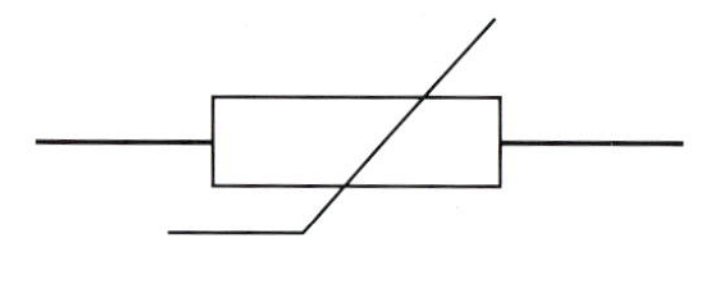

Figure 10.21
Symbol for a thermistor

The resistance of a **thermistor** changes with temperature. The resistance goes down as the temperature goes up.

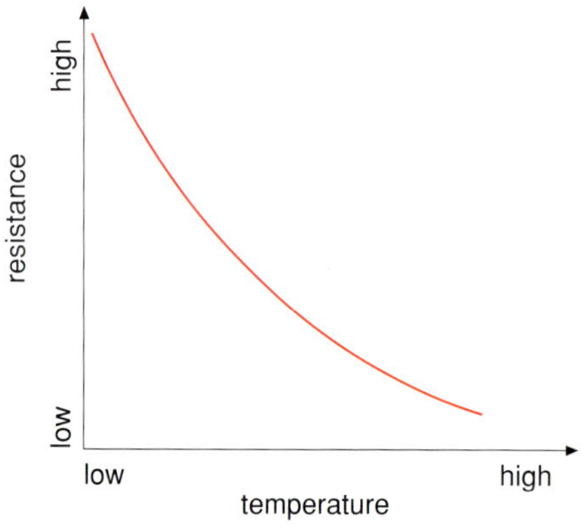

Figure 10.22
The thermistor resistance changes with temperature

The thermistor can be used as a sensor in temperature operated circuits, such as fire alarms.

Topic Questions

1 Copy and complete the following sentences.

a) The unit of resistance is the __________ .
b) The resistance of a ________________ goes down as its temperature goes up.
c) A light dependent resistor (_____) has its highest resistance in the __________ .

2 Calculate the potential difference across each of the following components.

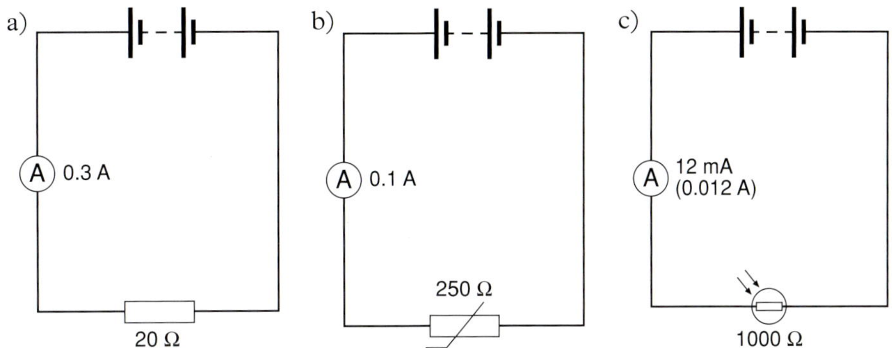

3 The circuit below shows two components joined in series. The potential difference across P is 4 volts. The potential difference across Q is 2 volts.

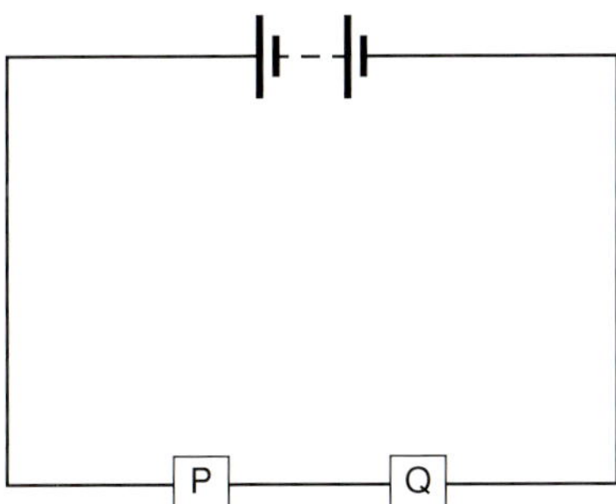

Which component, P or Q, has the largest resistance? Explain the reason for your choice.

4 A thermistor is in series with a high power lamp and is positioned close to it.

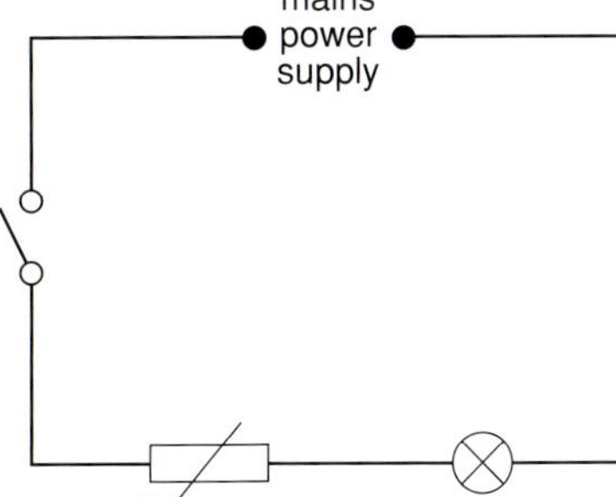

a) Explain why the lamp only shines dimly when the switch is closed.
b) Explain why the lamp gets slowly brighter (*Hint*: it also gets hotter).

10.5	
Co-ordinated	Modular
DA 12.1	DA 10
SA 12.1	SA 17

Current–voltage graphs

The circuit drawn in Figure 10.23 can be used to show how the current through a component depends on the voltage (potential difference) across it.

Figure 10.23

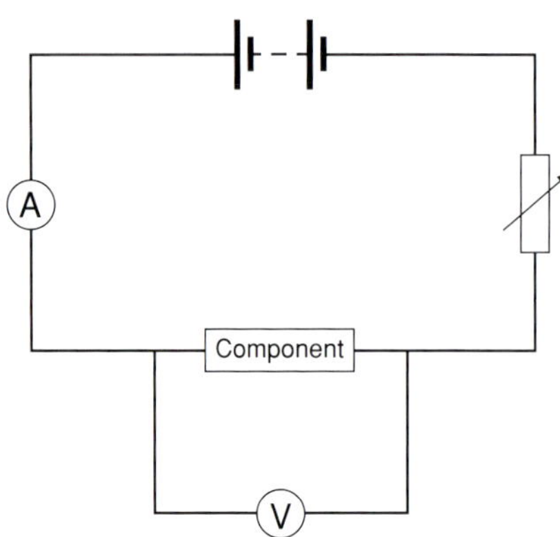

Adjusting the variable resistor changes the voltage across the component and the current through the component. In this way, a set of readings can be taken and a graph drawn.

When the component is a resistor

As long as the temperature of the resistor stays the same, the graph is always a straight line going through the origin (0,0).

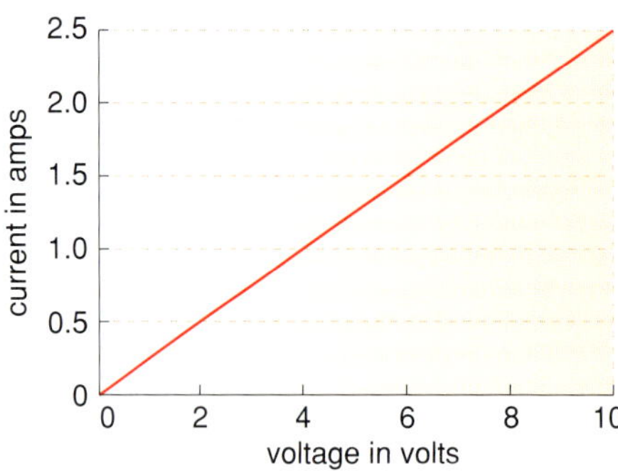

Figure 10.24 *Current–voltage graph for a resistor at constant temperature*

The graph shows that the current through the resistor is proportional to the voltage across the resistor. In other words, when the voltage changes the current changes in the same proportion.

For example: From the graph, a voltage of 4 V gives a current of 1 A. But when the voltage is doubled to 8 V, the current doubles to 2 A.

This means the resistance of the resistor is not changing, even though the voltage and current are changing. But remember, this is only true if the temperature of the resistor stays the same.

Connecting the resistor the other way round does not change its resistance. The graph just continues as a straight line.

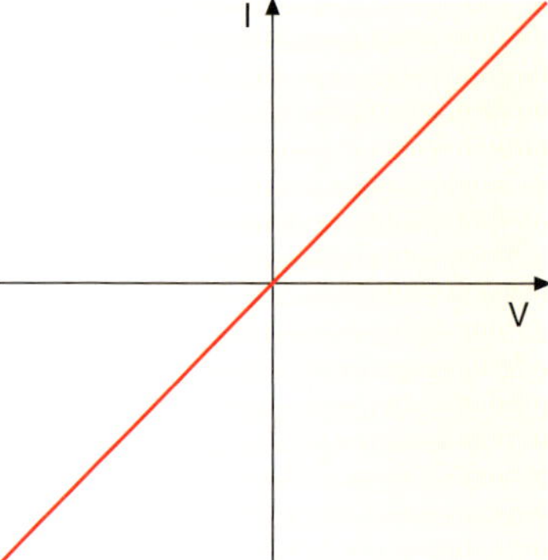

Figure 10.25 *Complete current–voltage graph for a resistor*

When the component is a filament lamp

A filament lamp uses a very hot wire (filament) to give light. As the voltage across the lamp increases, the light gets brighter. This is because the filament gets hotter. As the filament gets hotter its resistance increases. So the current does not increase in the same proportion as the voltage.

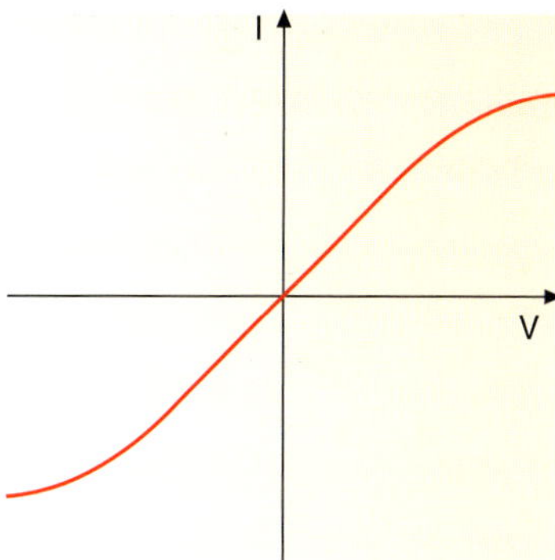

Figure 10.26 *Current–voltage graph for a filament lamp*

Connecting the lamp the other way round makes no difference to the way the resistance of the lamp changes. The resistance always goes up when the temperature of the filament goes up.

When the component is a diode

The pattern for a diode is different from both a resistor and a filament lamp. The resistance of a diode does depend on which way round it is connected in a circuit.

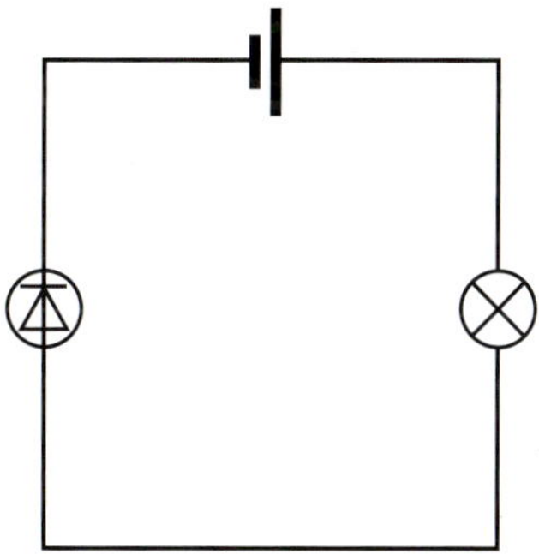

Figure 10.27 *Once the voltage goes above 0.7 volts, a current flows and the lamp lights*

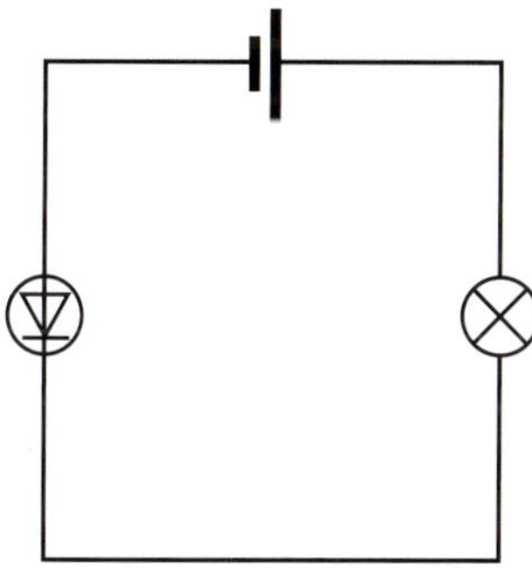

Figure 10.28 *With the diode reversed, no current flows and the lamp is off*

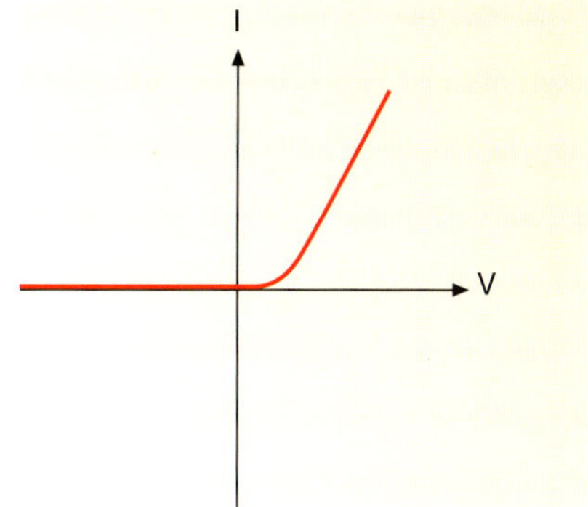

Figure 10.29 *Current and voltage for a diode*

Diodes let current flow in one direction only (called the forward direction). In the reverse direction, diodes have a very high resistance so no current flows.

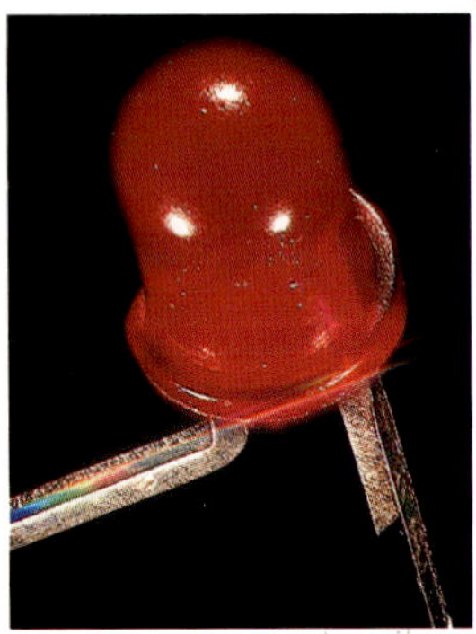

Figure 10.30
A light emitting diode (LED)

? Did you know?

Light emitting diodes (LEDs) produce light when they pass a current. The LEDs can give various colours of light. Because LEDs can be tiny, they are now used in miniature TV sets.

Topic Questions

1 Copy and complete the following sentences.

a) The resistance of a resistor does not ______________ provided the ______________ stays the same.
b) The resistance of a filament lamp ________________ as it gets __________ and brighter.
c) The ____________ of a diode depends on which way round it is connected.

2 Explain why a filament lamp might 'blow' (the filament melts) at the moment it is switched on.

3 An unknown component has the following current–voltage graph.

a) What is the component?
b) What happens to the resistance of the component as the voltage across it gets bigger?

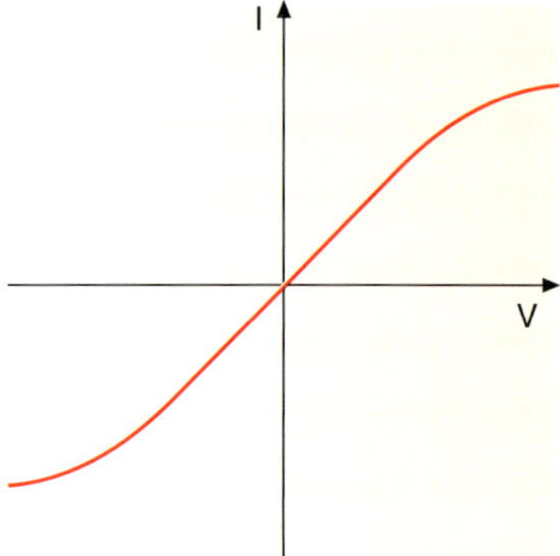

4 Which lamps are lit up when these circuits are switched on?

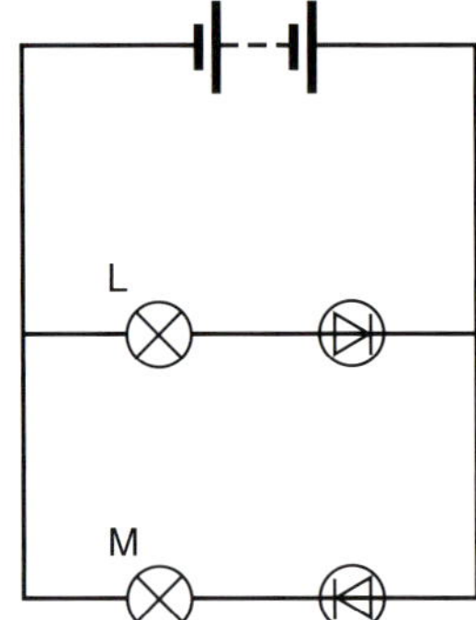

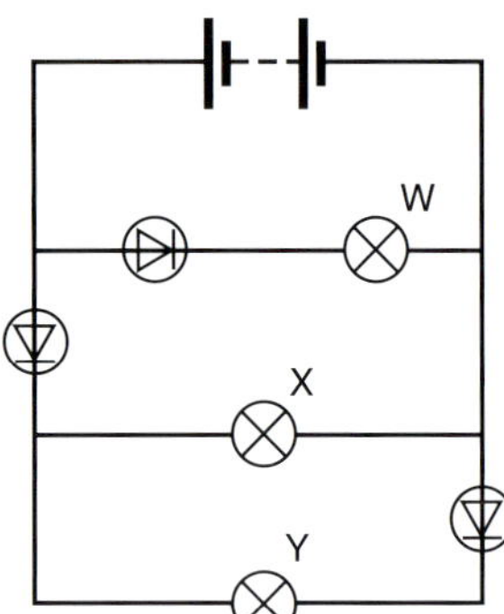

10.6 Electricity and magnetism

Co-ordinated	Modular
DA 12.2	DA 10
SA n/a	SA n/a

What you know

- Things made of iron or steel are attracted to magnets. This means there must be a force between the magnet and the iron or steel. If the force is big enough the magnet will stick to the iron or steel.
- Iron and steel are not the only magnetic materials. Nickel and cobalt are also strongly attracted to magnets.
- The space around a magnet where the force acts is called the **magnetic field**. The magnetic field lines show the direction of the magnetic force.

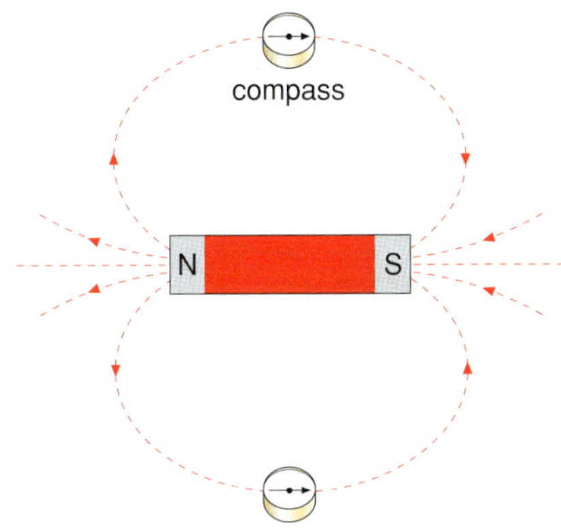

Figure 10.31
The magnetic field pattern around a bar magnet

- A small compass placed on a magnetic field line will point along the line. The compass shows the direction of the magnetic force. This is always from the north-seeking pole towards the south-seeking pole.
- Two magnets placed near each other will either push apart or pull together.
- Passing an electric current through a coil of wire makes a weak temporary magnet. The coil of wire is called a solenoid. This is a simple **electromagnet**. The coil of wire acts just like a bar magnet. One end of the coil is a north-seeking pole and the other end is a south-seeking pole.
- Reversing the direction of the current reverses the north and south poles of the electromagnet. Opening the switch turns the current off. This stops the coil being a magnet.
- Putting an iron bar in the coil makes the electromagnet much stronger as the iron bar becomes a temporary magnet.

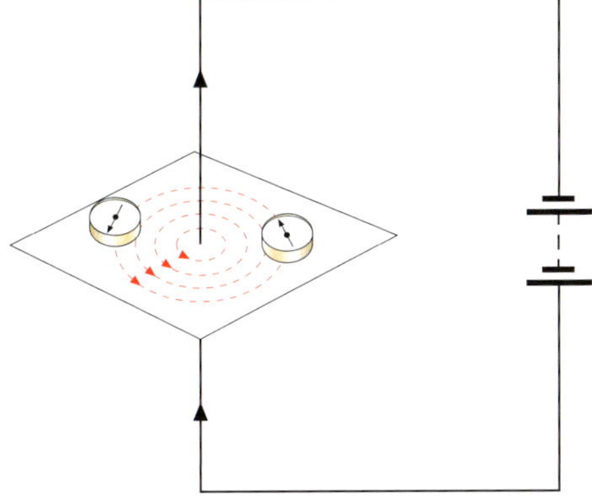

Figure 10.32
Magnetic field around a wire

When an electric current flows through a wire a magnetic field is created around the wire.

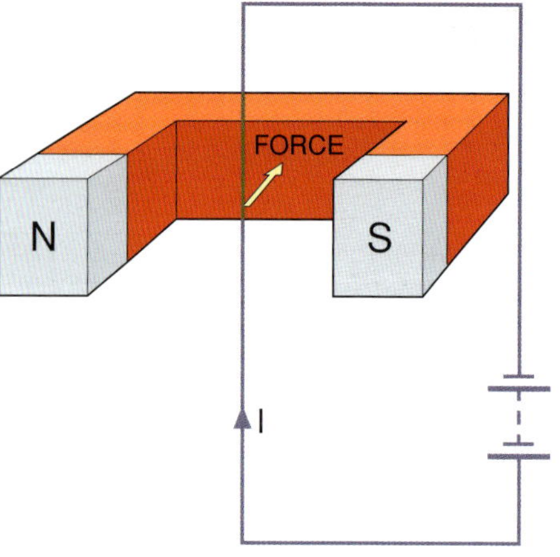

Figure 10.33
A force acts on the wire

If the wire is in the magnetic field of another magnet there may then be a force on the wire. This force will make the wire move.

The direction of the force depends on the direction of the current and the direction of the magnetic field.

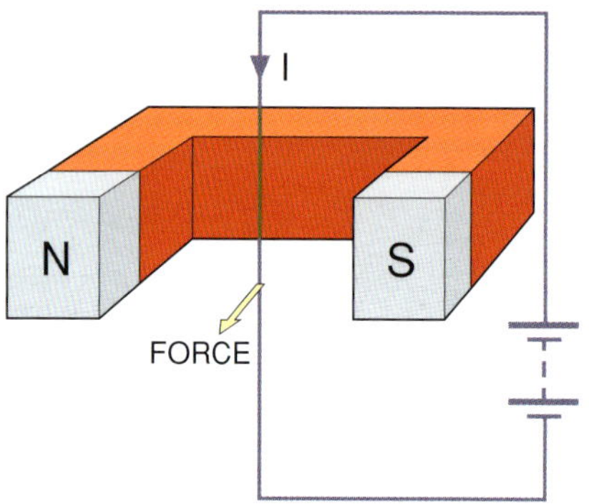

Figure 10.34
Reversing the direction of the current reverses the direction of the force

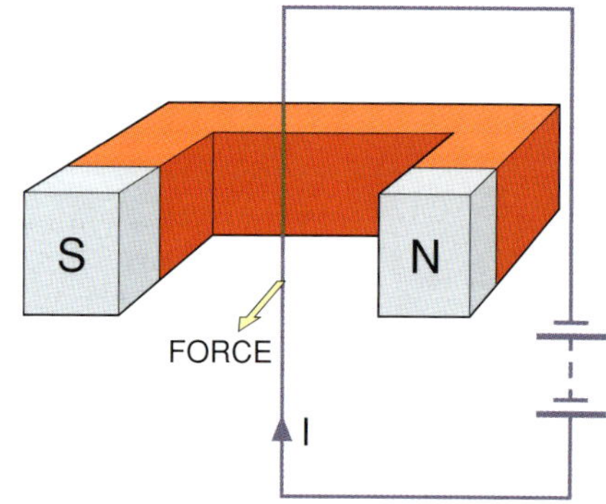

Figure 10.35
Reversing the direction of the magnetic field reverses the direction of the force

The force on the wire is made stronger if:

- The current flowing through the wire is made bigger.
- The magnetic field of the permanent magnet is stronger.

If the wire is in the same direction as the magnetic field lines of the magnet there is no force.

Did you know?

Doctors can use very strong magnetic fields to produce images of the brain or other parts of the body. This is called a magnetic resonance imaging (MRI) scan. During the scan the patient lies inside a tube that runs through the centre of a large electromagnet.

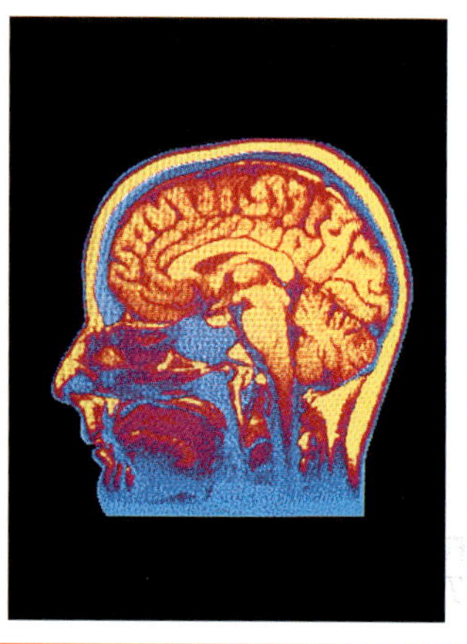

Figure 10.36
Image produced by a MRI scan

Topic Questions

1 Copy and complete the following sentences.

a) An electromagnet can be made by wrapping __________ around an __________ bar.

b) A ______________ field is created when a ____________ flows through a wire.

2 How could you show that there is a magnetic field around a wire that has a current flowing through it?

3 The diagram below shows a current flowing through a wire. The wire is in a magnetic field. There is a force on the wire.

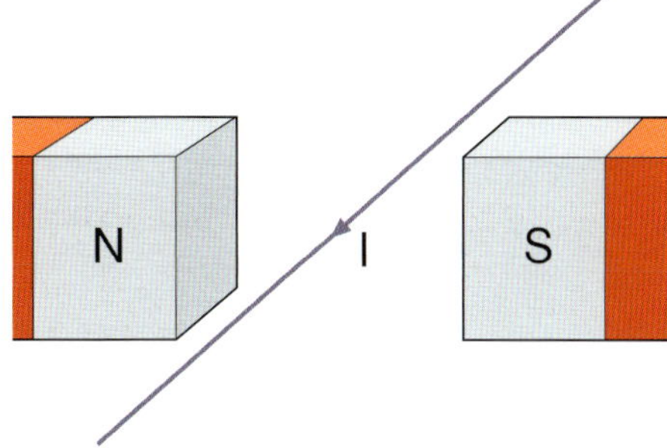

a) What happens to the size of the force if the current reverses direction?
b) What happens to the direction of the force if the current reverses direction?
c) What can be done to remove the force on the wire?

10.7	
Co-ordinated	Modular
DA 12.20	DA 10
SA n/a	SA n/a

Using the magnetic effect of an electric current

Electromagnets have many uses and applications.

They are used in motors and loudspeakers to produce movement. A doctor can use an electromagnet to remove a steel splinter from a patient's eye. Small electromagnets are used in buzzers, bells, relay switches and miniature circuit breakers.

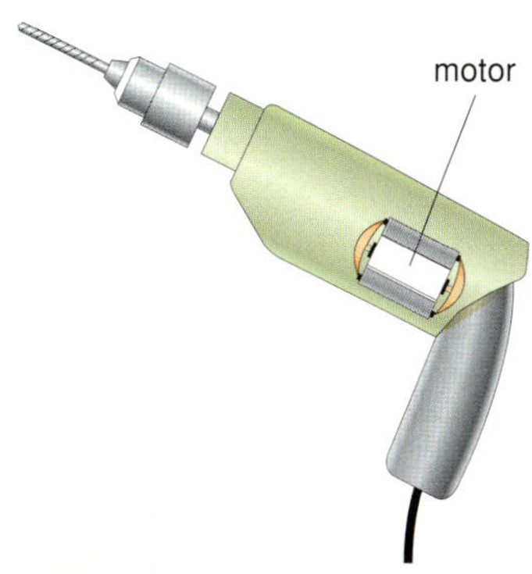

Figure 10.37
The motor inside an electric drill

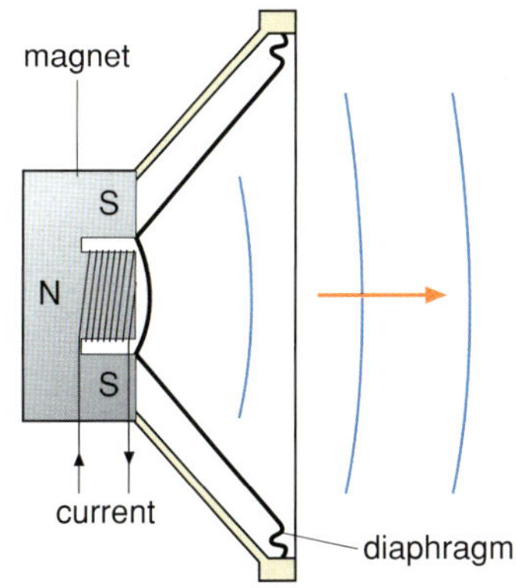

Figure 10.38
Inside a loudspeaker

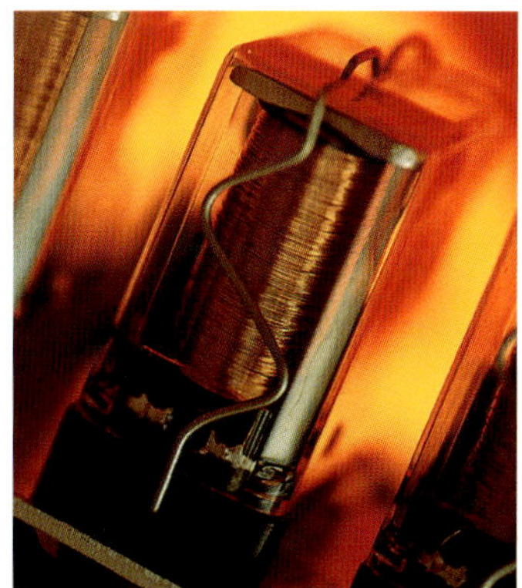

Figure 10.39
A relay switch

How a circuit breaker works

A circuit breaker is a switch that automatically breaks a circuit when the electric current is too big. Later the circuit breaker can be reset to switch the circuit back on.

Figure 10.40 shows one type of circuit breaker.

Figure 10.40
A circuit breaker

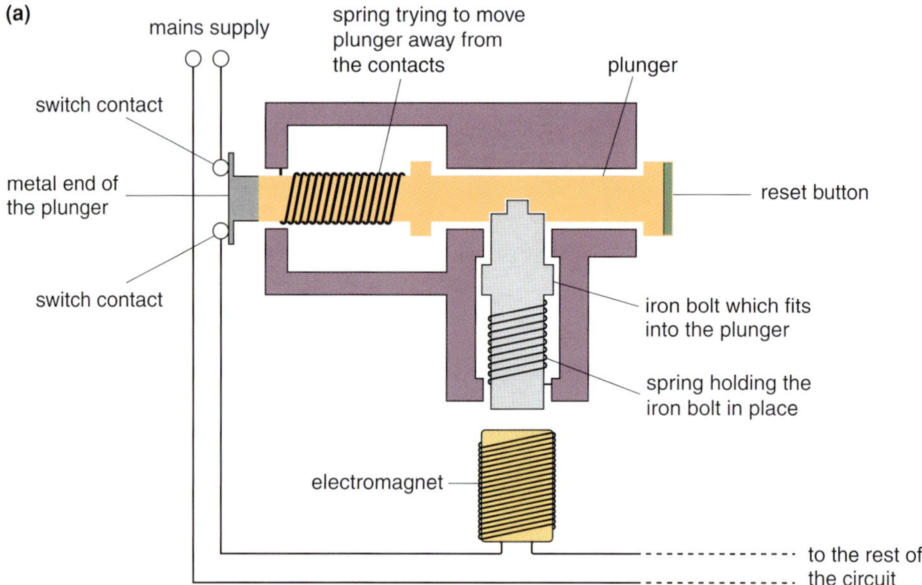

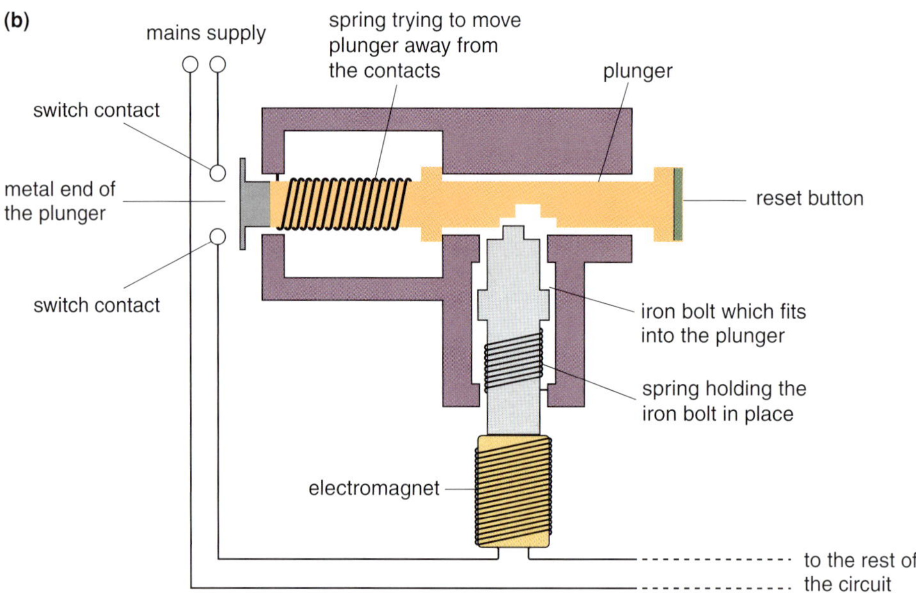

The current through the electromagnet is the same as the current through the circuit. With the right current flowing through the circuit, the electromagnet is not strong enough to attract and pull the iron bolt down. So the spring keeps the iron bolt in place.

A fault in the circuit may make the current go up. If this happens the electromagnet becomes stronger. The stronger electromagnet will pull the iron bolt down. The plunger is then pushed away from the contacts by its own spring. The contact switch is now open and the circuit is broken.

Once the fault is repaired the reset button can be pressed. This pushes the plunger in and closes the contact switch. The iron bolt will spring back to hold the plunger, keeping the contact switch closed.

The d.c. electric motor

A simple electric motor is made using a coil of wire suspended so that it can turn between the poles of a magnet.

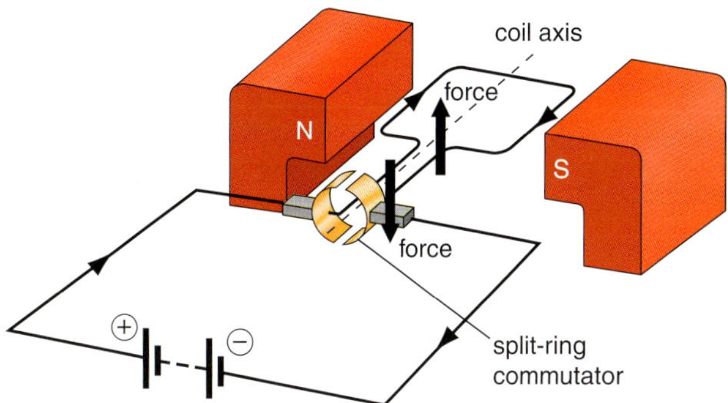

Figure 10.41 *A simple d.c. electric motor*

Closing the lower supply switch sends a current through the coil. This causes a force on each side of the coil. Since the current flows in opposite directions down each side of the coil the forces will also be in opposite directions. So as one side of the coil goes up, the other side goes down and the coil turns.

To keep the coil turning the current must change direction each time the coil becomes vertical. The split ring commutator is designed to make sure that the current does this.

The coil can be made to spin faster by using:

- a stronger magnet
- a coil with more turns of wire
- a bigger current through the coil.

Topic Questions

1 Write down the names of six appliances (not given in this section) that use an electric motor.

2 Copy and complete the following sentence by choosing the correct phrase.

The function (job) of a circuit breaker is to *reduce the current in a circuit / switch off the current when there is a fault / complete a circuit.*

3 Write down three ways to make the forces on the coil in an electric motor larger.

4 Copy and complete the diagram to show the energy change an electric motor is designed to produce.

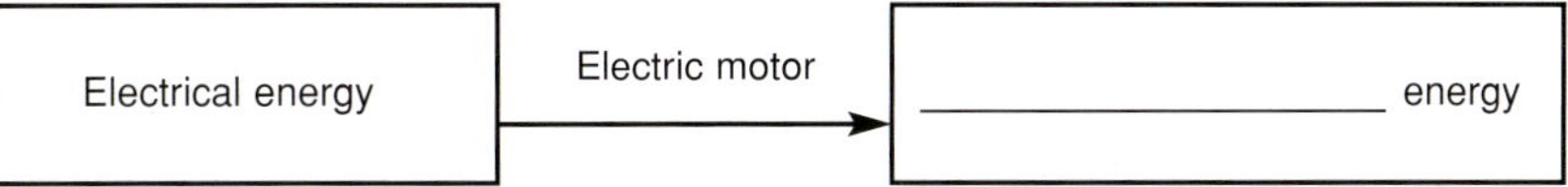

5 The diagram shows one type of circuit breaker.

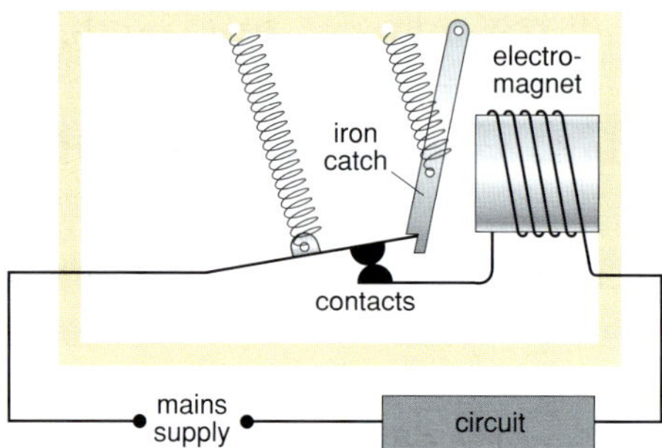

a) Write each of the following statements in the correct order.
- The strength of the electromagnet increases.
- The contacts spring apart breaking the circuit.
- A fault makes the current through the circuit increase.
- The iron catch is pulled across to the electromagnet.

b) Which important part of a circuit breaker is missing from the diagram?

10.8

Co-ordinated	Modular
DA 12.5	DA 10
SA n/a	SA n/a

Electrostatics

Rubbing different insulating materials together will usually make them become electrically charged. For example, rubbing a polythene rod with a cloth makes both the rod and the cloth charged. Once charged, the rod and cloth will stay charged for some time.

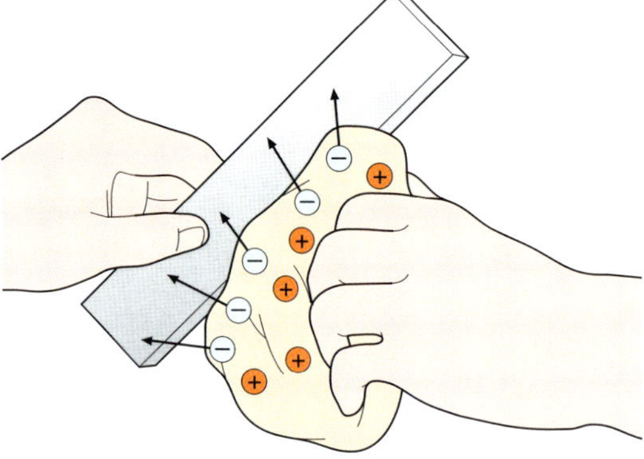

Figure 10.42
Charging a polythene rod

Where does the charge on the rod come from?

Like everything else, the rod and cloth are made of atoms. Atoms have the same number of negative charges (**electrons**) as positive charges (**protons**). (See Section 12.12.)

Rubbing the rod with the cloth doesn't make new charge. It moves and separates the charge that is already there.

Friction between the rod and the cloth removes electrons from some atoms in the cloth. These electrons move from the cloth onto the rod. The rod now has extra electrons so it is negatively charged. The cloth lost the electrons so it has a positive charge.

Only the negative charges (electrons) are free to move from one object to another.

When the rubbing stops no more electrons go onto the rod. The electrons that are already there will usually stay there and not move. We call this static electricity. But if the static charge does move someone could be in for a nasty shock!

A Perspex rod rubbed with the same cloth becomes positively charged.

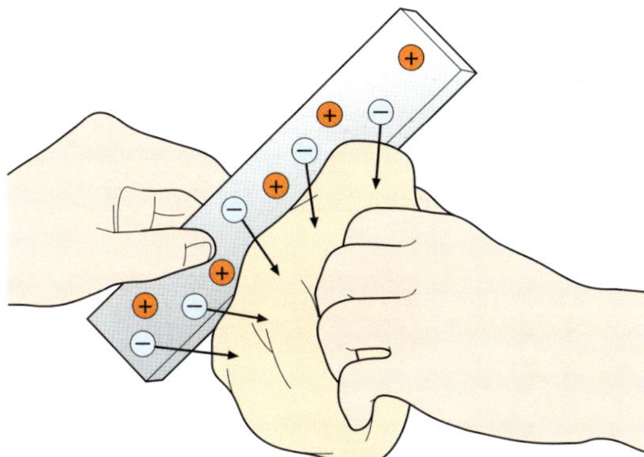

Figure 10.43
Charging a Perspex rod

This time electrons move from the Perspex rod onto the cloth. So while the Perspex becomes positive, the cloth becomes negative.

Electrostatic force

Charged objects exert forces on each other. If the objects are close together the forces may be large enough to make the objects move.

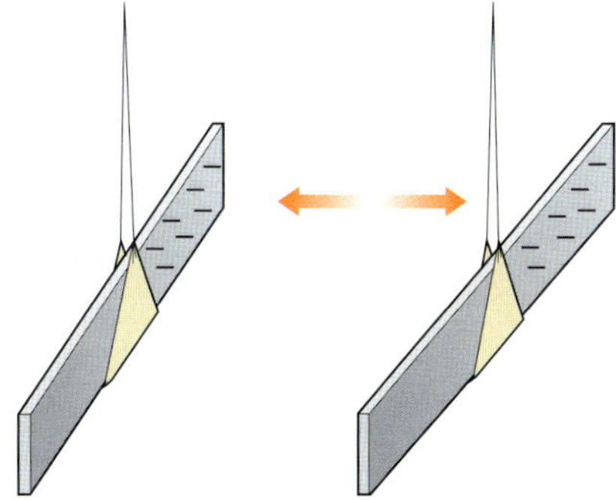

Figure 10.44
The forces between two charged polythene rods push the rods apart. The rods are repelling each other because they both have the same type of charge (negative)

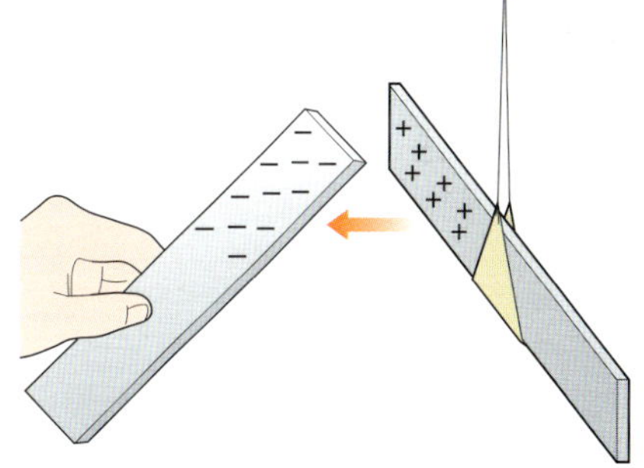

Figure 10.45
The forces between the polythene rod and the Perspex rod pull the rods together. The rods are attracting each other because they have opposite charge. One is positive and the other is negative

- Objects with the **same** charge **repel**.
- Objects with **opposite** charges **attract**.

Figure 10.46
Each hair has the same charge, so they repel each other

Charged objects can also attract small uncharged objects that are close by. For example, when you rub a balloon on your sleeve it becomes charged. The balloon can then attract your hair.

Topic Questions

1 Copy and complete the following sentences using the words in the box.

attract	charged	electrons	positively	repel	rubbed	small

A polythene rod is negatively ______________ when it is ____________ with a cloth. The cloth loses ____________ to the rod. The cloth is _____________ charged. The rod can attract ____________ uncharged objects. Opposite charges ____________ and charges of the same type _____________ .

2 A small polystyrene bead hangs on a nylon thread. The bead has a positive charge. Different rods, some of which are charged, are brought near the bead. The movement of the bead is recorded in the table. Copy the table. For each rod tick one of the columns to show the charge on the rod.

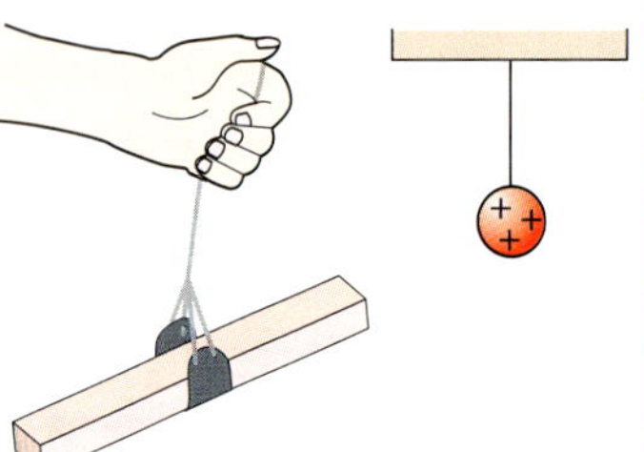

Type of rod	Bead movement	Charge on rod: Positive	Negative	No charge
Acetate	Repelled			
Ebonite	Attracted			
Polythene	Attracted			
Steel	None			

3 The diagram shows two charged balloons.

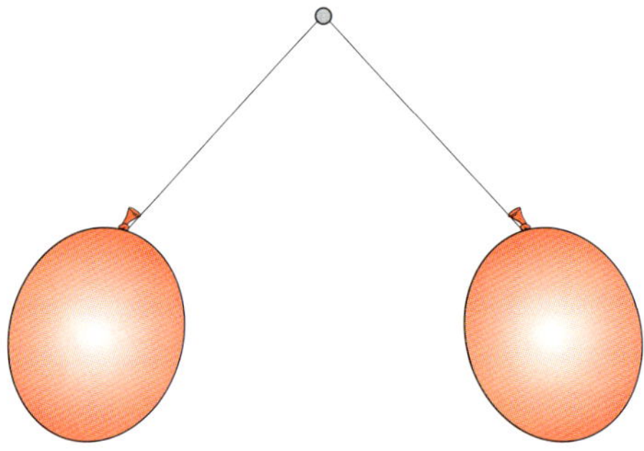

What type of charge is on each balloon? Explain the reason for your answer.

10.9	
Co-ordinated	Modular
DA 12.5 SA n/a	DA 10 SA n/a

Making use of electrostatic charge

The pictures in Figure 10.47 show three different ways that electrostatic charges can be used.

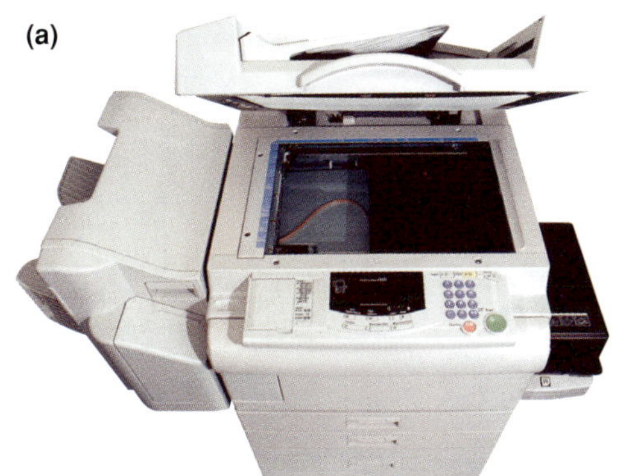

Figure 10.47 *Using electrostatic charge*

Photoconducting material

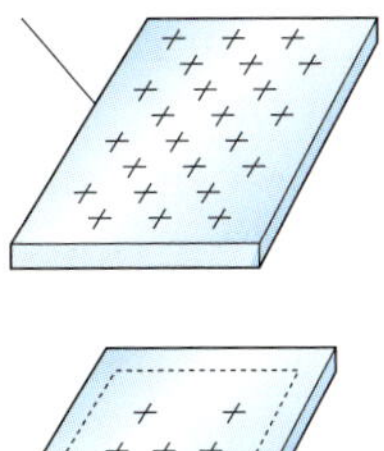

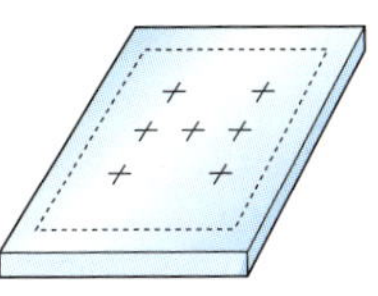

toner

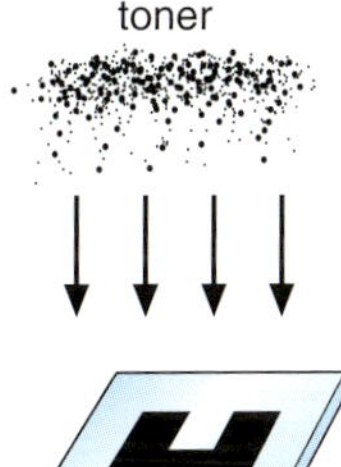

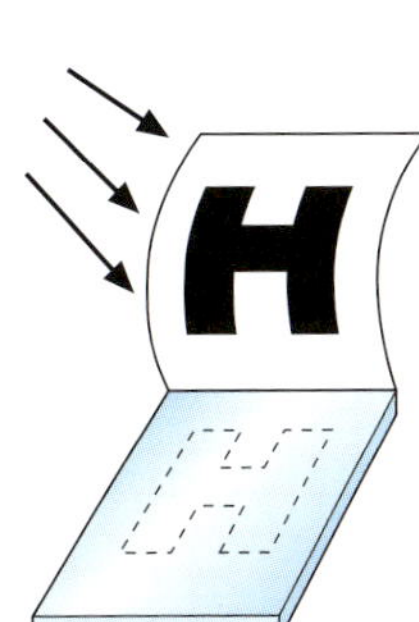

Figure 10.48 *Making a photocopy*

Making a photocopy

1 Inside the photocopier there is a light sensitive copying plate (or sometimes a roller). This plate is given a positive charge.

2 The page to be copied is lit with a strong light. This causes an image of the page to form on the charged copying plate. The bright areas of the plate lose their charge. The dark areas of the plate keep their charge. The pattern of the charge staying on the plate is the same as the dark parts of the original page.

3 Black negatively charged toner powder is spread over the plate. The toner powder is attracted to the positive charge that is still on the plate.

4 A blank sheet of paper is placed over the plate. The toner powder sticks to the paper.

5 The paper is heated and pressed by rollers against the plate. This makes sure the toner sticks to the paper. The sheet of paper is now a photocopy of the original page.

Electrostatic paint spray

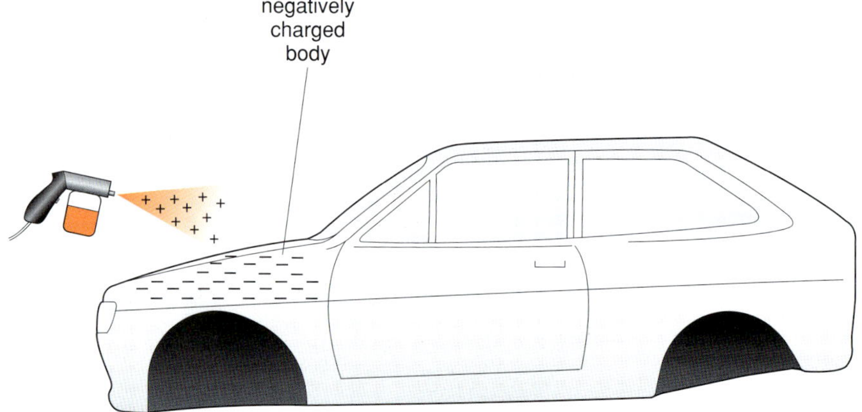

Figure 10.49
Spraying a car body

The tiny droplets of paint are charged as they leave the spray gun. Because the same charge is on each droplet they repel each other. This gives a fine even spray of paint. If the metal car body has the opposite charge to the paint, the paint droplets will be attracted to the car body. So the car body ends up covered in an even layer of paint.

Electrostatic smoke precipitator

Burning fuels like coal pollutes the air with waste gases and smoke. Smoke is made up of tiny particles of ash. The smoke can be taken from the waste gases before they go into the air by a smoke precipitator.

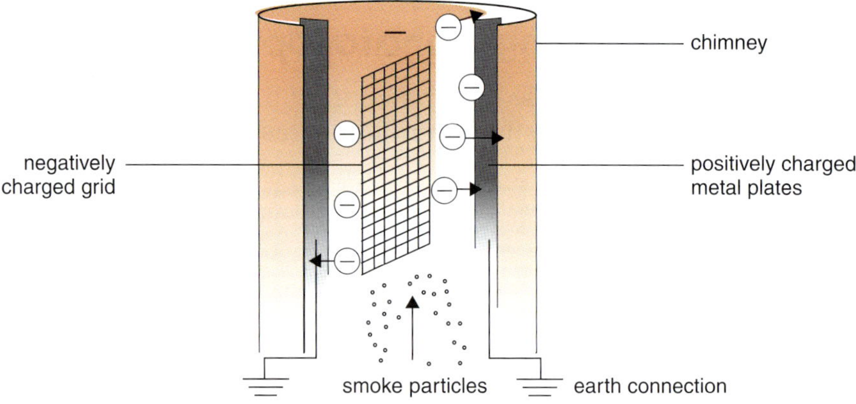

Figure 10.50
Electrostatic smoke precipitator

- The waste gases and smoke go up the chimney and through a negatively charged metal grid.
- As they go through the grid the smoke particles are given a negative charge.
- The grid repels the smoke particles.
- The metal collecting plates on the inside of the chimney have a positive charge.
- The smoke particles are attracted to the collecting plates and stick to them.

A coal-burning power station can produce up to 50 tonnes of fine ash every hour.

- The plates are knocked to make the smoke particles fall to the bottom of the chimney.
- The waste gases are now clean and free of smoke particles.

Topic Questions

1 Copy and complete the following sentences.

a) Smoke particles become ___________ when they pass through a charged metal grid. The smoke particles are then _______________ by the metal grid.

b) The copying plate in a photocopier is ___________ sensitive.

2 Why must the toner powder in a photocopier have a charge?

3 Why is a photocopy usually warm when it comes out of the photocopier?

10.10	
Co-ordinated	Modular
DA 12.5	DA 10
SA n/a	SA n/a

Charges on the move

Electrostatic charge can build up on an insulated object. After a while the charge may jump through the air causing a spark. In some situations even a small spark can be dangerous.

Delivering fuel

Figure 10.51 *Tanker delivering fuel safely*

Tankers deliver large amounts of fuel to garages and to aircraft. As the fuel flows through the delivery pipe, friction makes the pipe charge up. If the charge jumps to a nearby object the resulting spark may make the petrol vapour explode.

To stop charge building up, the tanker is **earthed**. This means the tanker is joined to the ground by a metal wire. Instead of building up on the pipe the charge flows along the wire to the earth.

Fine powders

In some industries, such as flour making, the friction between fast moving powders can cause the build up of charge. This could cause a spark that makes the powders explode.

Lightning

Figure 10.52 *Lightning between oppositely charged clouds*

If a cloud with a large charge passes close to tall objects on the ground, the charge may jump from the cloud to the ground causing a lightning strike.

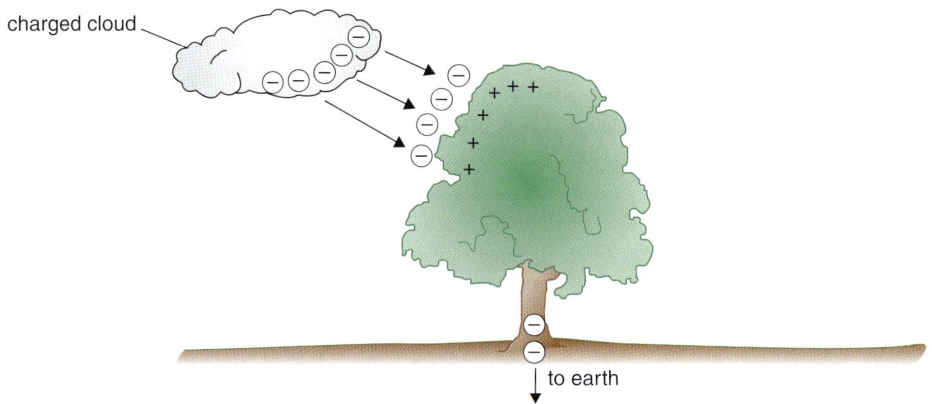

Figure 10.53 *A lightning strike between a charged cloud and a tree*

Tall buildings usually have a lightning conductor. This is a thick strip of copper – a good conductor. One end of the strip sticks up above the building and the other end is buried in the ground.

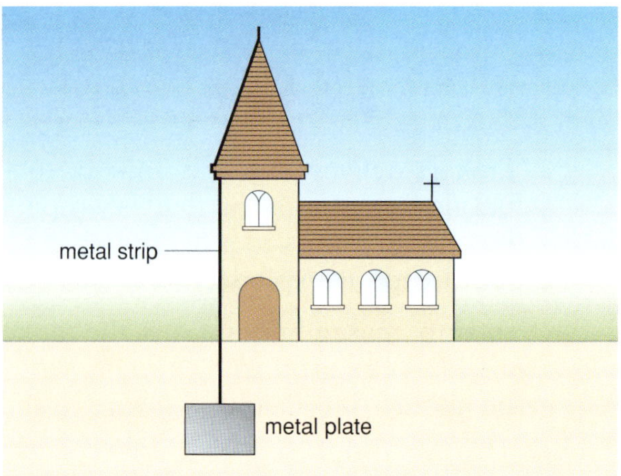

Figure 10.54 *Protecting a building with a lightning conductor*

Did you know?

Benjamin Franklin invented the lightning conductor. He got the idea after flying a kite during a thunderstorm. A Russian scientist copying Franklin's kite experiment died when charge from a thundercloud was conducted through him to the ground.

If lightning does strike, it is more likely to hit the top of the lightning conductor than the building. The charge would then flow down the conductor to the earth without harming the building.

Charge and current

An **electric current** is the movement of charged particles. In a metal wire, an electric current is a flow of electrons. When a wire is joined to the terminals of a battery, the electrons move from the negative terminal, along the wire to the positive terminal of the battery. As one electron arrives at the battery another sets off around the circuit.

Electrolysis

An atom has equal numbers of electrons and protons (see Section 12.12). So an atom is electrically neutral. If an atom loses or gains electrons it stops being neutral and becomes charged. The atom is now called an **ion**.

Some chemical compounds conduct electricity when they are melted or dissolved in water.

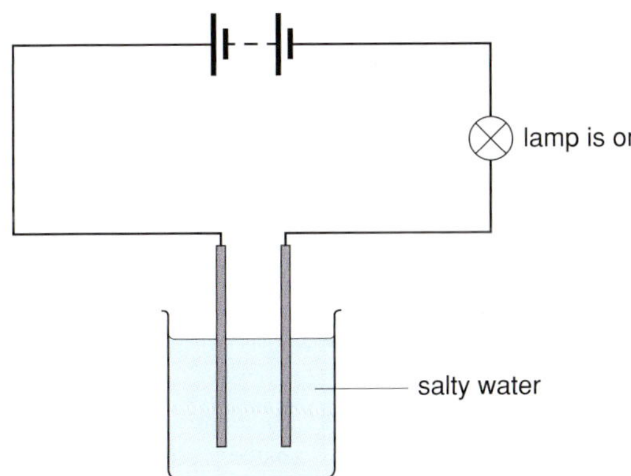

Figure 10.55
Sodium chloride dissolved in water (salty water) conducts electricity

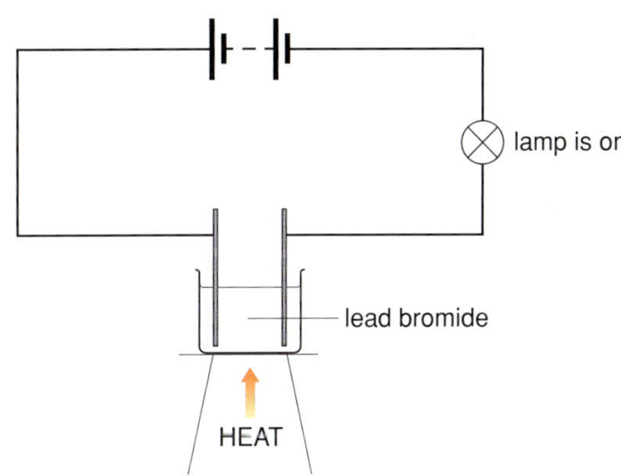

Figure 10.56
Lead bromide conducts only when it's melted

The compounds start to conduct because they are made up of ions. An electric current flows when the negatively charged ions move to the positive electrode, and the positively charged ions move to the negative electrode.

When the ions reach the electrodes, simpler substances, either gases or solids are released. This process is called **electrolysis**.

Figure 10.57 *Movement of ions in electrolysis*

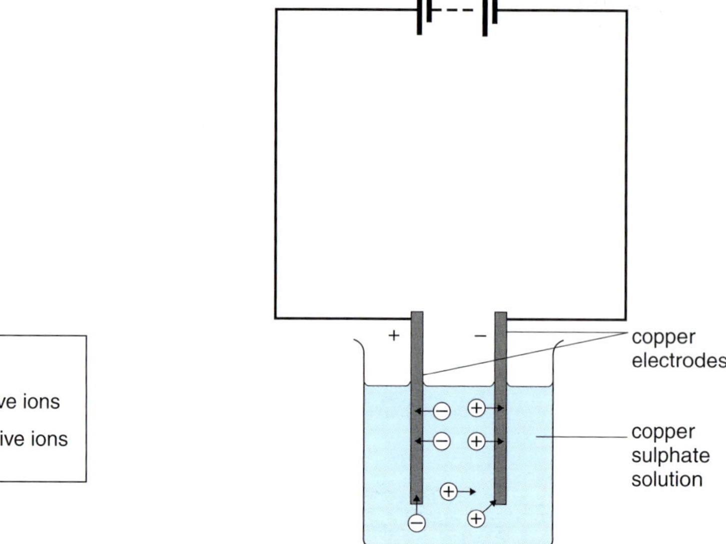

Did you know?

Electroplating uses electrolysis. A cheap metal spoon can be coated with a thin layer of silver.

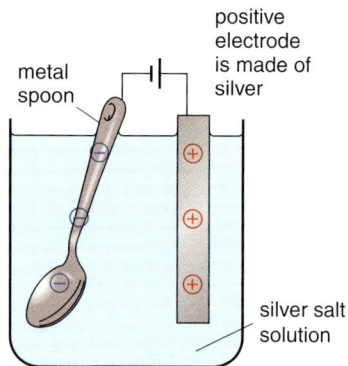

metal spoon

positive electrode is made of silver

silver salt solution

Figure 10.58

Topic Questions

1 Copy and complete the following sentences.

a) An atom that loses an ______________ is called an ion.

b) In electrolysis ____________ charged ions move to the negative ________________ .

c) In a solid conductor, an electric _____________ is a flow of electrons.

2 Why are the metal ends of fuel pipes earthed?

3 A current flows through copper sulphate solution.

Which colour circles represent the negative ions? Explain the reason for your choice.

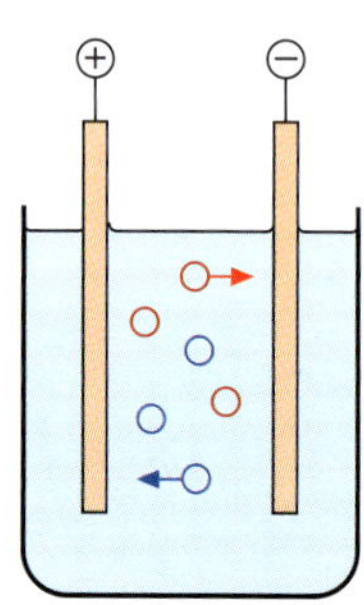

4 Explain how a conducting strip joined to the back of a car stops charge building up on the car body.

5 Why can molten lead bromide conduct electricity?

10.11

Co-ordinated	Modular
DA 12.3	DA 10
SA 12.3	SA 17

Mains electricity

Direct current (d.c.)

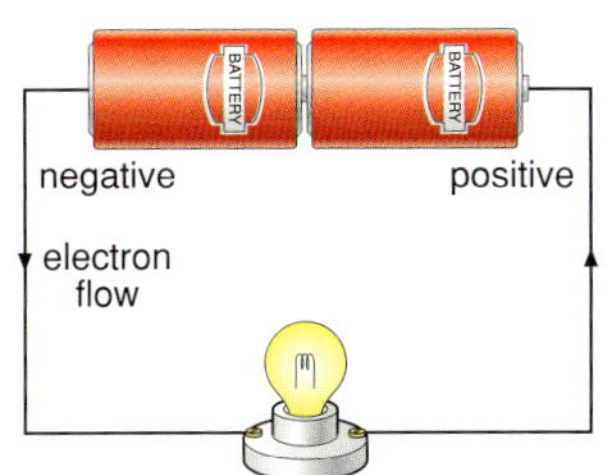

Figure 10.59
Electrons flow from negative to positive

A **direct current (d.c.)** always flows in the same direction. In circuit diagrams it is shown flowing from the positive to the negative terminal of the power supply.

A cell or battery gives a steady direct current.

We should remember that the current is actually a flow of electrons (see Section 10.10). So the electrons in a d.c. circuit always flow in the same direction.

Did you know?

The scientists who first investigated electricity did not know about electrons. They thought an electric current was a flow of positive charge. This idea is still used in circuit diagrams. It is called the conventional current.

An oscilloscope (CRO) can be used to show how the voltage (potential difference) of a d.c. supply changes with time (Figure 10.60).

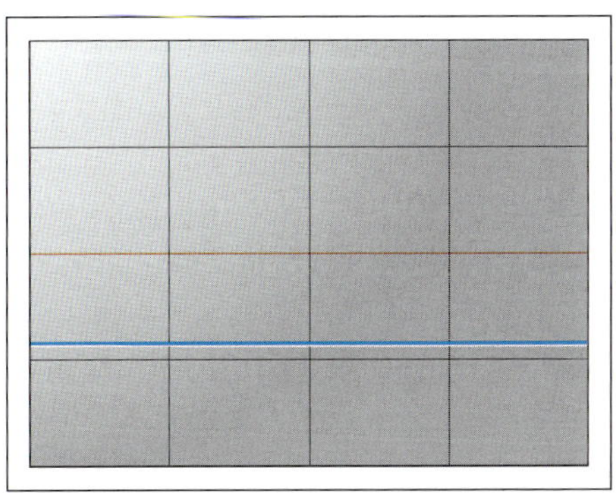

Figure 10.60
The voltage of a d.c. supply is steady and in the same direction. The zero voltage line is drawn in red

Alternating current (a.c.)

An **alternating current (a.c.)** in a circuit is constantly reversing its direction.

An oscilloscope (CRO) can be used to show how the voltage (potential difference) of an a.c. supply changes with time.

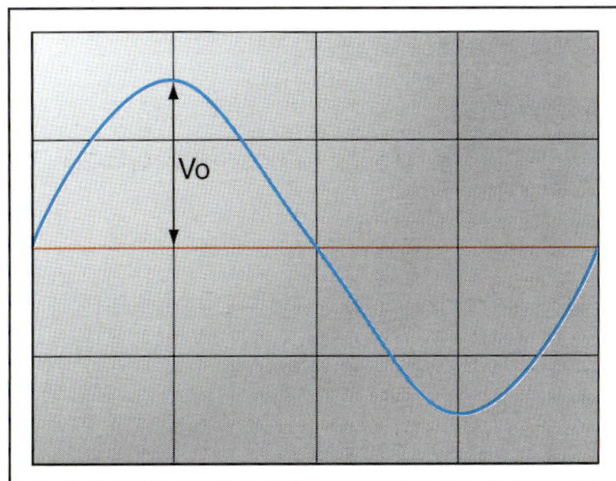

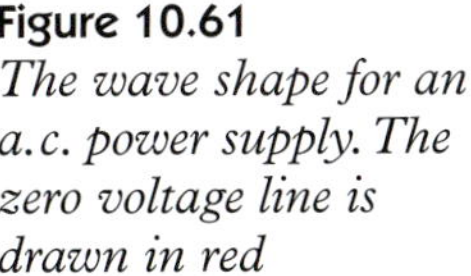

Figure 10.61
The wave shape for an a.c. power supply. The zero voltage line is drawn in red

The voltage of an a.c. supply is constantly changing. The voltage increases from zero to a peak (maximum) value, shown in Figure 10.61 as V_O. From the peak value, the voltage falls to zero, then changes direction. The voltage again increases to a peak value and then falls to zero. This changing voltage cycle is repeated many times a second.

Each time the line on the oscilloscope changes from above to below (or from below to above) the zero line, the current in the circuit changes direction.

The number of times the voltage (or current) changes direction in one second gives the frequency of the supply. Frequency is measured in cycles per second or **hertz** (Hz).

Comparing two different a.c. supplies

An oscilloscope can be used to compare different a.c. supplies.

Figure 10.62 shows the oscilloscope traces for two a.c. supplies. The controls on the oscilloscope were the same for both supplies.

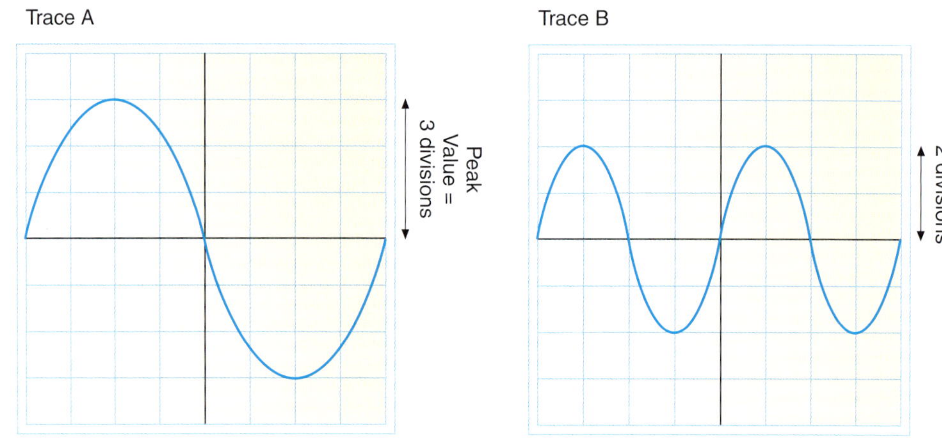

Figure 10.62
The two a.c. supplies have a different frequency and peak voltage

Trace A is 1½ times taller than trace B. So the peak voltage of supply A is 1½ times bigger than supply B.

Example: What is the peak voltage of supply A if the peak voltage of supply B is 4 V?

$$\text{peak voltage of supply A} = 1\tfrac{1}{2} \times \text{peak voltage of supply B}$$
$$= 1\tfrac{1}{2} \times 4 = 6\,\text{V}$$

Basically, as long as the controls on the oscilloscope stay the same, the taller the wave trace the bigger the peak voltage.

Supply B has twice as many waves on the oscilloscope as supply A. This means the voltage (and current) of supply B changes direction twice as often as the voltage (and current) of supply A. So the frequency of supply B is twice the frequency of supply A.

Example: What is the frequency of supply B if the frequency of supply A is 50 Hz?

$$\text{frequency of supply B} = 2 \times \text{frequency of supply A}$$
$$= 2 \times 50 = 100\,\text{Hz}$$

Basically, as long as the controls on the oscilloscope stay the same, the more waves that are shown on the screen, the higher the frequency.

Did you know?

An appliance like an electric toaster, designed to transform electrical energy to heat energy, would work with a d.c. supply. A steady 230 volt d.c. supply would give the same heating effect as the 230 volt a.c. mains supply.

The mains electricity supply

In the United Kingdom mains electricity is an a.c. supply. Although the voltage of the supply keeps changing we give its value as about 230 volts. This is not the peak voltage, it's a way of giving a special sort of average.

The frequency of the UK mains supply is 50 hertz. This means the current flows one way, then back again, 50 times each second.

Topic Questions

1 Copy and complete the following sentences.

a) The current from a _____________ is a _____________ current. This means it always flows in the same _____________.

b) Mains electricity is an _______________ current supply. It has a ___________ of 50 ___________ (Hz).

2 The diagram shows the oscilloscope trace for a battery.

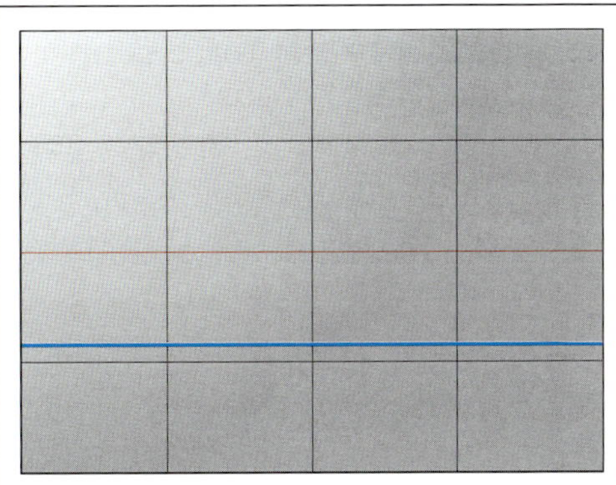

Copy the diagram. Draw a second line on your diagram to show the trace when the connection between the oscilloscope and battery are reversed.

3 The diagrams show the oscilloscope traces for different a.c. supplies.

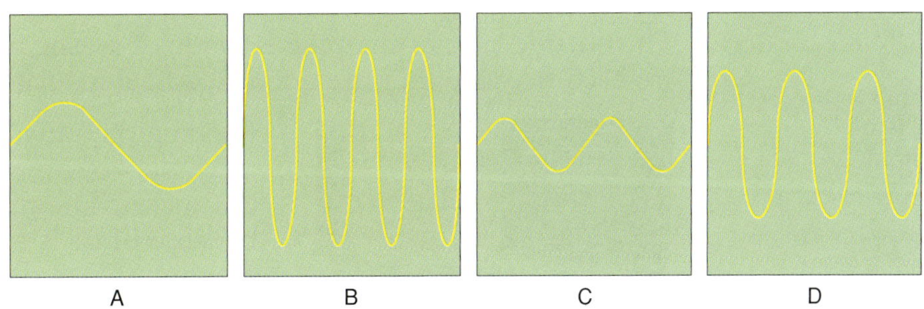

a) Which of the supplies has the largest peak voltage?
b) Which of the supplies has the highest frequency?
c) Which of the supplies has the lowest frequency?
d) Which of the supplies has the smallest peak voltage?

4 The diagram shows part of the information plates from two electric drills. One of the drills is from the UK the other is from the USA.

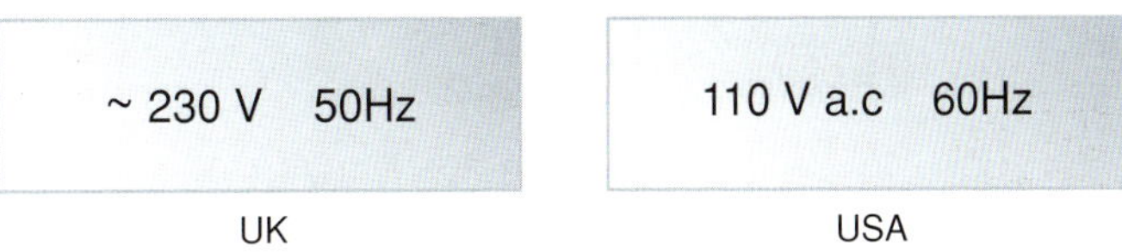

a) Write down two differences between the power supply used in the UK and the power supply used in the USA.
b) What is the same for both power supplies?

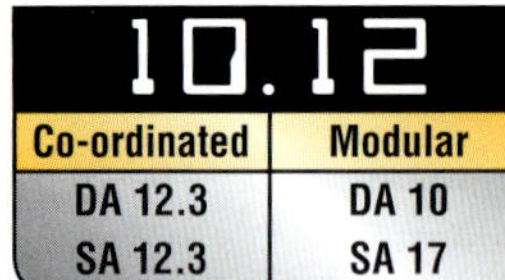

10.12

Co-ordinated	Modular
DA 12.3	DA 10
SA 12.3	SA 17

Using the mains electricity supply

Many electric appliances are designed to work from the mains electricity supply. Most of these appliances are connected to the mains using a cable and 3-pin plug. Both the cable and plug are designed to make the appliance safe to use.

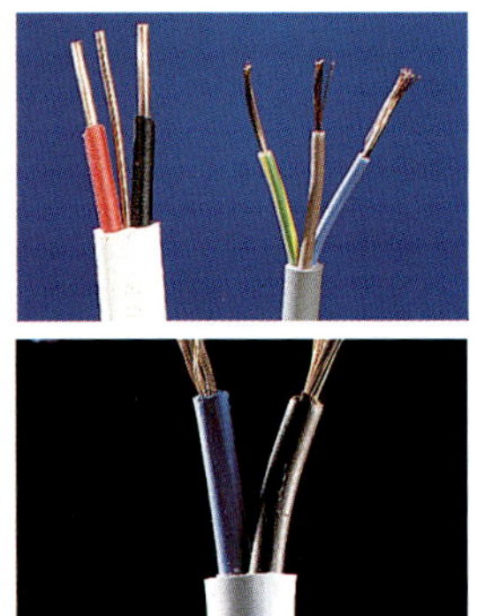

Figure 10.63
Three core and two core cables

The cable

Inside the cable are either two or three copper wires (the inner cores). Copper is used for the wires because it is a good electrical conductor. A layer of plastic covers each wire. The wires are held in place by an outside layer of flexible plastic.

Plastic is used to cover the inside wires and to form the outside part of the cable because it is a good insulator and is flexible.

Which type of cable is used will depend on the design of the appliance. Three core cable must be used with any appliance that has a metal case (see Section 10.13)

A 3-pin plug

The outside case of the plug is made of plastic or rubber. These materials are used because they are good electrical insulators.

Figure 10.64
The outside of a three-pin plug and its socket

Pushing the plug into a socket joins the three pins of the plug to the terminals of the socket. The pins are made from brass, because brass is a good electrical conductor.

Inside the plug there is a fuse and a cable grip.

The fuse joins the live pin to the live wire inside the plug. This means the current must flow through the fuse on its way to the appliance. To protect the appliance the fuse should always be the one recommended by the maker of the appliance (see Section 10.13).

The cable grip is there to hold the cable firmly in place. If the cable is pulled the cable grip will stop the copper wires inside the plug being pulled loose.

Connecting an appliance to a 3-pin plug

When you buy a new electrical appliance it will come with a plug already fitted. But at some time a new plug may need to be fitted to an older appliance. When this is done care must be taken to make sure the correct wire from the appliance is joined to the correct pin in the plug. The wires are colour coded to help make sure they go to the right place.

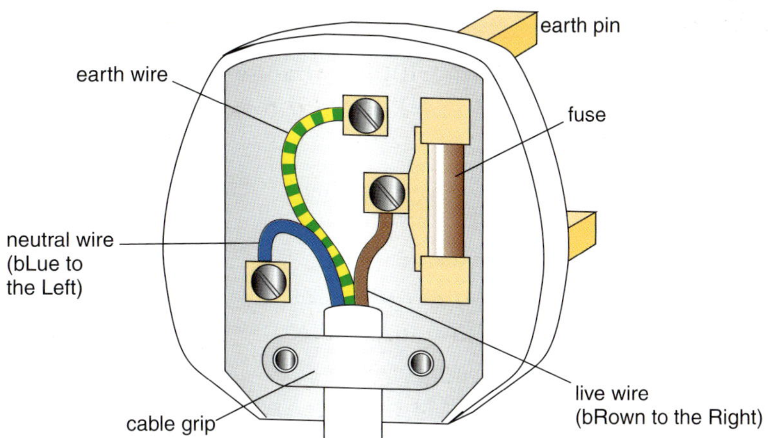

Figure 10.65 *Correct wiring in a 3-pin plug*

- The brown covered wire is joined to the fuse, which is joined to the live pin.
- The blue covered wire is joined to the neutral pin.
- The green/yellow covered wire is joined to the earth pin.

Topic Questions

1 Copy and complete the following sentences.

a) When wiring a 3-pin plug, the __________ covered wire goes to the live pin, the __________ covered wire to the neutral pin and the green/yellow covered wire to the __________ pin.

b) Appliances with a metal case must be connected to a plug using __________ core cable.

c) A __________ joins the live pin to the live wire.

2 Which of the following materials are electrical insulators – brass, copper, plastic or rubber?

3 Say why each plug shown below is not safe to use.

a)

b)

c)

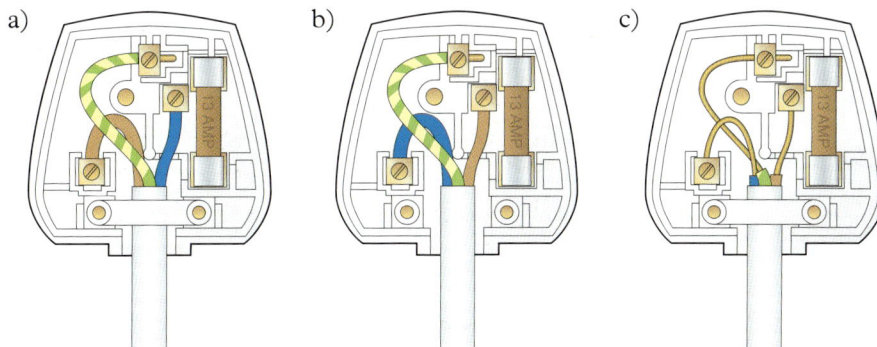

10.13	
Co-ordinated	Modular
DA 12.3	DA 10
SA 12.3	SA 17

Safety devices and electrical appliances

Voltages larger than 50 volts and currents as small as 0.03 amps (30 milliamps) can kill. So it is very important electricity is used correctly and safely.

Safety measures include:

- using insulation (see Section 10.12)
- devices that automatically switch off the current
- earthing an appliance.

A fault in an appliance or circuit may cause a larger than normal current to flow. If this happens the circuit can be automatically switched off by a fuse or circuit breaker.

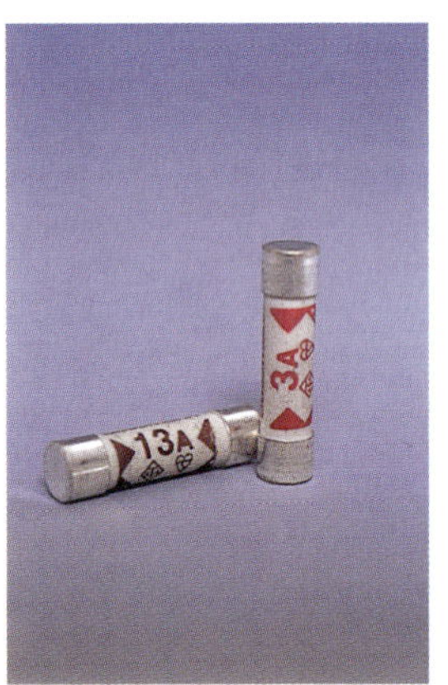

Figure 10.66
Fuses

Fuses

A fuse is a thin piece of wire that lets a current up to a certain value flow through it. Above this value the fuse will overheat and melt. We often say that the fuse has 'blown'. For example, a 5-amp fuse will melt if a current bigger than 5 amps flows through it.

Most plugs are fitted with a 3-amp, 5-amp or 13-amp fuse. It is important that the right fuse is used with an appliance.

The fuse must have a higher value than the normal current that flows through the appliance. But it should be as close to the normal current as possible.

Most fuses take 1–2 seconds to melt. This is enough time for an electric shock to kill someone. So fuses protect the appliance more than the person using the appliance.

Did you know?

Many household fires are caused by electrical faults.

Figure 10.67

Circuit breakers

Circuit breakers are fast acting. If there is an electrical fault they can switch the current off in less than 0.05 seconds. This is fast enough to stop an electric shock killing someone.

There are two main types of circuit breaker. One type is described in Section 10.7. The other type is the residual current circuit breaker (RCCB).

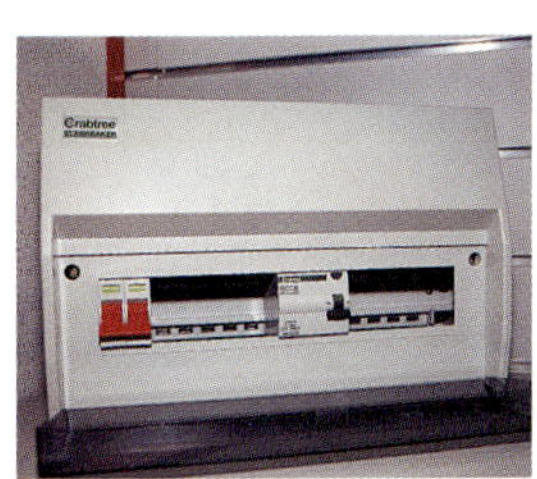

Figure 10.68
A RCCB plugged into a socket ready for use

A RCCB works by detecting any difference between the current in the live wire and the current in the neutral wire of the supply cable. If everything is working properly the two currents are the same. If something happens to make the current in the wires different, the RCCB switches the circuit off.

The person in Figure 10.69 has accidentally cut through the cable insulation. This causes a current to flow through them to earth. So the current in the live wire is now different from the current in the neutral wire. The RCCB detects this difference and switches the circuit off.

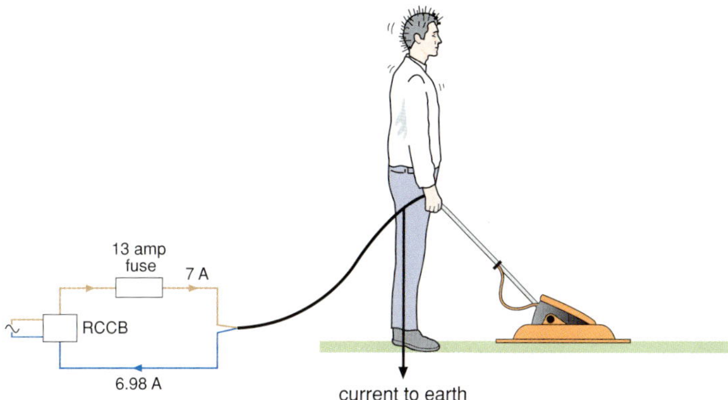

Figure 10.69
Using a RCCB may save your life

Earthing

Any electrical appliance which has metal parts that are easily touched should be earthed.

Figure 10.70 shows an electric kettle. It's been earthed by joining the earth wire from the three core cable to the metal casing. A fault may cause the live wire to touch the metal casing. If this happens while the kettle is switched on, a large current will flow from the live wire to the earth wire. This large current causes the fuse to melt, which switches the kettle off.

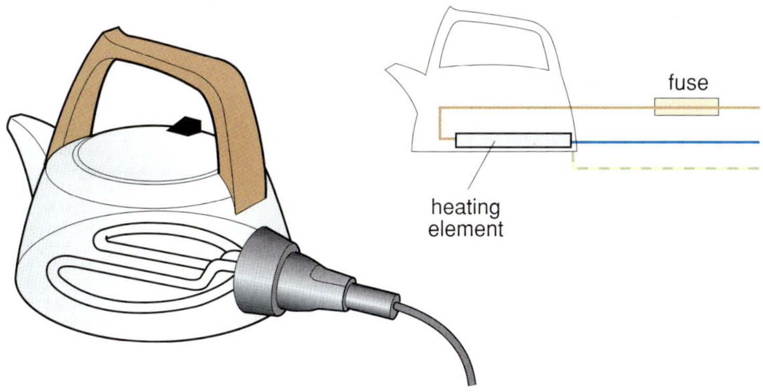

Figure 10.70
An electric iron should always be earthed

Without the earth wire the metal casing would become live. Any person touching the metal casing would then get an electric shock as the current flows through the person to earth.

Did you know?

The sign ⧈ is used to show an appliance that is double insulated. This means that the appliance does not need an earth wire. The outside case is made from an insulating material, often plastic.

Topic Questions

1 Copy and complete the following sentences.

a) A fuse is connected to the __________ wire. It will _________ if the ____________ in the circuit is higher than the fuse value. This will ___________ the circuit off.

b) Appliances with metal cases must be ________________ .

c) A RCCB will switch off an appliance if the ___________ in the live wire is different to the current in the ______________ wire.

2 Copy and complete the following table.

Appliance	Normal current in amps (A)	Correct fuse (3 A, 5 A or 13 A)
Toaster	4	
Fan heater	9	
Table lamp	0.5	
Hairdryer	6	

3 An electric coffee maker takes a current of 2 amps from the mains supply. Explain why the plug of the coffee maker should not be fitted with a 13-amp fuse.

4 Explain how a fuse works.

5 The following pictures show electricity being used dangerously. Explain the danger in each picture.

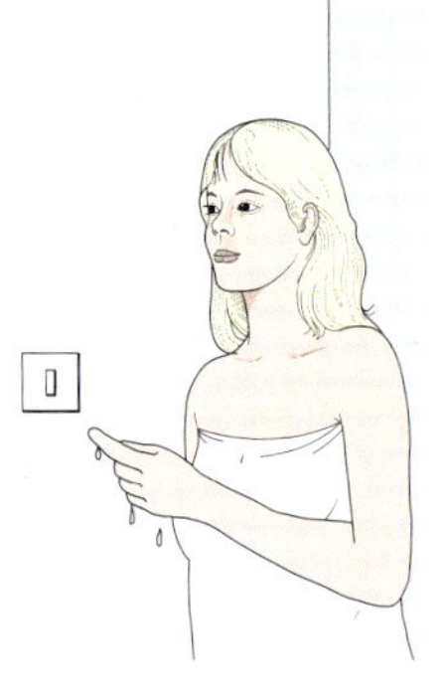

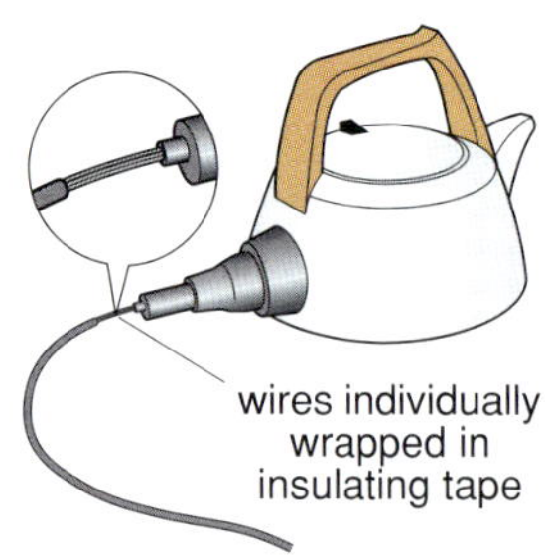

10.14	
Co-ordinated	Modular
DA 12.2	DA 10
SA n/a	SA n/a

Energy and power in a circuit

What you know

- An electric current will flow through a circuit if:
  - the circuit is complete
  - the circuit contains a power supply.
- The power supply can be:
  - a single cell (|ı)
  - a battery (|ı··|ı)
  - the mains supply (∿).
- When an electric current flows energy is transferred from the power supply to the components or appliances in the circuit.

An electric current is a flow of charge. In a metal wire, an electric current is a flow of negatively charged electrons (see Section 10.10).

In any circuit, when charge flows through a resistor, some electrical energy is transferred as heat. Sometimes this is useful, but sometimes it leads to energy being wasted.

Electric power

The two televisions in Figure 10.71 are both designed to transfer electrical energy as light and sound. Both televisions waste some energy as heat.

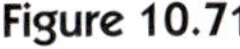

Figure 10.71

The portable television uses (transfers) less energy each second than the large television. The portable television has less power than the large television.

Electrical power measures how much electrical energy is transferred by an appliance or component in a circuit every second.

An appliance or component in a circuit that transfers 1 joule of energy in 1 second has a power of 1 watt (see Section 9.3).

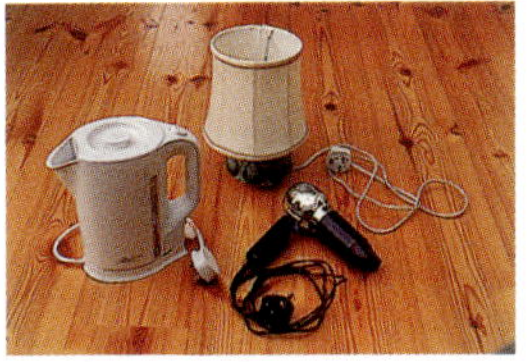

Figure 10.72 *Electrical appliances*

The kettle shown in Figure 10.72 transfers 2000 J of electrical energy as heat energy every second. So the kettle has a power of 2000 W. This is very high. It means that the kettle heats the water quickly.

The power of an appliance or the power in a circuit can be worked out using this equation:

$$\text{power (W)} = \text{potential difference (V)} \times \text{current (A)}$$

Remember you may see the word voltage used instead of potential difference.

Example: A TV and video take a current of 2 A from the 230 V mains supply. What is the power of the TV and video?

$$\begin{aligned} \text{power} &= \text{potential difference} \times \text{current} \\ &= 230 \times 2 \\ &= 460\text{ W} \end{aligned}$$

Topic Questions

1 The table gives the power of three different electric heaters.

Heater	Power
A	2000 W
B	1500 W
C	2500 W

Which heater would warm a room the fastest? Explain the reason for your choice.

2 Copy and complete the following table.

Appliance	Potential difference (in volts)	Current (in amps)	Power (in watts)
Kettle	230	10	
Oven	230	15	
Radio	9	0.6	
Motor	450	7	

3 The table shows how long it takes three different electric kettles to boil some water. The starting temperature of the water in each kettle is the same.

Kettle	Mass of water in the kettle	Time to boil the water
X	1.0 kg	4 minutes
Y	0.5 kg	2½ minutes
Z	1.2 kg	4 minutes

Which one of the kettles has the highest power? Explain the reason for your choice. (You can assume the kettles are equally efficient.)

10.15	
Co-ordinated	Modular
DA 12.21	DA 10
SA 12.13	SA 17

Induced current

Just as an electric current causes a magnetic field (see Section 10.6), so a changing magnetic field around a wire produces a current. This is called electromagnetic induction.

Electromagnetic induction can be shown by moving a wire through the magnetic field of a strong magnet. The wire must move so that it cuts across the magnetic field lines.

Moving the wire through the magnetic field produces a small voltage (potential difference) between the ends of the wire. We say that a voltage has been induced across the wire. Since the wire is part of a complete circuit, the induced voltage will cause a current to flow. An electric current made in this way is called an **induced current**.

A current only flows while the wire is moving in the magnetic field. If the wire is held still the induced current is lost.

The size of the induced voltage and current can be increased by:

- using a stronger magnet
- moving the wire faster.

An induced current is also produced when a magnet is moved into or out of a coil of wire. The direction of the induced current depends on which way the magnet is moving.

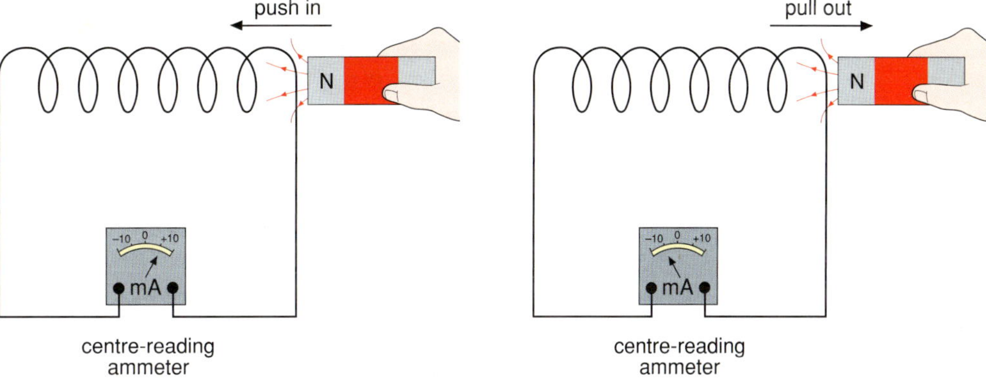

Figure 10.73
Electromagnetic induction in a coil

Pushing the magnet into the coil makes the current flow one way. Pulling the magnet out of the coil makes the current flow the opposite way.

Did you know?

Michael Faraday discovered the electromagnetic induction effect in 1831. The effect is used in power stations to generate electricity.

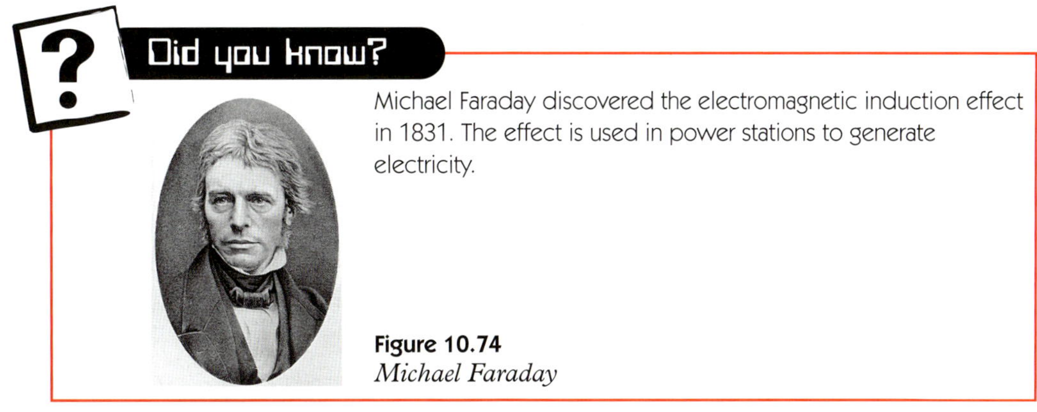

Figure 10.74
Michael Faraday

The direction of the induced current also depends on which pole (north or south) of the magnet moves into the coil.

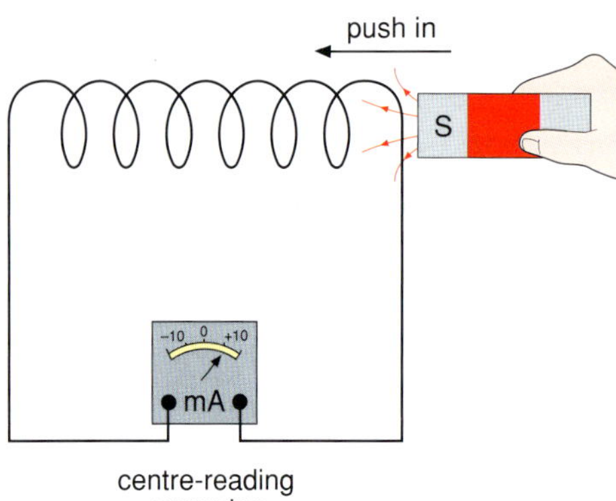

Figure 10.75
The effect of changing the pole on the direction of the current

Moving the south pole into the coil makes the current flow the opposite way to when the north pole moves into the coil.

No current is induced when the magnet is not moving.

The size of the induced voltage and current can be increased by:

- moving the magnet faster
- using a stronger magnet
- having more turns on the coil (the turns should be close together).

So it doesn't matter if it's the wire that moves or the magnet. The effect is the same; magnetic field lines are being cut. When this happens a voltage is induced across the wire. Because the wire is part of a complete circuit an induced current will flow in the wire.

Topic Questions

1 Copy and complete the following sentences.

a) Moving a wire so that it cuts through ____________ ____________ lines will ____________ a voltage across the wire.
b) An induced voltage will cause an induced ____________ to flow in a wire only if the wire is part of a ____________ circuit.
c) To induce a bigger voltage across a wire, a ____________ magnet should be used.

2 The diagram shows a coil of wire joined to a sensitive ammeter. A magnet is about to be pushed into the coil.

motion
N
centre-reading ammeter

What happens to the ammeter needle when the magnet:

a) moves into the coil?
b) stays still in the coil?
c) moves out of the coil?
d) moves faster into the coil?

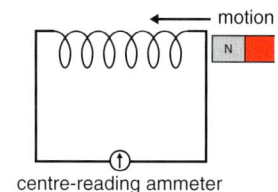

3 Explain why a wire that is held still between the poles of a strong magnet does not have a voltage induced across it.

4 Describe an experiment to show how the size of an induced current can be increased.
Your answer should include a diagram.

10.16	
Co-ordinated	Modular
DA 12.21	DA 10
SA 12.13	SA 17

Generating and transmitting electricity

Generators

Electricity is generated using the electromagnetic induction effect. This can be done in two ways.

- A coil of wire can be rotated in a magnetic field.
- A magnet can be rotated inside a coil of wire.

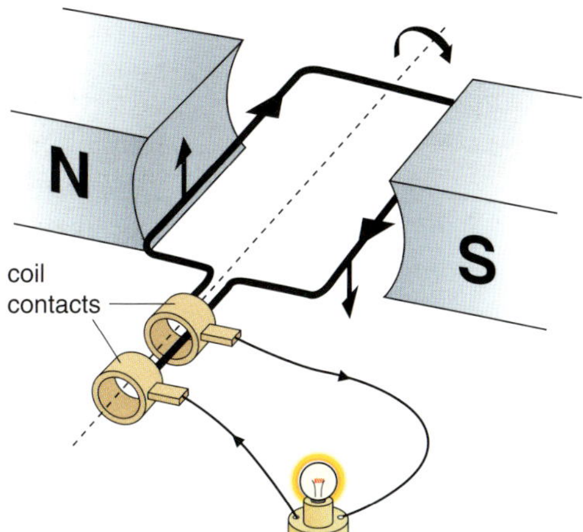

Figure 10.76
An a.c. generator

As the coil rotates it cuts through the magnetic field lines. This induces a voltage in the coil, which causes a current to flow.

The size of the induced voltage and current can be increased by:

- rotating the coil faster
- using a stronger magnet
- having more turns on the coil
- making the area of the coil bigger.

Electricity generators like those used in a power station use large stationary coils of wire. The magnetic field of a powerful electromagnet, which is rotating inside the coils, cuts through the coils.

Did you know?

A bicycle dynamo generates electricity by electromagnetic induction. A small magnet inside the dynamo spins as the back wheel of the bicycle rotates.

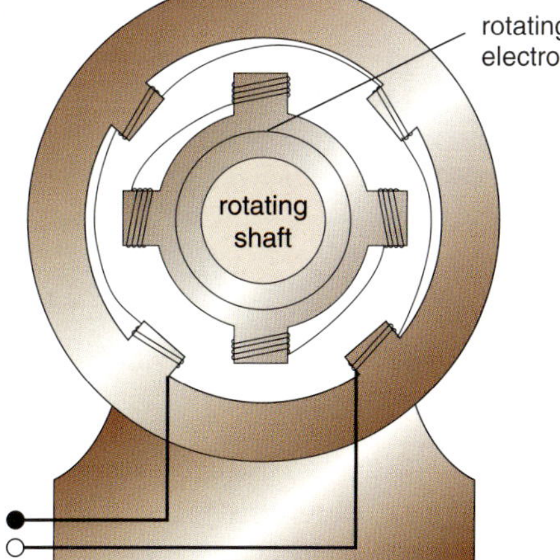

Figure 10.77
A simplified mains electricity generator

Figure 10.78
An actual mains electricity generator

Transformers

A transformer is used to increase or decrease the voltage of an a.c. supply.

- A step-up transformer increases voltage.
- A step-down transformer reduces voltage.

Transformers do not work with a d.c. supply.

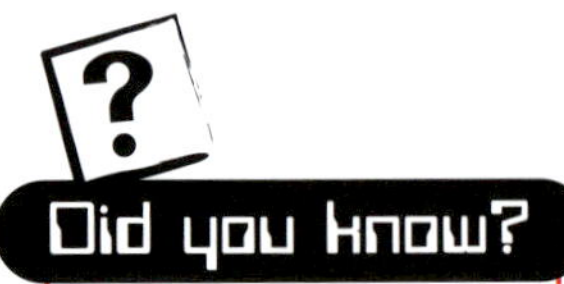

Did you know?

The circuit inside a mobile phone charger includes a small step-down transformer. It reduces the voltage from 230 V to about 6 V.

Figure 10.79
A small laboratory transformer

Figure 10.80
A large transformer used in the National Grid

Transmitting mains electricity

Power stations generate electricity at voltages of about 25 000 V (25 kV). A step-up transformer increases this voltage to about 400 000 V. The electric power is then transmitted (sent) through the cables of the National Grid to all parts of the country.

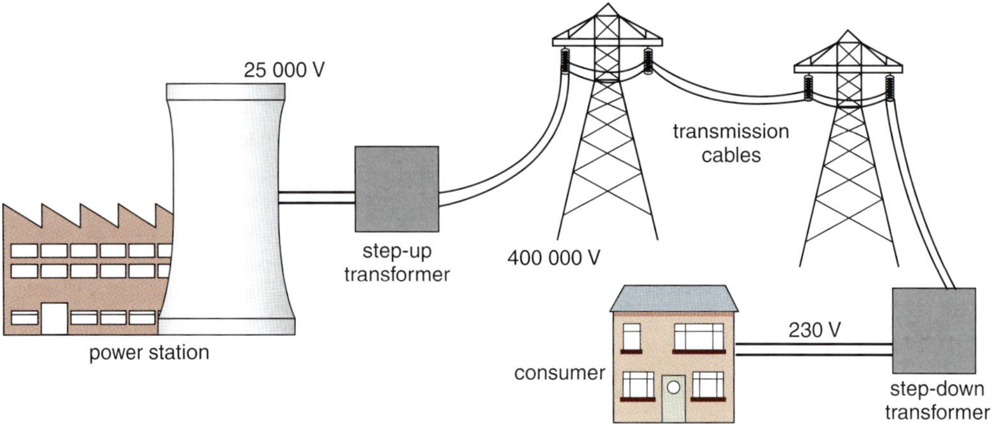

Figure 10.81
Simplified diagram of power transmission

At the consumer end, the voltage is reduced to a safer level by a series of step-down transformers. Mains electricity used at home is stepped-down to 230 V.

This system reduces the amount of energy lost from the cables as heat. It makes the transmission of mains electricity more efficient.

Did you know?

Before the National Grid was set up in 1926, each area had its own power station. The voltage, frequency and the choice of a.c. or d.c. varied from one power station to the next.

Topic Questions

1 Copy and complete the following sentences.

a) A bicycle dynamo is a type of _______________ generator.
b) A _______________ is used to change the voltage of an a.c. supply.
c) The National Grid is a system of ___________ that link __________ stations with consumers.

2 A coil of wire with 20 turns rotating once each second between the poles of a magnet induces a current to flow in the coil. What changes can be made to increase the size of the induced current?

3 Why is the mains electricity voltage reduced to 230 V for use in homes?

Summary

- For components connected in series:
  - the total resistance is the sum of their resistances
  - the same current flows through each component
  - the potential difference of the power supply is shared between the components.
- For components connected in parallel:
  - the total current in the circuit is the sum of the currents in each parallel branch of the circuit
  - there is the same potential difference (voltage) across each parallel branch.
- Current, potential difference (voltage) and resistance are linked by the equation:

potential difference = current × resistance

- For a filament lamp the resistance increases as the temperature of the filament increases.
- Diodes only let current flow in one direction.
- The resistance of a light dependent resistor (LDR) goes down as the intensity of the light falling on it goes up.
- The resistance of a thermistor goes down as the temperature goes up.
- There are two types of electrical charge – negative and positive.
- An object becomes negatively charged when it gains electrons, and positively charged when it loses electrons.
- Two objects with a similar charge repel; two objects with opposite charges attract.
- A charged object will discharge when it is connected by a conductor to earth (earthing).
- An electric current is due to the movement of charged particles. In a solid the charged particles are electrons. In a gas or liquid the charged particles are ions.
- Power is the rate at which energy is transferred.
- The power of an appliance is measure in watts (W).
- Current, potential difference (voltage) and power are linked by the equation:

power = potential difference × current

- Alternating current (a.c.) changes direction but direct current (d.c.) flows in one direction only.
- In the UK, the mains electricity supply is at 230 volts with a frequency of 50 hertz.
- An earth wire should be connected to the metal casing of an appliance.
- When a current flows through a wire it produces a magnetic field.

- A wire carrying a current in a magnetic field experiences a force.
- If a magnet is moved into or out from a coil of wire, or if a coil of wire moves in a magnetic field, a current is induced in the wire.
- Transformers increase or reduce voltages.

Examination Questions

1 Two thin strips of plastic are held close to each other. One strip of plastic is electrically charged. The other strip of plastic is negatively charged.
Which two of the following statements are true?
- The strips will attract each other.
- The strips will repel each other.
- The strips will have no effect on each other.
- If the uncharged strip is given a negative charge, the strips will attract each other.
- If the uncharged strip is given a negative charge, the strips will repel each other.

2 A current is passed through a solution of sodium chloride.
Which two of the following statements are correct?
- Positive ions move to the positive electrode.
- Negative ions move to the positive electrode.
- Negative electrons move to the positive electrode.
- Negative ions move to the negative electrode.
- Positive ions move to the negative electrode.

3 The diagram shows a device labelled P connected in parallel with a lamp labelled Q.
The voltage across the battery is 12 volts.
The current from the battery is 4 amps.

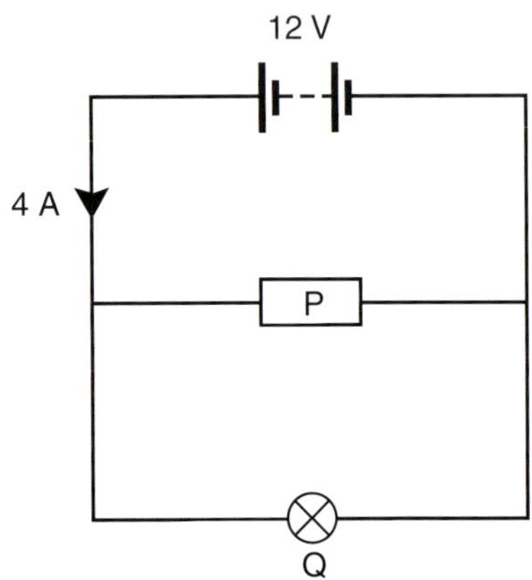

a) The current through P is 3 A.
What is the current through the lamp?
A 1 A
B 3 A
C 4 A
D 8 A

b) What is the voltage across the lamp?
A 3 V
B 4 V
C 6 V
D 12 V

c) What can you say about the resistance of lamp Q?
A It is the same as the resistance of P.
B It is bigger than the resistance of P.
C It is less than the resistance of P.
D We do not know if it is larger or smaller than the resistance of P.

d) The graph shows how the current through P varies when the voltage across it is changed.

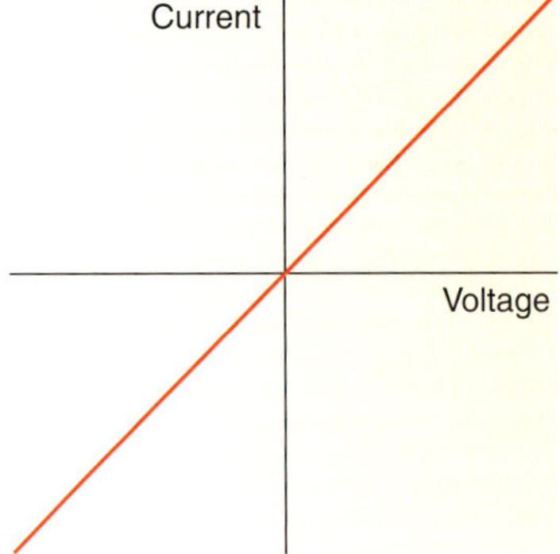

What is device P?
A A resistor at constant temperature.
B A transformer.
C A diode.
D A filament lamp.

4 Different electrical devices have different jobs. Match the words from the list with each of the numbers 1–4 in the table.

fuse	lamp	switch	transformer

Device	What the device does
1	Breaks the circuit when a fault develops
2	Changes the voltage of a mains supply
3	Lets us break the circuit when we want
4	Transfers electrical energy as light

5 A Van de Graaf machine is used to produce very high voltages. When the machine is switched on, a charge collects on the dome.

a) A girl stands on a thick plastic block and she touches the dome. When the machine is switched on, her hairs stand on end.
Her hairs stand on end because
- A each hair has the same charge that is the same as the charge on her head.
- B each hair has the same charge that is opposite to the charge on her head.
- C the hairs have opposite charges to each other.
- D the hairs are charged but her head has no charge.

b) Why must the girl stand on a plastic block?
- A So that the charge on the dome does not become too great.
- B To stop both her and the dome from discharging.
- C Plastic is a good conductor of static electricity.
- D So that her voltage does not become too great.

c) A strip of paper is held close to the charged dome. The strip is attracted to the dome.
What can you say about the strip of paper?
- A It might have the same charge as the dome or it might be neutral.
- B It might have the opposite charge to the dome or it might be neutral.
- C It has the same charge as the dome.
- D It has the opposite charge to the dome.

d) The charge on the dome is positive. The dome can be discharged by connecting it to earth.
Which of these happens as the dome is discharging?
- A Positive charges flow from the dome to earth.
- B Positive charges flow from earth to the dome.
- C Negative charges flow from the dome to earth.
- D Negative charges flow from earth to the dome.

6 A student did an experiment with two strips of polythene. She held the strips together at one end. She rubbed down one strip with a dry cloth. Then she rubbed down the other strip with the dry cloth. Still holding the top ends together, she held up the strips.

a) (i) What movement would you expect to see?
(ii) Why do the strips move in this way?

b) Copy and complete the four spaces in the passage.
Each strip has a negative charge. The cloth is left with a ______ charge. This is because particles called ______ have been transferred from the ______ to the ______.

7 The diagram shows a 13 amp plug.

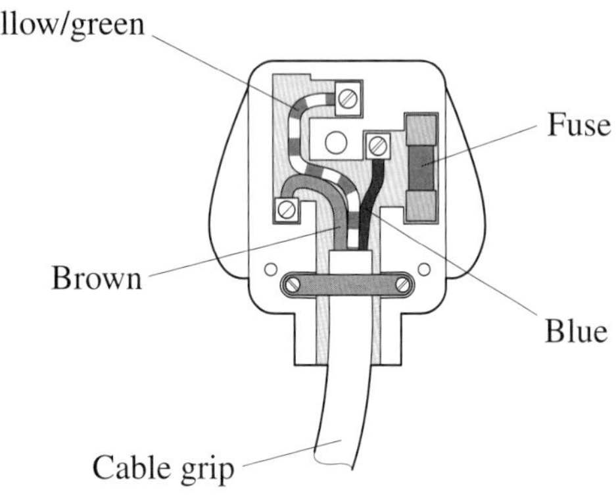

(i) What is wrong with the way this plug has been wired?
(ii) Why do plugs have a fuse?

8 A fault in an electrical circuit can cause too great a current to flow. Some circuits are switched off by a circuit breaker.

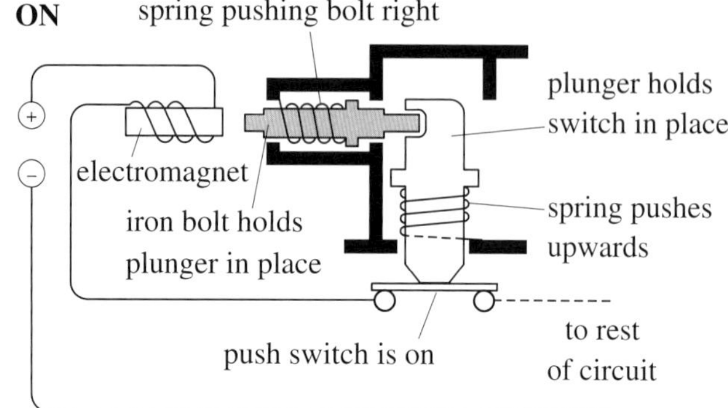

One type of circuit breaker is shown above. A normal current is flowing.
Explain, in full detail, what happens when a current which is bigger than normal flows.

Chapter 11

Forces and motion

Key terms

acceleration • air resistance • artificial satellite • braking distance • elastic potential energy • galaxy • geostationary satellite • gravity • mass • Milky Way • orbit • planet • red giant • satellite • star • stopping distance • terminal velocity • thinking distance • Universe • velocity • weight • white dwarf • work

11.1 Speed, velocity and acceleration

Co-ordinated	Modular
DA 12.6	DA 11
SA n/a	SA n/a

What you know

If you travel at a fast speed, it takes less time to finish a journey than when you travel at a slow speed. Some passenger trains are so fast they can travel 78 metres in just one second, which is a speed of 78 metres per second (78 m/s). This is the average speed of the train. During a one hour journey the train will sometimes go faster and sometimes slower.

Speed can be worked out using this equation:

$$\text{speed (m/s)} = \frac{\text{distance travelled (m)}}{\text{time taken (s)}}$$

Example: In a race, a horse travels 1280 metres in 80 seconds. What is the average speed of the horse in metres/second?

distance travelled = 1280 metres

time taken = 80 seconds

$$\text{speed} = \frac{\text{distance travelled}}{\text{time taken}} = \frac{1280}{80} = 16 \text{ m/s}$$

Figure 11.1
Racehorses in action

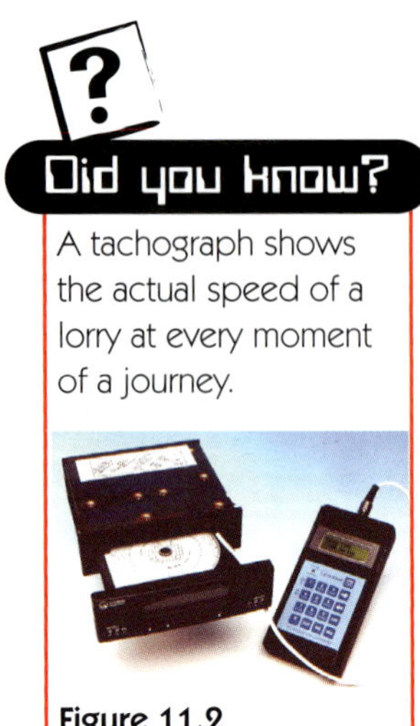

Did you know?

A tachograph shows the actual speed of a lorry at every moment of a journey.

Figure 11.2

Velocity

On a journey it's not just speed that is important, direction also counts. Figure 11.3 shows the route taken by a pupil going to school.

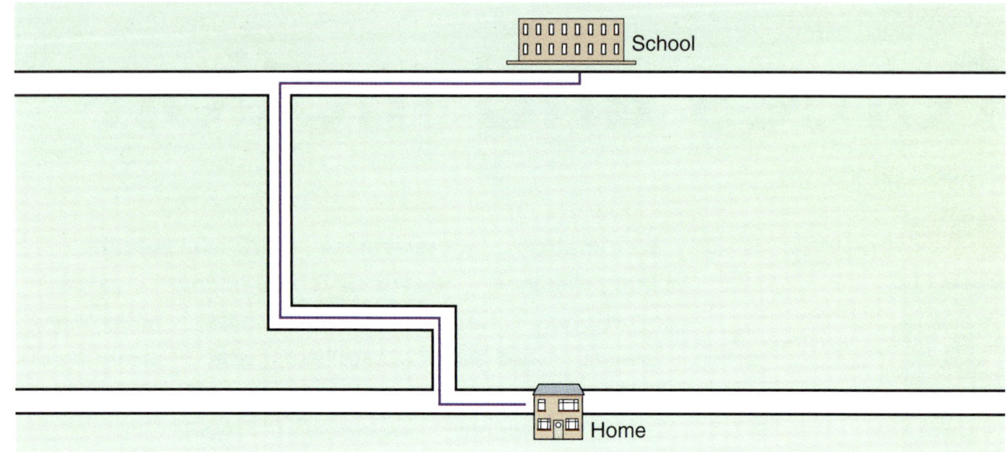

Figure 11.3

Each time the pupil changes direction, the **velocity** of the pupil changes. This is because velocity is speed in a given direction. If direction changes, velocity changes, even though speed may stay the same.

To give a velocity, both speed and direction must be given. Direction can be shown in different ways. You can use:

- an arrow
- a compass direction
- a + or − sign.

In Figure 11.4 the two joggers are moving at the same speed but have different velocities.

Figure 11.4

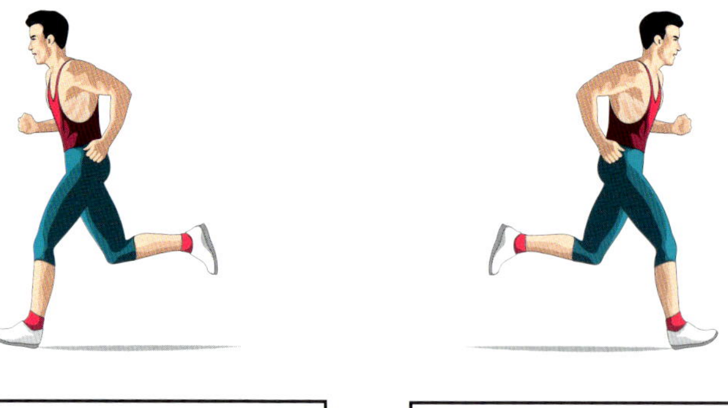

velocity = 3 m/s (←)
or 3 m/s west
or −3m/s

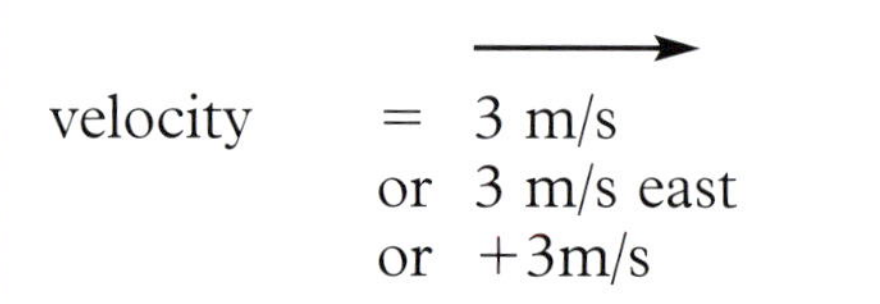

velocity = 3 m/s (→)
or 3 m/s east
or +3m/s

Acceleration

An object is accelerating when its velocity is changing. The faster the velocity changes, the larger the **acceleration**.

Figure 11.5
A rocket has a rapid acceleration – it reaches a high speed in a short time

Acceleration can be worked out using this equation:

$$\text{acceleration (m/s}^2\text{)} = \frac{\text{change in velocity (m/s)}}{\text{time taken for change (s)}}$$

When objects move in a straight line, 'change in velocity' and 'change in speed' are the same thing.

Example: At the start of a 100 metre race, an Olympic runner accelerates to 12 metres/second in 2 seconds. What is the acceleration of the runner?

$$\text{starting velocity} = 0 \text{ m/s}$$
$$\text{final velocity} = 12 \text{ m/s}$$
$$\text{time taken} = 2 \text{ s}$$

$$\text{acceleration} = \frac{\text{change in velocity}}{\text{time taken}} = \frac{12 - 0}{2} = \frac{12}{2} = 6 \text{ m/s}^2$$

An object that is slowing down is decelerating. The object has a negative acceleration.

Figure 11.6
Olympic runners accelerating off the blocks

Topic Questions

1 Copy and complete the following sentences.

a) To work out speed you need to know the ________________ travelled and the ______________ taken.
b) Velocity is the speed of an object in a particular ________________ .
c) Acceleration is measured in ______________ per __________ __________ .

2 Copy and complete the following sentence by choosing the correct word or words.

The speed of a decelerating car is *increasing / not changing / decreasing*.

3 The diagram shows the speed and direction of four animals.

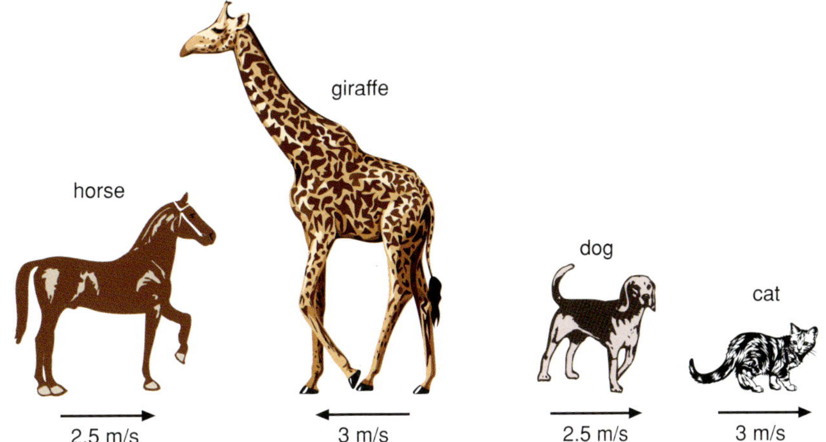

Which two animals have the same velocity? Give a reason for your answer.

4 The speed of a train travelling along a straight track increases from 10 m/s to 35 m/s in 100 seconds. Calculate the acceleration of the train.

11.2

Co-ordinated DA 12.6 SA n/a | Modular DA 11 SA n/a

Motion graphs

Motion graphs can be used to describe the movement of an object.

Distance–time graphs

A distance–time graph can be used to describe the speed of an object.

Figure 11.7 shows the distance–time graph for a short car journey.

Figure 11.7

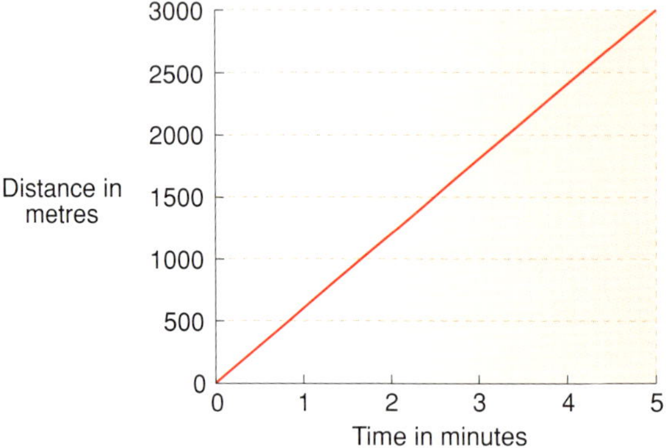

The car is travelling the same distance, 600 metres every minute. So the car is moving at a steady speed. A straight line means a steady (or zero) speed.

Figure 11.8 shows the distance–time graph for a horse and rider. The graph has been divided into three parts.

Figure 11.8

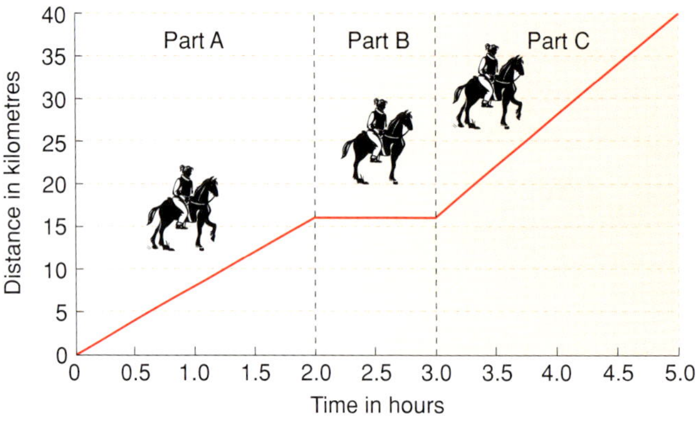

Part A – The horse and rider move at a steady speed. They travel 16 kilometres in 2 hours.

Part B – The distance moved does not change. The horse and rider have stopped to take a rest. When an object is not moving (stationary), the distance–time graph is flat.

Part C – The horse and rider move at a steady speed. They travel 24 kilometres in 2 hours. This means the speed of the horse and rider was the greatest during this part of the journey.

The steeper the slope of a distance–time graph, the greater the speed of the object.

Figure 11.9

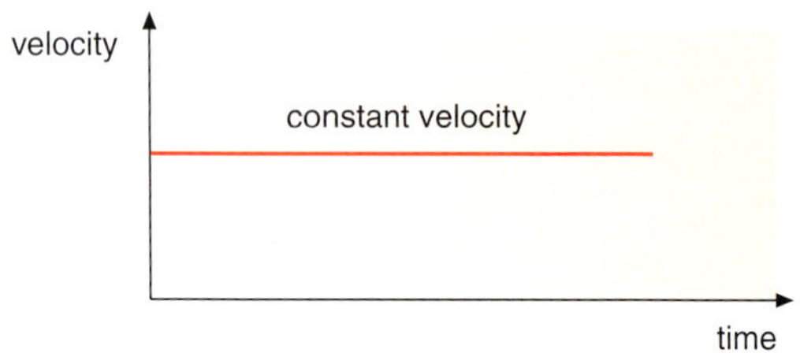

Velocity–time graphs

A velocity–time graph can be used to describe the velocity and acceleration of an object.

Figure 11.9 shows a velocity–time graph for a car moving at a constant speed along a straight road. This means the car is moving with a constant velocity. Remember when objects move without changing direction, velocity and speed are the same thing.

When an object moves with a constant velocity, the velocity–time graph is flat.

Figure 11.10

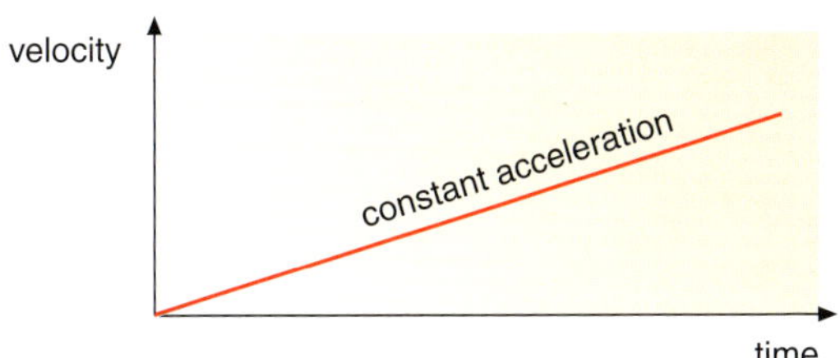

Figure 11.10 shows the velocity–time graph for a car accelerating along a straight road. The slope of the graph shows how fast the velocity is changing, which means the slope shows the acceleration of the car. A straight line means the acceleration was constant.

The steeper the slope of a velocity–time graph the greater the acceleration of the object.

Figure 11.11

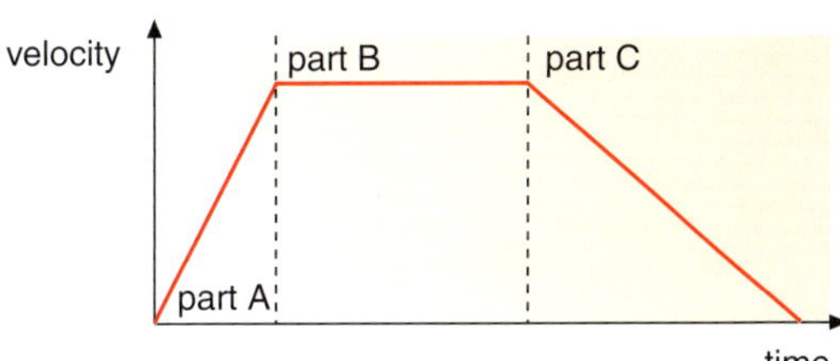

Figure 11.11 shows the velocity–time graph for a lift moving up between two floors in a hotel. The graph has been divided into three parts.

Part A – The lift has a constant acceleration; the velocity is increasing.

Part B – The lift is not accelerating; the velocity is constant.

Part C – The lift has a constant deceleration; the velocity is decreasing.

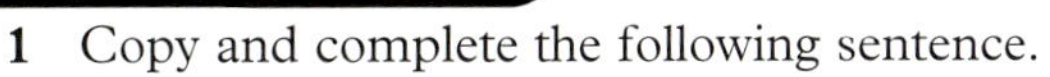

Topic Questions

1 Copy and complete the following sentence.

The ____________ the slope of a velocity– __________ graph the greater the ______________________ .

2 Copy and complete the following sentence by choosing the correct word or words.

This graph shows an object which is moving at *zero / a constant / an increasing speed.*

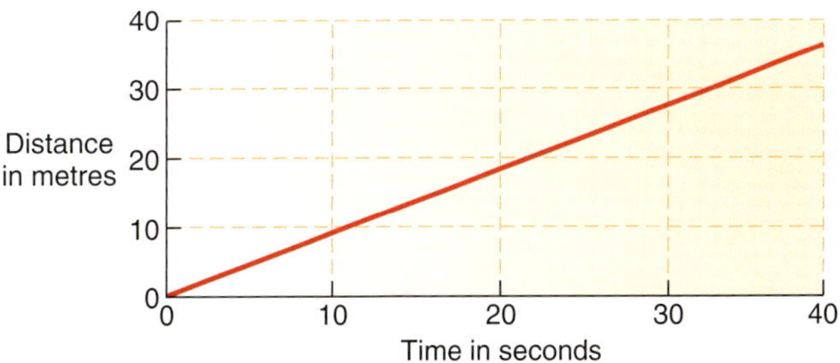

3 A pupil cycles to school. The journey is shown on the distance–time graph below.

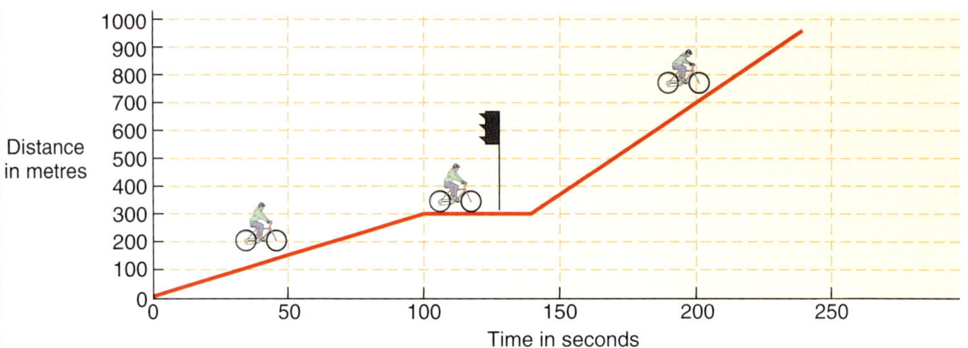

a) How far does the pupil live from school?
b) How long does it take the pupil to cycle to school?
c) How do you know that the pupil had to stop at the traffic lights?
d) During which part of the journey was the pupil cycling the fastest?

4 Two students, Bill and Vneeta, took part in a sponsored run. The distance–time graph for Bill's run is shown. Vneeta did not start the run until 500 seconds after Bill. She completed the whole run at a constant speed in 3000 seconds.

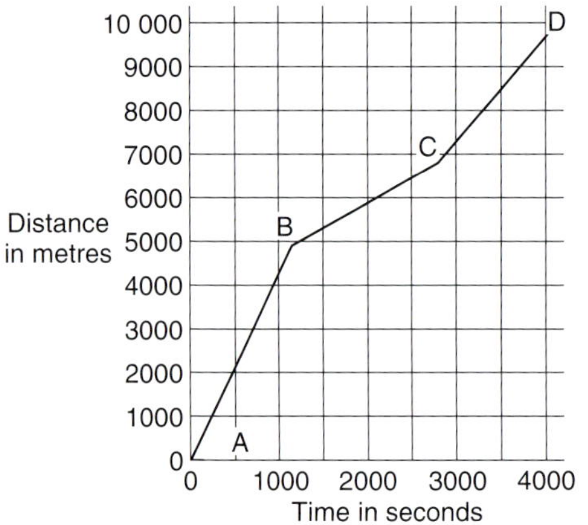

a) Copy the graph and on the same axes, draw a distance–time graph for Vneeta's run.
b) How far had Bill run when Vneeta overtook him?

5 Describe the motion of the cheetah whose velocity–time graph is shown below.

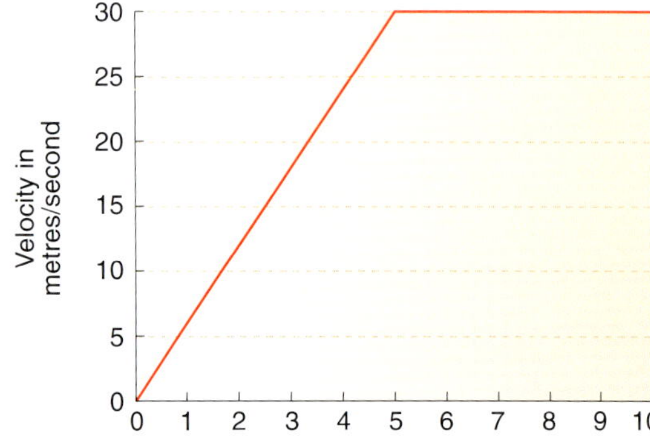

11.3	
Co-ordinated	Modular
DA 12.7 SA n/a	DA 11 SA n/a

Standing still, moving steadily – balanced forces

What you know

- A force has both size and direction. The size of a force is measured in newtons (N). In diagrams, a force is represented by an arrow. The longer the arrow, the larger the force.

Figure 11.12 shows two of the forces that act when a person sits on a chair.

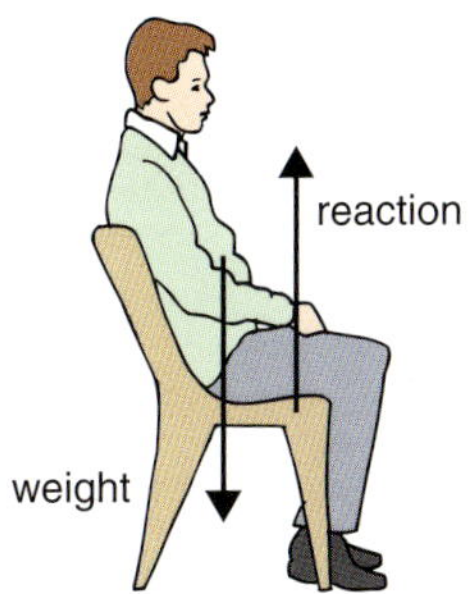

Figure 11.12

Because of gravity, the person's weight pulls him downwards against the chair. The chair, which supports the person, must be pushing upwards. The two forces on the person are the same size but in opposite directions. This means the forces are balanced.
In effect they have cancelled each other out.

- Friction is a force that will:
  - try to stop a stationary object from starting to move
  - slow down a moving object.

Whichever way an object moves, or tries to move, the force of friction will be in the opposite direction.

A force of friction also acts when an object moves through the air or through water.

Friction is sometimes called the drag force.

The pictures shown in Figure 11.13 show the forces exerted on different objects. The objects are all stationary (not moving). So the forces on them must be equal in size and opposite in direction. The forces are balanced.

Figure 11.13

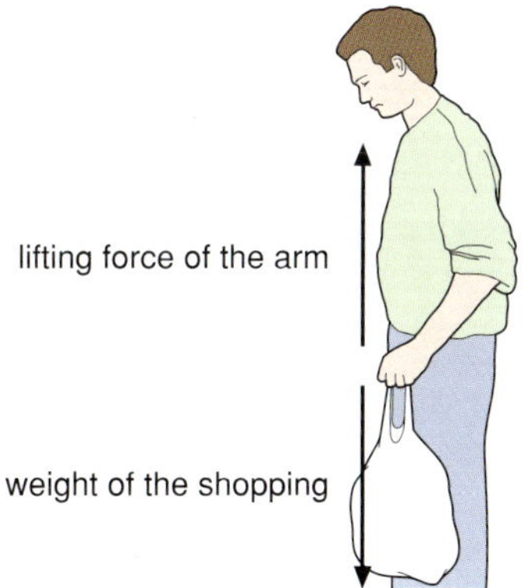

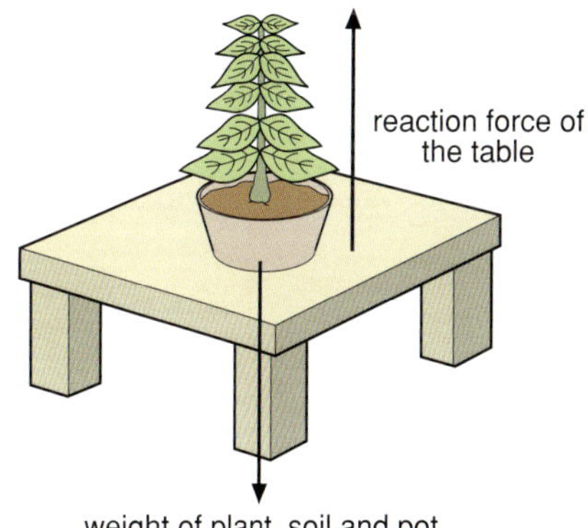

Balanced forces do not change the movement of an object. If an object is stationary, it will remain stationary. But what if an object is moving? If the forces are balanced it will keep moving at the same speed and in the same direction.

Figure 11.14 shows the forward and backward forces acting on a cyclist.

Figure 11.14

The cyclist is riding along a straight road, keeping the same constant speed. This means that there is no change in the movement of the cyclist. So the two forces must be balanced.

Figure 11.15 shows the forces acting on a flying aircraft.

Figure 11.15

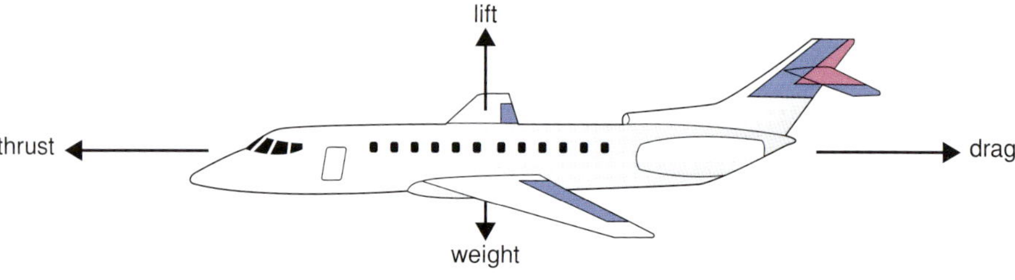

Did you know?

When working outside, a rope joins astronauts to their spacecraft. With no friction to stop them, a small force could send the astronaut drifting into space.

When the aircraft is flying at a constant height and speed, the forces acting up and down balance and the forces acting forwards and backwards balance. Lift is equal to **weight**. Thrust is equal to **air resistance**.

Topic Questions

1 Copy and complete the following sentences.

a) Two forces that are balanced must be ___________ in size and _______________ in direction.

b) When the forces on an object are balanced, the object could be _________________ or moving at a ______________ speed in a _________________ line.

c) When a book rests on a shelf its ____________ pushes downwards. The shelf pushes ______________ on the book, with the same __________ force.

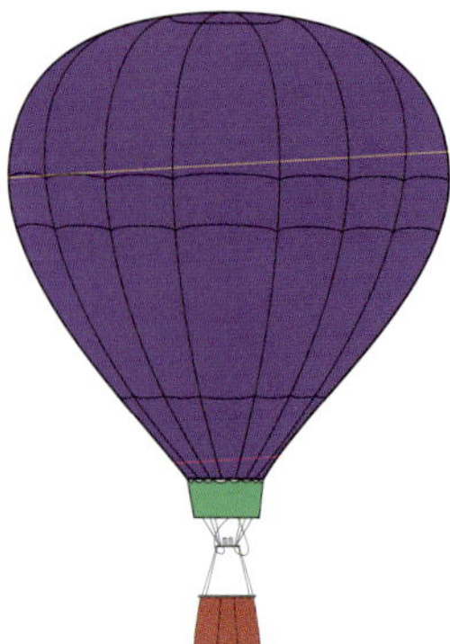

2 Copy and complete the following sentence by choosing the correct words.

When a car moves at a constant velocity along a level road, the force from the engine is *bigger than / the same as / smaller than* the total frictional forces on the car.

3 The diagram shows an air balloon. The balloon is rising at a steady speed.

Copy the diagram. Add arrows to show the size and direction of the forces acting on the balloon.

4 The diagrams show how the size of the forces acting on an aircraft at different points during a flight change.

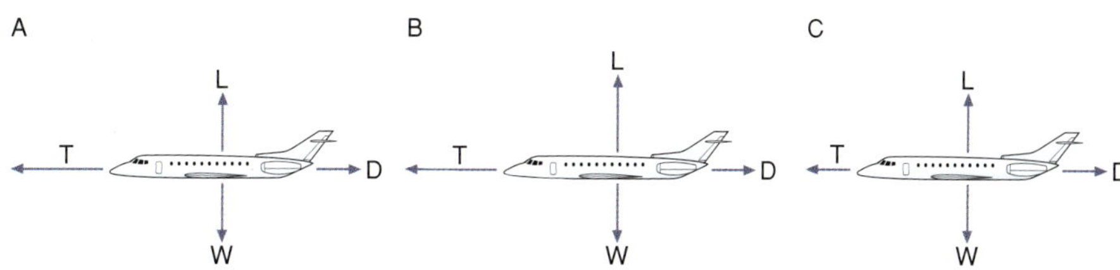

In which one of the diagrams, A, B or C, would the aircraft be flying at a constant height and speed? Explain the reason for your choice.

11.4

Co-ordinated	Modular
DA 12.7	DA 11
SA n/a	SA n/a

Speeding up, slowing down – unbalanced forces

In a tug of war, two teams pull against each other. When both teams pull equally hard, the forces are balanced and the rope does not move. But when one team starts to pull with a larger force the rope moves. The two forces do not cancel each other out. The forces are no longer balanced. The rope and both teams will start to move in the direction of the unbalanced force.

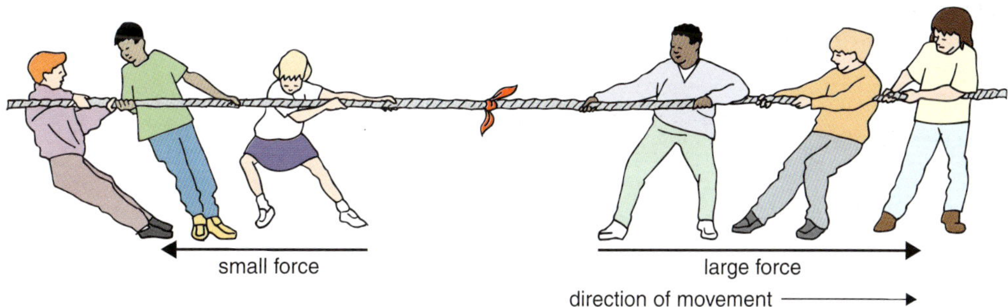

Figure 11.16
Unbalanced forces

Unbalanced forces affect the movement of an object. An unbalanced force can make an object speed up or slow down.

An engine produces the driving force needed to keep a car moving forwards. When the car moves at a steady speed the driving force forwards is balanced by the frictional forces backwards. So the faster the speed of the car the bigger the driving force forwards and the bigger the frictional forces backwards.

Figure 11.17

If the car engine stops, the car will slow down (decelerate). This happens because the force of friction is in the opposite direction to the car's movement.

An object moving in the opposite direction to the unbalanced force will slow down.

If the driver pushes on the accelerator pedal, the forward driving force increases. The forces on the car are now unbalanced. The car will speed up (accelerate) in the direction of the unbalanced force.

An object will only accelerate when an unbalanced force acts on it. It then accelerates in the direction of the unbalanced force.

A car with a flat battery can usually be push started. With only one person pushing, the acceleration of the car is small. The more people that push, the greater the acceleration.

The greater the force on an object the greater the acceleration of the object.

Figure 11.18

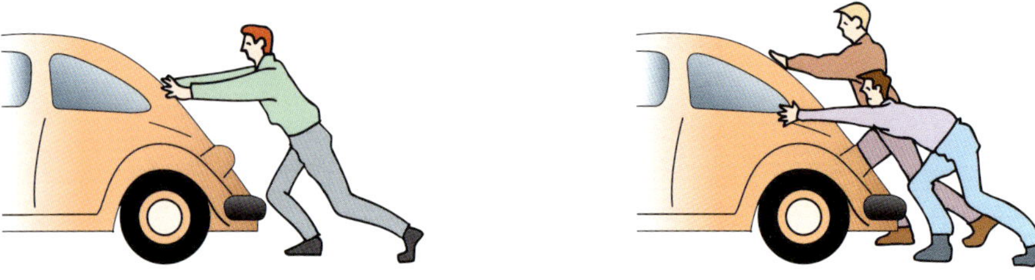

A much bigger force is needed to give a van the same acceleration as a car. This is because the **mass** of the van is far bigger than the mass of the car.

The bigger the mass of an object the bigger the force needed to make it accelerate.

Figure 11.20 *The force needed to accelerate this mass is enormous*

Figure 11.19

Topic Questions

1 Copy and complete the following sentences.

a) An object will ________________ when an ________________ force acts on it.

b) The faster a car moves, the bigger the ____________ ____________ trying to stop it.

2 The diagram shows two of the forces acting on a cyclist.

Is the cyclist speeding up, slowing down or moving at a steady speed? Explain the reason for your choice.

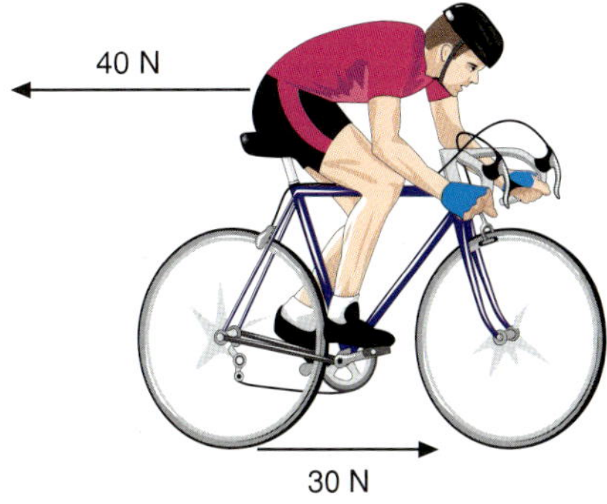

3 Explain why canoeists slow down when they stop paddling.

4 Why does a lorry need more powerful brakes than a car?

5 The diagram shows a ball which is about to be kicked.

Describe the two different effects that the force of the kick will have on the ball.

11.5	
Co-ordinated	Modular
DA 12.8 SA n/a	DA 11 SA n/a

Stopping safely

What you know

The brakes on a vehicle rely on friction. Pushing a brake pedal or pulling a brake lever causes friction between the brakes and the wheels. The friction slows the wheels and stops the vehicle.

If a dog runs out in front of a car, it takes the average driver about three quarters of a second to react and put the brakes on. This is called the driver's reaction time. During the reaction time the car carries on moving at a steady speed. The distance the car moves during the reaction time is called the **thinking distance**. Once the brakes have been put on the car slows down and comes to a stop. The distance the car moves once the brakes are put on is called the **braking distance**.

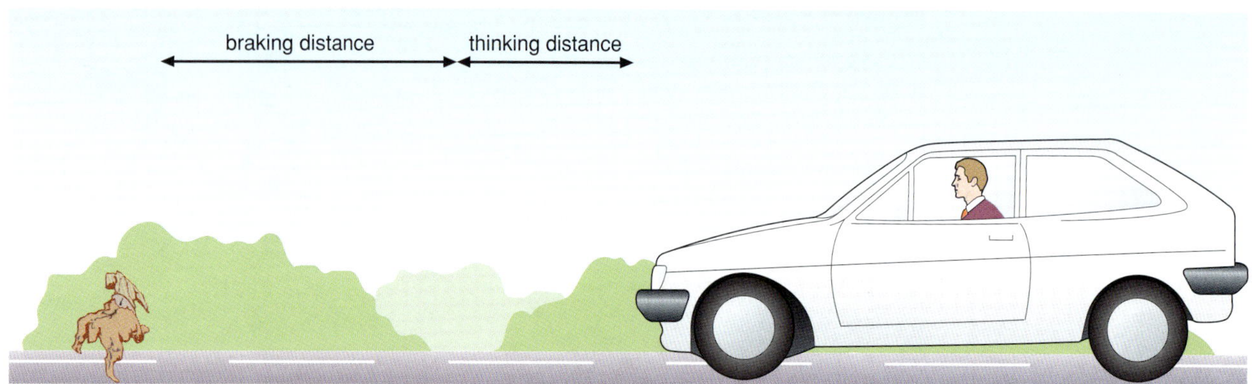

Figure 11.21 *Thinking distance and braking distance*

The total **stopping distance** of a car is made up of the two parts: the thinking distance and the braking distance.

The slower a driver's reaction time, the greater the thinking distance becomes. A driver's reactions are much slower if they:

- have been drinking alcohol
- have been taking certain types of drug
- are tired
- are driving in fog or other bad visibility conditions
- are talking on a mobile phone.

Speed affects both the thinking distance and the braking distance. Figure 11.22 shows the stopping distances for a car at different speeds. The distances are average values for a family car in good condition on a dry road. The force applied by the brakes is the same for each of the speeds.

Figure 11.22

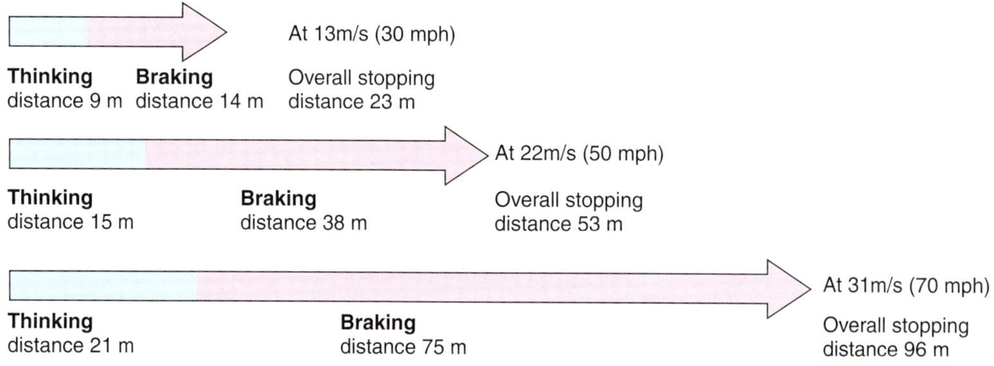

For a certain braking force, the faster the car, the greater the total stopping distance. The faster the car, the greater the braking force needed to stop the car quickly. But beware, if the braking force is too big the car may skid.

Tyres are designed to grip the road. The more friction between the tyres and the road the smaller the chance of skidding. But a strong braking force may make the wheels 'lock' and stop turning. If this happens the tyres will not be able to grip the road and the car will skid.

Did you know?

Figure 11.23
A racing car fitted with 'slicks'

Figure 11.24
A 'slick' tyre

The tread on a tyre is designed to throw the water out from between the tyre and the road. So on a dry day, a racing car uses tyres called 'slicks', these have no tread at all. If it starts to rain the tyres must be changed to ones with a tread or the driver risks loosing control of the car.

Did you know?

Modern cars are designed with crumple zones. In an accident, these parts crush slowly, causing the car to slow down gradually. The longer it takes for the car to stop the less chance there is that the passengers will be injured.

Friction between the car tyres and the road is smaller when the road is wet or icy. This gives the tyres less grip, increasing the braking distance. On an icy road the friction is so small that it is very difficult to stop quickly without skidding.

The braking distance is also increased if the car has worn brakes or tyres.

Topic Questions

1 Copy and complete the following sentences.

a) The distance a car travels during the driver's ____________ time is called the ________________ distance.
b) A driver's reactions are ______________ when they are tired.
c) Applying a large braking ___________ may make a car __________ .

2 Explain the effect on the stopping distance of a car when there is ice on the road.

3 People should not drive a car after drinking alcohol. Why?

4 Figure 11.22 gives the braking distance for a car travelling at 22 m/s as 38 m. Give two reasons why the braking distance at this speed could be bigger than 38 m.

5 What effect does driving slower have on:

a) a driver's reaction time?
b) thinking distance?
c) stopping distance?

6 Drivers should leave a gap between their vehicle and the one in front. A learner driver said to his instructor 'If I go twice as fast, I should leave a gap twice as big'. Explain to the learner driver why he is wrong.

11.6	
Co-ordinated	Modular
DA 12.8 12.19 SA n/a	DA 11 SA n/a

Falling bodies

If you wanted to know your weight, you would probably stand on some bathroom scales.

If the bathroom scales measure in kilograms, they show your mass not your weight.

Mass and weight are not the same thing.

Remember, mass is the amount of matter that makes up an object; it is measured in kilograms. Weight is the force which **gravity** exerts on a mass; it is measured in newtons.

Figure 11.25

But more mass does mean more weight, because there is more mass for the force of gravity to pull on.

On Earth, gravity pulls on every 1 kilogram of mass with a force of about 10 newtons. This is called the gravitational field strength (g).

$$g = 10 \text{ N/kg}$$

So someone with a mass of 50 kg will weigh:

$$50 \times 10 = 500 \text{ N}$$

You can use this equation to work out the weight of an object.

$$\text{weight (N)} = \text{mass (kg)} \times \text{gravitational field strength (N/kg)}$$

Gravity is a force of attraction. A ball thrown into the air is pulled back to the ground by gravity. As the ball falls downwards, gravity will make it accelerate. If there were no frictional forces (air resistance), the ball would accelerate at 10 m/s^2.

Usually air resistance does act on a falling object. Air resistance can only be ignored if the force it exerts is very small.

When sky-divers jump from a plane, the forces on them are unbalanced. The downward force due to gravity (weight) is bigger than the upward force of air resistance. So the sky-diver accelerates towards the ground.

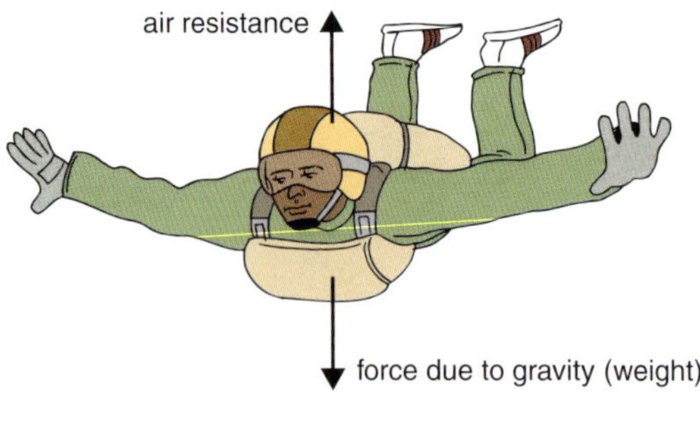

Figure 11.26 *Sky-diver at terminal velocity*

As the speed of the sky-diver increases so does the air resistance. Eventually the weight of the sky-diver downwards and the air resistance upwards will be balanced. The sky-diver will stop accelerating but will carry on falling at a constant speed. The sky-diver is falling at his **terminal velocity**.

Opening the parachute increases air resistance. The sky-diver will slow down until the two forces are again balanced. This gives the sky-diver a new slower terminal velocity.

The velocity–time graph for the sky-diver is shown in Figure 11.27.

Figure 11.27

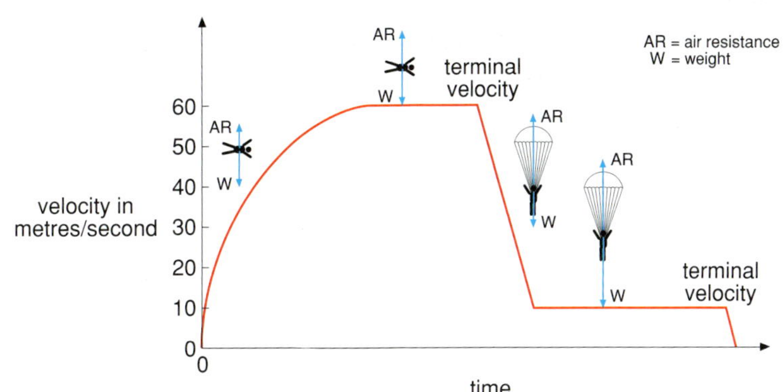

Did you know?

Vesna Vulovic, an air hostess, fell over 10 kilometres without a parachute and survived.

Topic Questions

1 Copy and complete the following sentences.

a) Air resistance is a ____________________ force.
b) All falling objects, if they fall far enough, will reach their __________ ___________.
c) Someone with a ____________ of 80 kg has a ______________ of 800 N.

2 An astronaut standing on the Moon lets go of a hammer and a feather.

What will happen, and why?

3 Someone drops two sheets of paper. Each sheet has the same weight but one has been screwed up into a ball.

a) What force causes the paper to fall to the ground?
b) Explain why the sheets of paper fall at different speeds.

4 Three people decide to go on a diet. The table shows the mass of each person at the start and end of the month.

	Start of the month	End of the month
Kathy	72 kg	69 kg
Jon	115 kg	107 kg
Cary	68 kg	66 kg

How much **weight** did each person lose?

5 Aid parcels are dropped from an aircraft. Each parcel has a parachute that opens straight away.

a) Copy the diagram. Add labelled arrows to show the forces acting on the parcel and parachute.

b) Explain why the parachute slows down the falling parcel.

11.7 Work and energy

Co-ordinated	Modular
DA 12.19	DA 11
SA n/a	SA n/a

What you know

When two objects rub against each other friction will make the objects heat up.

Only a small force is needed to lift a book off a table. But to lift the table, a much bigger force is needed. In both cases the forces are making something move. Forces that make something move are doing **work**. The bigger the force, and the further it moves, the more work is done.

Work, like energy, is measured in joules (J).

The work done by a force in moving an object can be calculated using the equation:

$$\text{work done (J)} = \text{force applied (N)} \times \text{distance moved (in the direction of the force) (m)}$$

Figure 11.28 *A weightlifter doing work*

Example: Calculate the work done by a weightlifter, lifting a mass of 75 kg to a height of 2.4 m.

weight lifted = mass lifted × gravitational field strength (see Section 11.6)

weight lifted = 75 × 10 = 750 N

work done = force × distance moved

work done = 750 × 2.4

= 1800 J

When the builder in Figure 11.29 starts pushing the wheelbarrow, it moves, so work is being done. But to make the wheelbarrow move energy must be transferred. In this case chemical energy in the builder's muscles is transferred to the movement (kinetic) energy of the wheelbarrow.

Figure 11.29

Whenever a force moves an object, work is done and energy must be transferred.

Work and energy are linked by the equation:

work done = energy transferred

This means that if 100 J of work is to be done, then 100 J of energy must be transferred.

The builder must keep pushing to keep the wheelbarrow moving at a steady speed. So the builder is doing work. But the energy of the wheelbarrow is not changing (the wheelbarrow is not going any faster or any slower). The work done by the builder is against the frictional forces that are trying to stop the wheelbarrow moving. Chemical energy from the builder's muscles is being transferred to heat.

When work is done against frictional forces most energy is transferred as heat.

Example: A force of 120 N is used to pull a heavy box across a rough wooden floor. The box moves at a steady speed for 5 metres. Calculate the work done against the frictional forces.

work done = force × distance moved

work done = 120 × 5 = 600 J

So, 600 J of energy will be transferred as heat to the floor, box and air.

Elastic potential energy

An elastic object will change shape when a force is applied. But it will go back to the way it was when the force is taken away. To stretch, squash, twist or bend an elastic object work must be done and energy transferred. **Elastic potential energy** is the energy stored in an elastic object when work is done to change its shape. In Figure 11.30 chemical energy from muscles is transferred to the stretched elastic cords.

Figure 11.30
Using a chest expander

? Did you know?

A bungee rope is designed to stretch. The jumper accelerates until the rope becomes tight. The rope then starts to stretch and work is done. Energy is transferred from the jumper to the rope. Some work is also done against air resistance (a frictional force), so some of the jumper's energy is transferred as heat. This is just as well or the bungee jumper would bounce up and down forever.

Topic Questions

1 Copy and complete the following sentences.

a) When a ______________ moves an object, ______________ is done and ______________ is transferred.
b) When work is done against friction, most energy is transferred as ______________.
c) When an elastic object changes ______________, it stores energy as elastic ____________________ energy.

2 Calculate the work done in each of these scenarios.

a) A shopper pushing a trolley with a force of 40 N for 75 m.
b) A pole-vaulter lifting her own weight of 520 N a height of 3.6 m.
c) A car using a braking force of 700 N to stop in 23 m.
d) A crane lifting a crate of mass 1400 kg a height of 12 m.

3 Is the pole used by a pole-vaulter elastic? Give a reason for your answer.

4 The picture shows a person pulling a heavy sack up a ramp.

Explain why the energy transferred to the sack is less than the work done pulling the sack up the ramp.

11.8	
Co-ordinated	Modular
DA 12.14 SA 12.8	DA 11 SA 18

The solar system

What you know

In our solar system there is one star (the Sun) and nine planets. The Earth is one of the planets. Some planets have moons that move around them.

The Earth turns on its axis. It takes 24 hours for the Earth to make one complete turn. For places facing towards the Sun, it is daytime. For places facing away from the Sun, it is night-time. As the Earth turns, places in the light move into the dark, so day becomes night.

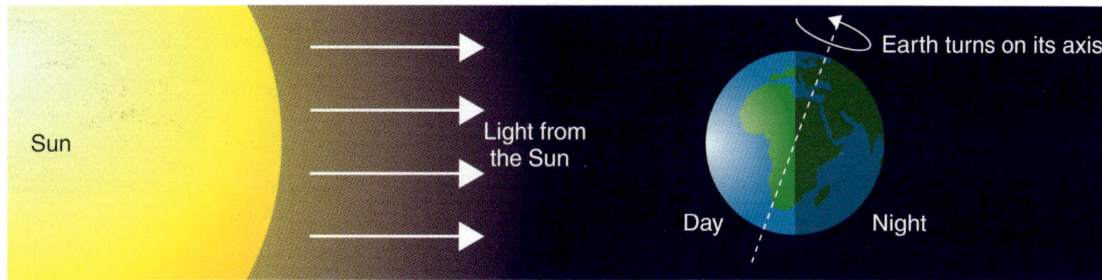

Figure 11.31

The Earth moves around (orbits) the Sun. It takes just over 365 days to make one orbit.

The other planets, like the Earth, orbit the Sun. In the night sky some planets look like very bright stars. But planets do not give out their own light. We can see the planets because they reflect light from the Sun towards us. Figure 11.32 shows how this happens.

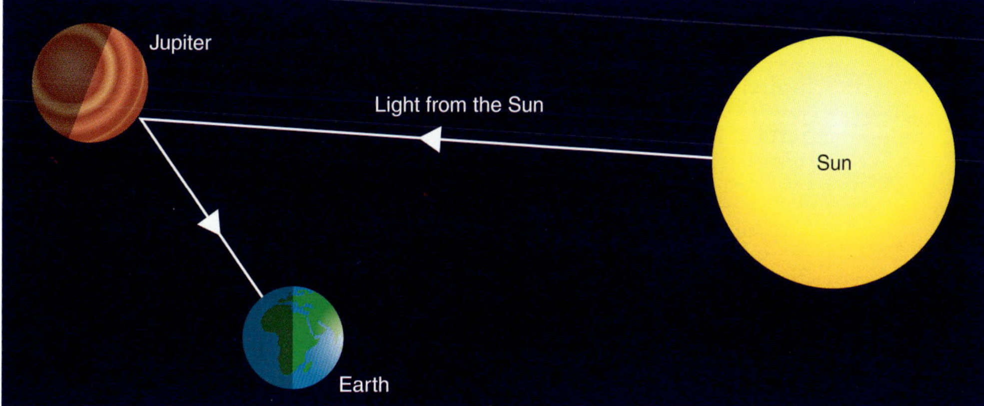

Figure 11.32

The planets

The **orbits** of the **planets** are not perfectly round. The orbits are slightly squashed circles with the Sun close to the centre. They are called elliptical orbits. The orbit of Pluto is so elliptical that it is sometimes closer to the Sun than Neptune.

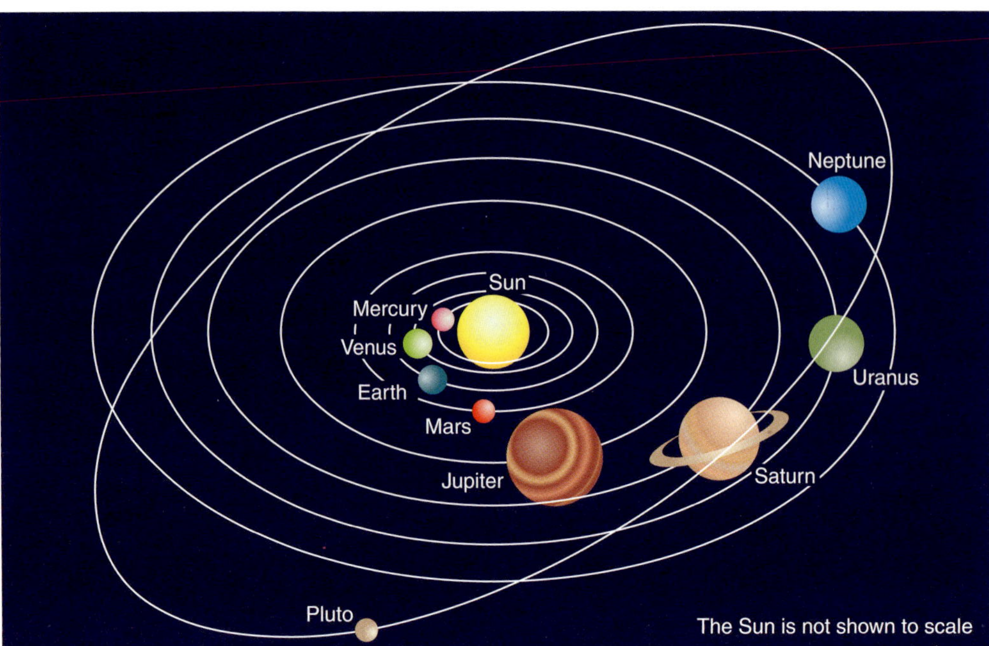

Figure 11.33
Parts of the solar system

Did you know?

Orbiting the Sun between Mars and Jupiter are thousands of pieces of rock, called asteroids. The largest, Ceres, has a diameter of about 1000 kilometres!

The further a planet is from the Sun, the longer it takes to complete one orbit.

Figure 11.34

Planet	Average distance from the Sun in million km	Time taken to orbit the Sun once
Mercury	58	88 days
Venus	108	225 days
Earth	150	$365\frac{1}{4}$ days
Mars	228	687 days
Jupiter	780	12 years
Saturn	1430	29 years
Uranus	2800	84 years
Neptune	4500	165 years
Pluto	5900	248 years

The Moon

The Moon is the Earth's natural **satellite**. As the Earth orbits the Sun, so the Moon orbits the Earth. Most of the other planets are orbited by at least one moon. Saturn has 24 moons.

Did you know?

The spacecraft Cassini launched in 1997 will reach Saturn in 2004. A probe will then explore the atmosphere and surface of Titan, Saturn's largest moon.

Comets

Comets orbit the Sun. But compared to the planets the orbits are very elliptical. So most of the time a comet will be a huge distance from the Sun. We can only see a comet when its orbit passes close to the Sun. Then heat from the Sun turns some of the comet's solid material into a long tail of gases. We can see the tail because it reflects light from the Sun.

Figure 11.35
The path of a comet

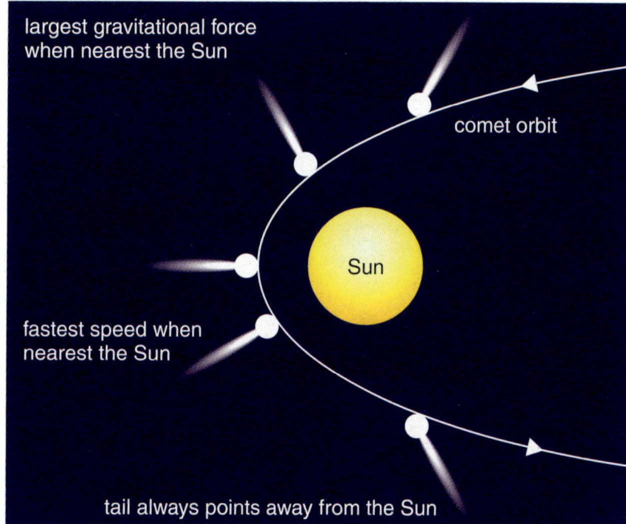

Figure 11.36
The comet Hyakutake showing its glowing tail

Did you know?

In 1682, Edmund Halley predicted that a comet he had been studying would be seen again in 1758. He was right. The comet, which can be seen every 76 years, was named Halley's Comet.

Topic Questions

1 Copy and complete the following sentences.

a) Planets orbit a ________________ .
b) A moon orbits a ____________________ .
c) We can see Venus because it __________________ light from the Sun.
d) The orbit of Pluto is very ____________________ .
e) It takes Saturn _________ years to travel once around the _________ .

2 Using words from the box, copy and complete the following sentences. Each word may be used once or not at all.

comet	galaxy	moons	star

The solar system has one ________________ called the Sun. There are nine planets in orbit around the Sun. Some planets have one or more ______________ in orbit around them.

3 The box contains the names of eight of the nine planets in the solar system.

Earth	Jupiter	Mars	Mercury	Neptune	Pluto	Saturn	Uranus

a) Which planet has **not** got its name in the box?
b) Which planet has the shortest orbit?
c) Which planet has a year that is almost twice as long as the Earth's year?

4 a) Why can comets only be seen during a small part of their orbit?
b) Why can a comet's tail be seen?

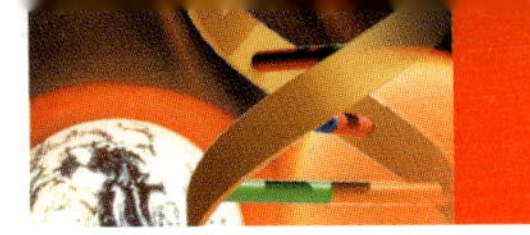

11.9 Gravity in space

Co-ordinated	Modular
DA 12.14	DA 11
SA 12.8	SA 18

Gravity is a force which tries to pull objects together. It is a force of attraction (see Section 11.6). Gravity acts between all objects, no matter how big or how small. But the bigger the mass of the object, the bigger the gravity force. The force is only big enough to feel if one of the objects has a very big mass, like a planet or a star.

Figure 11.37 *Small masses, small forces of attraction*

The gravitational pull of the Sun affects all objects in the solar system. The force of gravity between the Sun and a planet keeps the planet in its orbit around the Sun. If the Sun's gravitational pull suddenly stopped pulling, the Earth and all the other planets would shoot off into space.

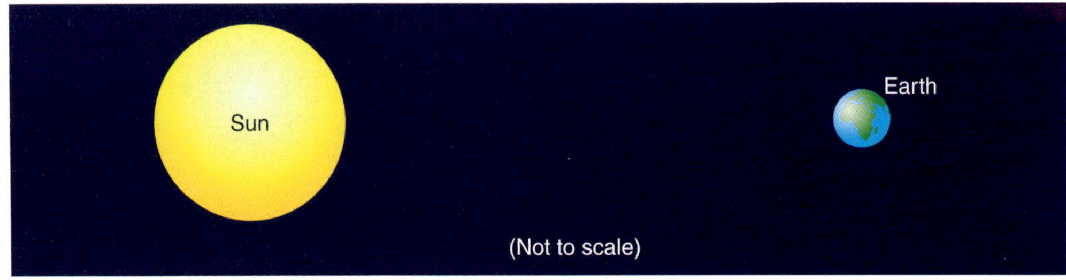

Figure 11.38 *Large masses, large forces of attraction*

The orbit of a moon around a planet is due to the gravitational pull between the moon and the planet.

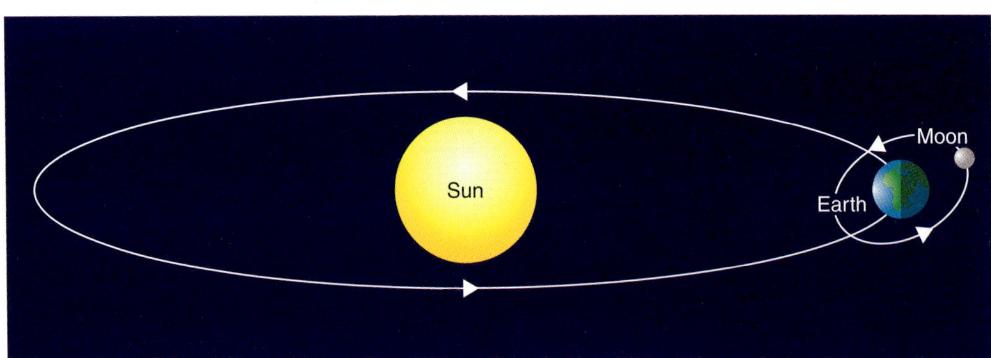

Figure 11.39 *Orbit of the Moon*

As the distance between objects gets bigger, the gravitational pull between them gets smaller. The Earth's gravitational pull on a rocket gets smaller and smaller as the rocket gets further and further away from the Earth. When the distance between the rocket and the Earth doubles the gravitational pull goes down by more than half.

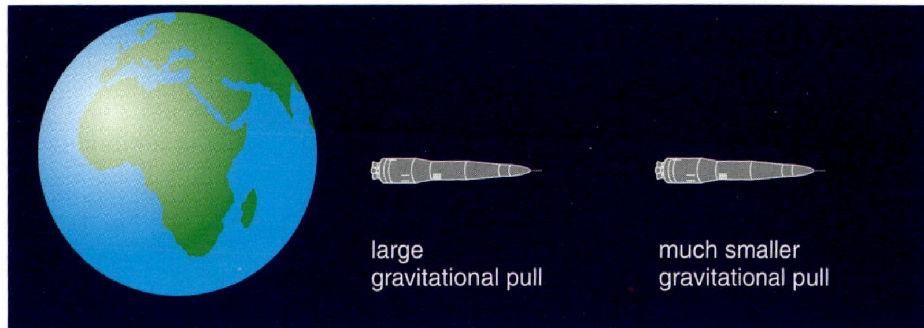

Figure 11.40 *Gravitational pull on a space rocket*

The force of gravity between planets is very weak because the planets are a long way apart. But the force of gravity between a planet and the Sun is much stronger because the Sun has such a huge mass. This force together with the high speed of the planet keeps the planet in its orbit.

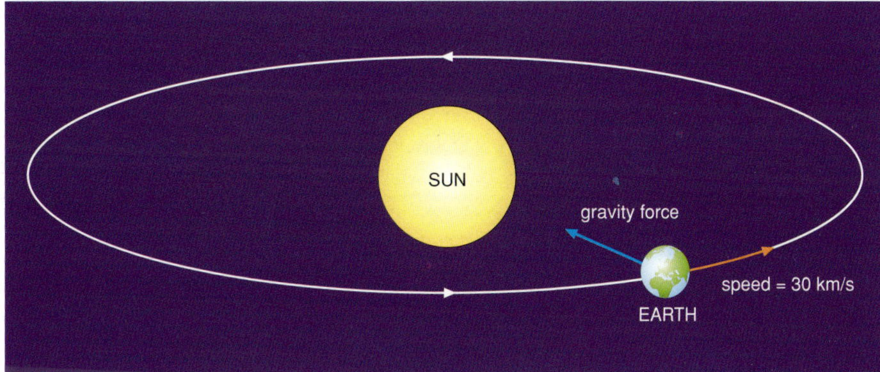

Figure 11.41

Did you know?

Gravity can be used to change the direction of a spacecraft. Figure 11.42 shows the path of the Cassini spacecraft. Each time Cassini gets close to a planet, the gravitational pull of the planet swings it round for the next part of its journey.

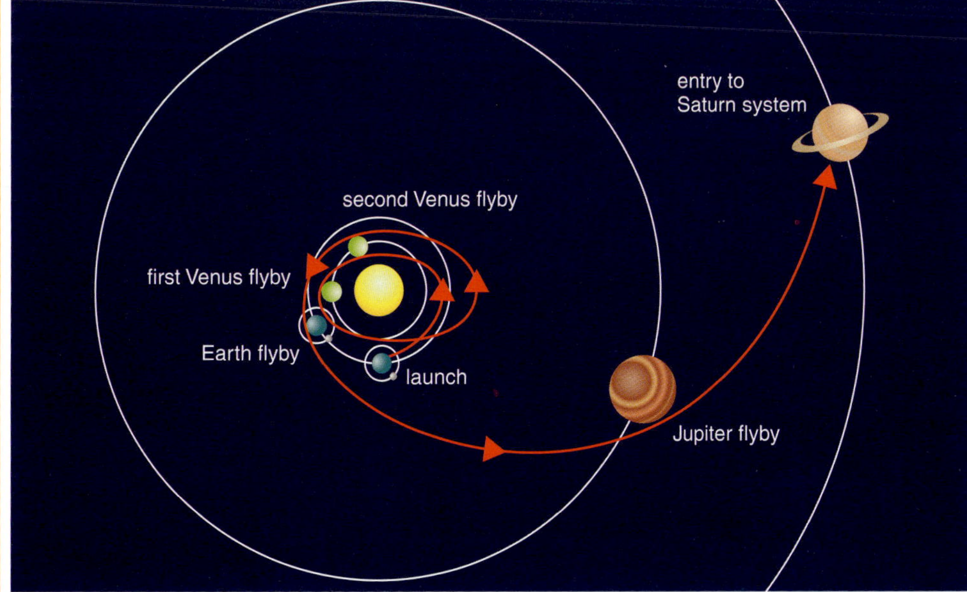

Figure 11.42

Topic Questions

1 Copy and complete the following sentences.

a) As the distance between two objects gets bigger, the force of gravity between them gets ________________ .

b) The force of gravity between the ____________ and the moon, keeps the moon in its orbit.

c) The force of gravity between a planet and the Sun is ______________ than the force of gravity between two planets.

2 The diagram shows the orbit of Venus around the Sun.

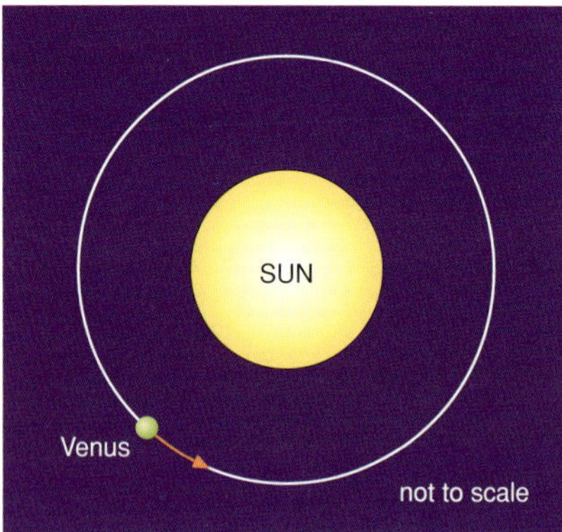

Copy the diagram and draw an arrow to show the gravity pull of the Sun on Venus.

3 The diagram shows the orbit of Mercury around the Sun.

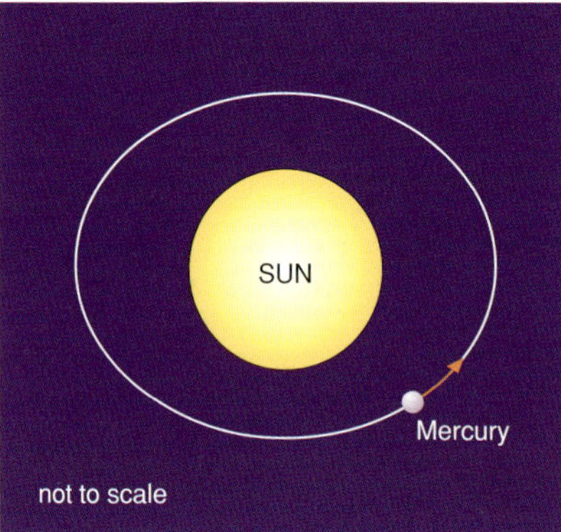

The speed of Mercury changes as it orbits the Sun. Why is this?

11.10

Co-ordinated	Modular
DA 12.14	DA 11
SA 12.8	SA 18

Satellites

What you know

There are many artificial satellites in orbit around the Earth. Each satellite has a job to do. Satellites can be used to:

- send information from one place to another
- monitor weather conditions
- look into space.

To stay in orbit a satellite must be given a forward speed parallel to the Earth's surface. So, as the force of gravity pulls the satellite towards the Earth, its forwards speed makes it move in a curve. To stay in orbit at a particular distance from the Earth, the satellite must move forwards at the right speed.

Figure 11.43

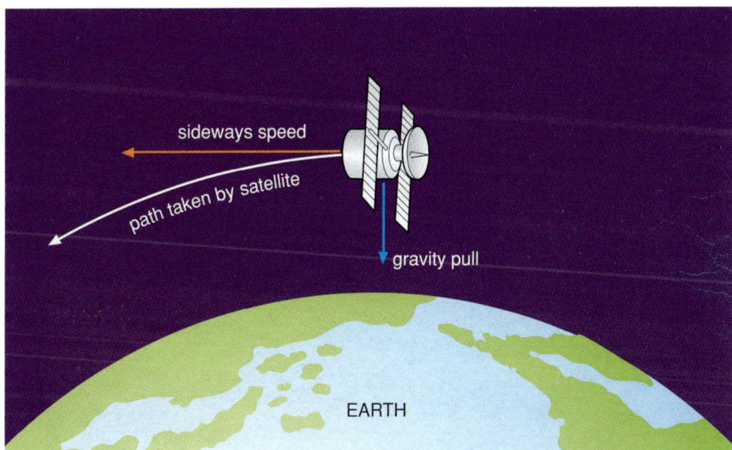

Did you know?

The first artificial satellite, Sputnik 1, was launched on 4th October 1957. It orbited the Earth every 96 minutes for three months.

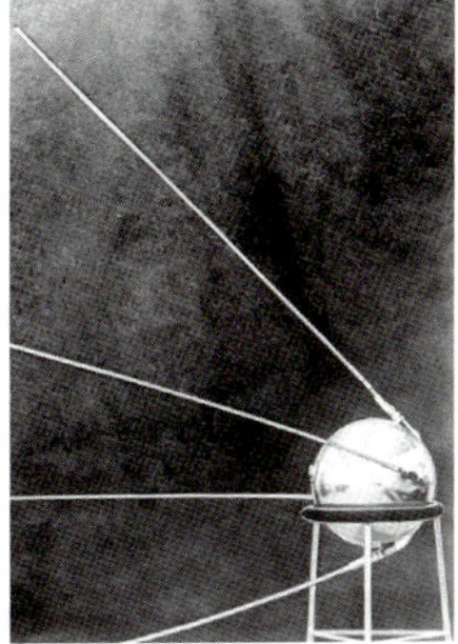

Figure 11.44
Sputnik 1

If the forward speed is too low the gravitational pull will make the satellite fall out of its orbit and drop towards Earth.

Artificial satellites have many different uses.

Looking at the Earth

Satellites that look at the Earth are usually put into a low orbit that passes over the North and South poles – a polar orbit. As they pass over the Earth they scan the surface sending back detailed pictures. Such things as the position of an oil slick, the path of a hurricane or the eruption of a volcano can all be watched and monitored. Because the Earth spins around its axis, the whole of its surface can be scanned each day.

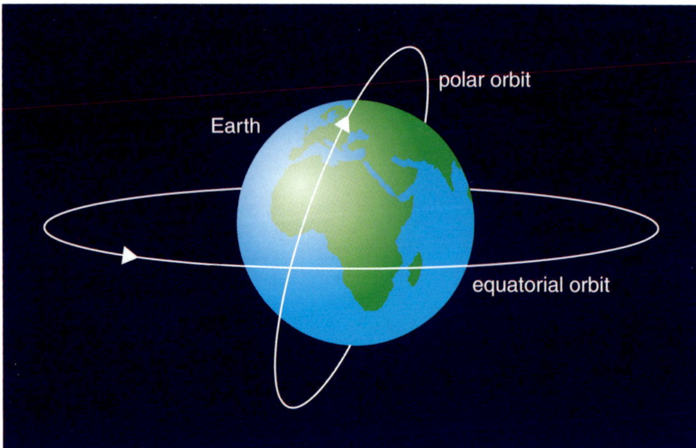

Figure 11.45
Polar and equatorial orbits

Figure 11.46
A satellite picture of the River Thames in London

Weather monitoring

Satellites, such as Meteostat, are put into a high orbit above the equator – an equatorial orbit. They monitor the weather patterns over one part of the Earth's surface. With this information, weather forecasts can be made more accurate.

Figure 11.47
Meteostat in orbit

Figure 11.48
The weather pattern seen from a satellite

Looking into space

From a satellite the view of the solar system and distant stars is much clearer than from the Earth. This is because the satellite is above the Earth's atmosphere.

Figure 11.49
The Hubble Telescope sends pictures back to Earth

Communications

Satellites used to send television and telephone signals around the world are put into a **geostationary orbit**. They are high above the equator taking 24 hours to orbit the Earth. This means the satellite moves around the Earth at the same rate as the Earth spins. So the satellite will always be above the same point on the Earth's surface.

TV signals are sent to the satellite using microwaves. The satellite then sends the signals back to an area of the Earth as big as Europe. A dish fixed to point in the right direction is used to pick up the signals.

Figure 11.50
Satellite dishes receive the TV signal from a satellite

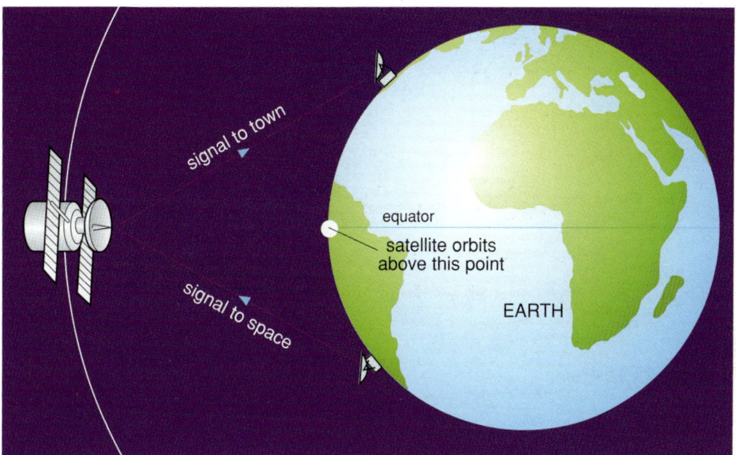

Figure 11.51
Using a communications satellite

Topic Questions

1 Copy and complete the following sentences.

a) The path taken by a satellite is called its ____________ .
b) ____________ satellites are used to send TV ____________ around the world.
c) A ____________________ satellite orbits the Earth once every 24 hours above the ____________________ .

2 Copy and complete the following sentence by choosing the correct phrase.
If the forwards speed of a satellite goes down, the satellite will *drift into space / stay in its orbit / fall towards the Earth.*

3 What is the difference between a polar orbit and an equatorial orbit?

4 Explain why the solar system can be seen better from a satellite than from the ground.

5 Write down two different uses for a satellite placed in a polar orbit.

6 Find out what a global positioning system (GPS) does.

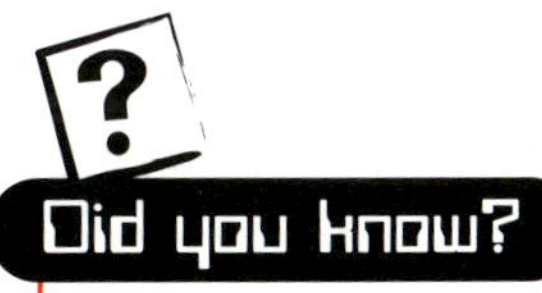

Did you know?

To orbit the Earth once every 24 hours, a geostationary satellite must be 35 900 km above the equator, moving with a forwards speed of 11 088 km/hour.

11.11	
Co-ordinated	Modular
DA 12.15	DA 11
SA 12.9	SA 18

Outside the solar system

A **galaxy** is a vast number of **stars** held together by gravity. The Sun is just one of the 100 000 million stars in the **Milky Way** galaxy. Figure 11.52 shows what the Milky Way would look like if you could see it from above and from the side.

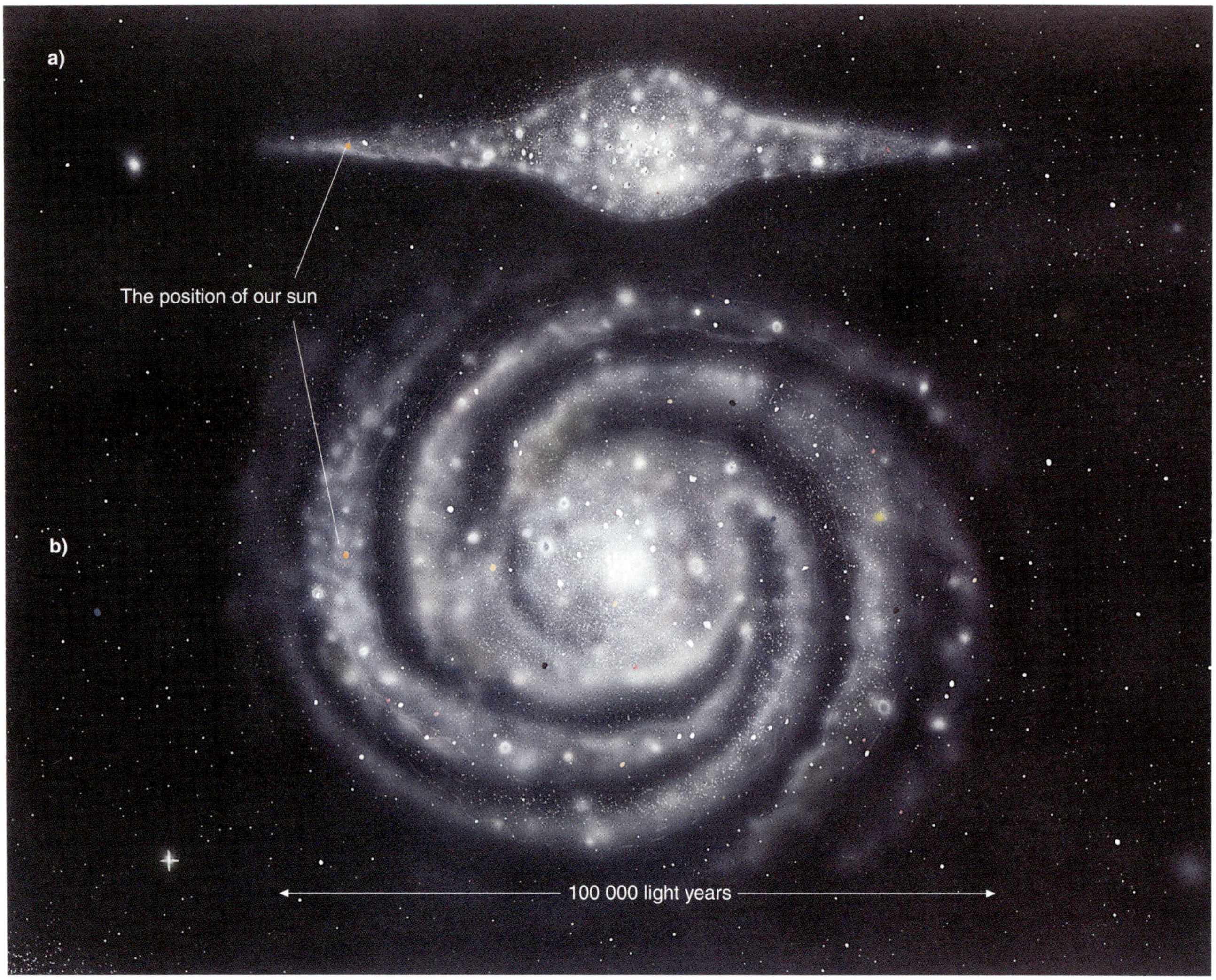

Figure 11.52

Did you know?

The Sun is our nearest star. Light takes 8 minutes to travel from the Sun to the Earth. Proxima Centauri is our next nearest star. The light from Proxima Centauri takes 4.2 years to reach the Earth.

The distance between stars in a galaxy is often millions of times bigger than the distance between the planets in the solar system. The Milky Way is so huge that it takes light 100 000 years to go from one side to the other.

The Milky Way is not the only galaxy. Using large telescopes, millions of similar galaxies can be seen in all directions in space. The **Universe** is made up of at least a billion (a thousand millions) galaxies. Each galaxy has about as many stars as the Milky Way.

The distance between galaxies is often millions of times bigger than the distance between the stars in a galaxy.

Figure 11.53
The Andromeda Galaxy

Did you know?

Distances in space are so huge that special units are used. One of these is the light year. A light year is how far light travels in one year.

1 light year is about 9 500 000 000 000 kilometres

Some stars are 15 000 million light years away from the Earth.

The Andromeda galaxy is one of the closest galaxies to the Milky Way. The light from Andromeda takes 2 million years to reach the Earth.

Topic Questions

1 Copy and complete each sentence by choosing the correct word from the box. Each word can be used once or not at all.

Milky Way	planet	solar system	star	Universe

The Sun is the nearest ________________ to the Earth. The Sun is in the galaxy called the _____________________ . In the ________________ there are millions of galaxies.

2 Rewrite the following list in order of size. Put the smallest at the top of the list.

- galaxy
- planet
- solar system
- Universe
- star

3 The solar system is in which galaxy?

4 What is a galaxy?

5 Find out what a constellation is.

11.12 The life cycle of a star

Co-ordinated	Modular
DA 12.15	DA 11
SA 12.9	SA 18

Stars do not last forever. They go through a life cycle from birth to death. The Sun is half way through its life cycle.

Did you know?

The Sun has been changing 4 million tonnes of its mass into energy every second for the last 5000 million years. It will carry on doing this for another 5000 million years.

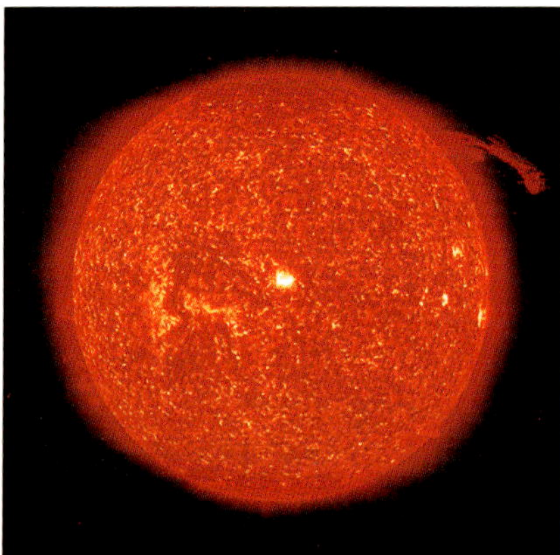

Figure 11.54

The birth of a star

Like all stars, our Sun was formed from a huge cloud of gas (nearly all hydrogen) and dust called a nebula.

Figure 11.55
Stars are born in the swirling cloud of dust and gas in the Great Nebula in Orion

Over millions of years the force of gravity pulled the dust and gas particles together. As more particles are pulled together enough heat is produced for nuclear reactions to start. The compressed cloud starts to glow and a star is born.

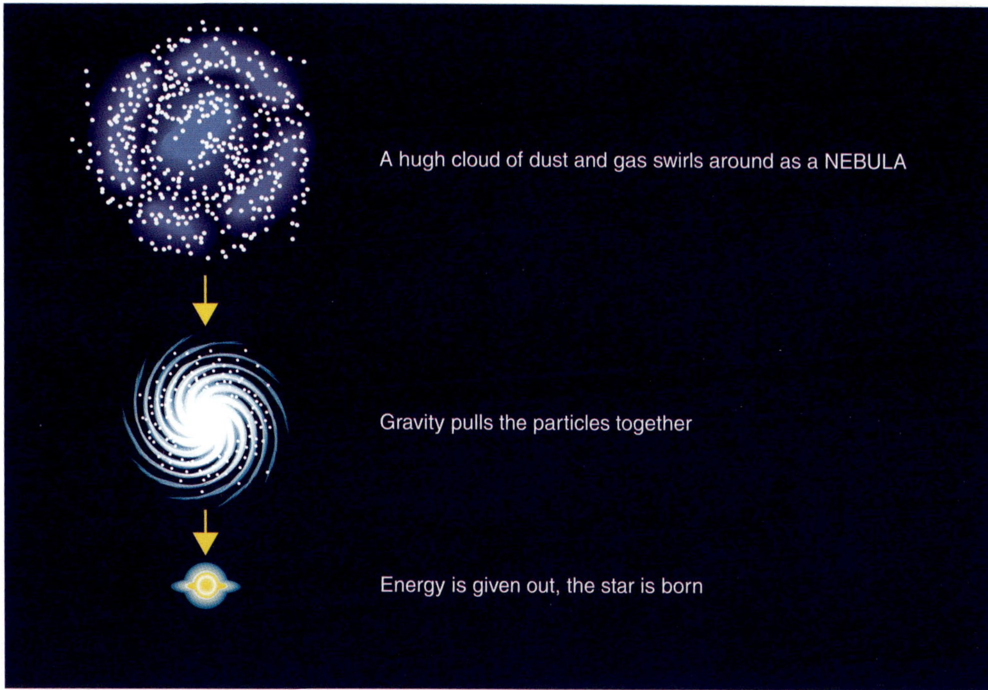
A hugh cloud of dust and gas swirls around as a NEBULA
Gravity pulls the particles together
Energy is given out, the star is born

Figure 11.56
The birth of a star

Scientists think that at the same time as the Sun was born smaller bodies were formed from the same nebula. These bodies were too small to produce their own heat and light. They became the planets and moons of the solar system.

For most of its life a star is stable. The gravitational forces pulling inwards balance the outward forces caused by the high temperature of the star. The Sun is in the stable part of its life cycle.

Figure 11.57
A stable star

The death of a star

Eventually the reactions causing the star to give out huge amounts of energy begin to stop. The star will then expand to become a **red giant**. When our Sun reaches this stage it will be so large that the Earth will end up destroyed inside it!

What happens next depends on the size of the star.

For a small star like the Sun, gravitational forces will make the star contract. The star gives out more energy; it is now a **white dwarf**. The matter in a white dwarf may be millions of times more dense than anything on Earth.

If a red giant is big enough, it may contract rapidly and explode as a supernova. The centre of the exploding star that is left behind becomes a very dense neutron star.

> **Did you know?**
>
> If the ball on the end of a ball-point pen were made from material with the same density as a neutron star, the ball would have a mass of about 91 000 tonnes.

Figure 11.58
The death of a star

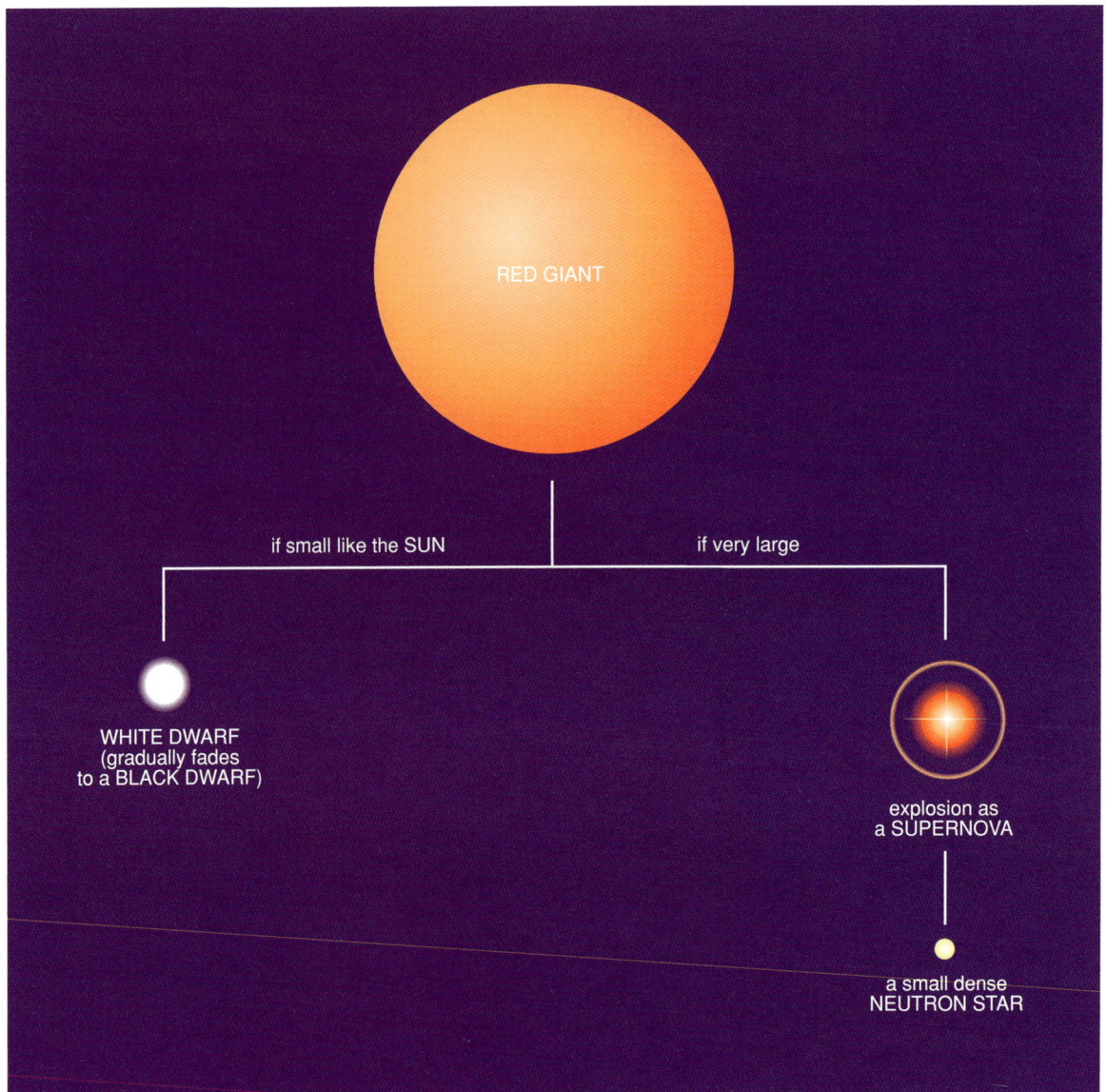

The dust and gas thrown into space by the explosion make a new nebula. So the life cycle starts again with new stars being born.

Figure 11.59
A supernova

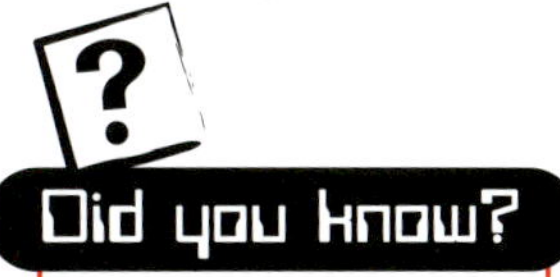

Did you know?

In 1987 the star Sandulek-69 exploded into a supernova. It was brighter than the whole of the Magellan Cloud, Sandulek's galaxy.

Topic Questions

1 Copy and complete the following sentences.

a) A star is formed from ____________ and _____________ particles pulled together by the force of ______________________ .

b) The _________________ were formed from the same _______________ and at the same time as the Sun.

c) When a red giant explodes as a ___________________ a _____________ star is formed.

2 What happens to a star at the end of the stable part of its life cycle?

3 What will happen to our Sun after it has been a red giant?

4 Find out what the word supernova means.

11.13 Life on other planets

Co-ordinated	Modular
DA 12.15	DA 11
SA 12.9	SA 18

The Earth is a special planet. It has water, an atmosphere containing oxygen and a temperature just right for plants and animals to live. These are the conditions for life as we know it. But things have not always been like this. Living organisms have helped to change the Earth. For example, because of plants there is much more oxygen in the atmosphere.

In the search for life on other planets, unmanned space probes have been sent to Mars and to Venus.

Figure 11.60
An image of the surface of Mars taken by the Mars Pathfinder space probe

Scientists looking for clues have designed and set up experiments:

- to test for water and oxygen
- to measure changes in the atmosphere
- to see if there are fossils of organisms in the rocks.

But the experiments have not been successful. Maybe the scientists have been looking for the wrong things or in the wrong places.

Recent photographs of Jupiter's moon Europa show that its surface could be an icy crust several kilometres thick. Below the ice may be liquid water. Scientists are thinking about sending a spacecraft to Europa in order to drill through the ice. A probe called a hydrorobot could then investigate the contents of the water.

Figure 11.61
An artist's impression of a hydrorobot

On Earth, organisms have now been found that live in conditions once thought impossible.

- Microbes have been found in the sulphurous atmosphere of a volcano.
- Other microbes have been found deep underground, getting their energy from hydrogen.

Perhaps life does exist on other planets and in other galaxies – but in forms we may not know and conditions once thought impossible.

Did you know?

In 1984, a meteorite was discovered that is thought to have come from Mars thousands of years ago. The meteorite seems to contain microscopic worm-like structures.

Figure 11.62
The Mars meteorite

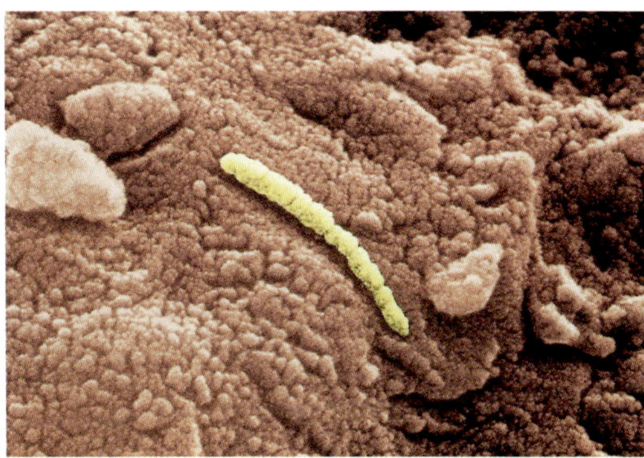

Figure 11.63
Are these worm-like structures fossilized Martian bacteria?

SETI

Frank Drake was the first scientist to start a careful search for intelligent radio signals from space. He was unsuccessful, but the search goes on. However, after 40 years the SETI (Search for Extra-Terrestrial Intelligence) project has not found anything.

Topic Questions

1 Copy and complete the following sentences.

a) Plants have ______________ the amount of oxygen in the Earth's ______________.

b) The Earth has the right ___________________ , _______________ and _____________________ for plants and animals to live.

2 What conditions exist on Jupiter's moon Europa that might be able to support life?

3 What is the SETI project? Has it been successful?

4 Radio waves travel at the same speed as light. Imagine aliens are living in the Andromeda galaxy and sending radio messages to Earth. How long would it be before the messages were received?

5 Find out why a spacecraft could not land on Jupiter.

Summary

- The velocity of an object is its speed in a given direction.
- The acceleration of an object is the rate at which its velocity changes.
- Distance–time graphs can be used to show when an object is not moving or moving at a steady speed.
- Velocity–time graphs can be used to show when an object is moving with constant velocity or with constant acceleration.
- Balanced forces do not change the movement of an object.
- Unbalanced forces can make an object speed up or slow down.
- The bigger the mass of an object, the bigger the force needed to make it accelerate.
- The stopping distance of a car will change if the braking force, the driver's reaction time, the condition of the road or car, the weather conditions or the speed of the car change.
- The force of gravity acts to pull objects together.
- An object moving as fast as possible is moving at its terminal velocity.
- When work is done energy is transferred.
- The joule (J) is the unit of work and energy.
- Elastic potential energy is the energy stored in an elastic object when work is done to change its shape.
- The orbit of a planet around the Sun is like a squashed circle (it's an elliptical orbit).
- Comets have very elliptical orbits.
- A planet is kept in its orbit by the force of gravity between the planet and the Sun.
- Communication satellites are put into a geostationary orbit.
- The Universe is made up of at least a billion galaxies.
- Stars are formed when gas and dust are pulled together by gravity.
- Stars do not stay the same, they go through a life cycle from birth to death.
- So far, life has not been found anywhere else in the Universe.

Examination Questions

1 a) A shopping trolley is being pushed at a constant speed. The arrows represent the horizontal forces on the trolley. Which one of the distance–time graphs, K, L or M, shows the motion of the trolley? Draw a circle around your answer.

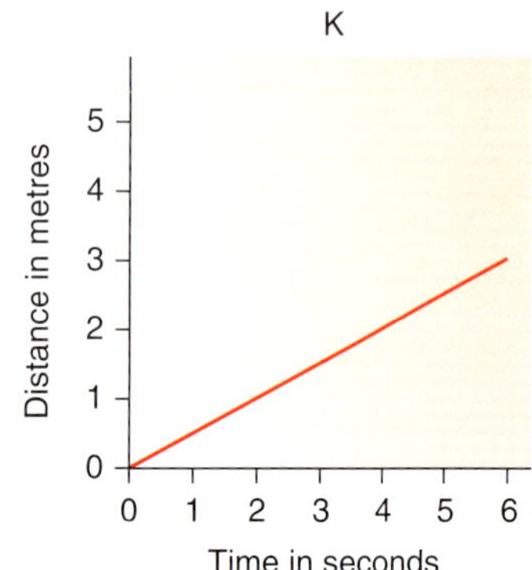

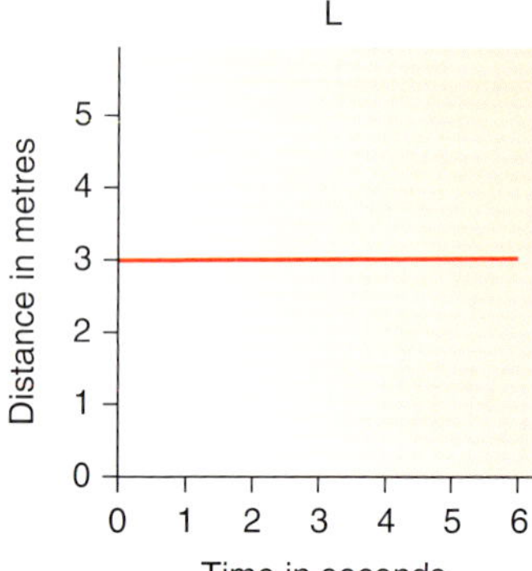

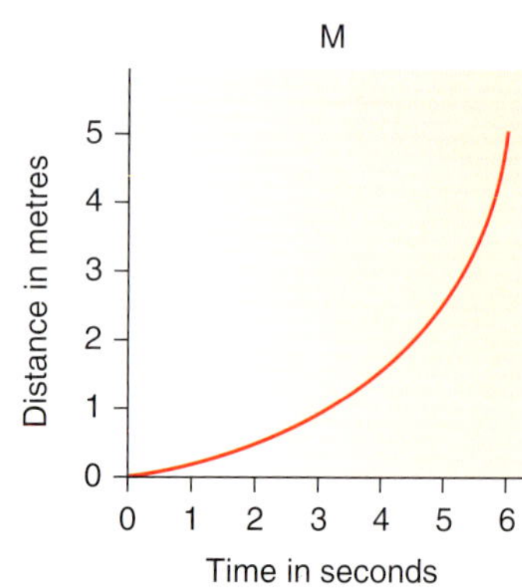

b) Complete the sentence by crossing out the two words in the box that are wrong. Acceleration is the rate of change of energy / speed / velocity.

2 A satellite is in orbit around the Earth.

a) Below are the names of three types of force.

drag	friction	gravitation

Which one of these forces keeps the satellite in its orbit?

____

b) Give one reason why some satellites are used to take pictures of the Earth.

____

3 a) Two sky-divers jump from a plane. Each holds a different position in the air. Copy and complete the following sentence. Sky-diver ______ will fall faster because ______.

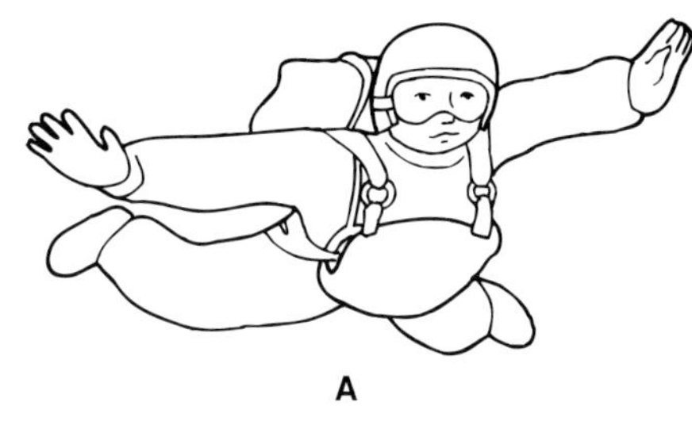

A

B

b) The diagram shows the direction of the forces acting on one of the sky-divers.

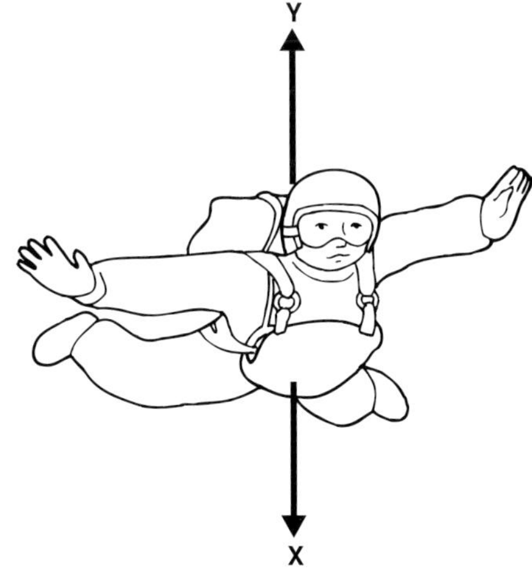

Copy the following sentences and complete them by choosing the correct endings.

i) Force X is caused by air resistance / friction / gravity.

ii) Force Y is caused by air resistance / gravity / weight.

iii) When force X is bigger than force Y, the speed of the sky-diver will go up / stay the same / go down.

iv) After the parachute opens, force X goes up / stays the same / goes down.

c) How does the area of an opened parachute affect the size of force Y?

4 Two students, Anna and Graham, took part in a sponsored run. The distance–time graph for Graham's run is shown. Four points have been labelled A, B, C and D.

a) Between which pair of points was Graham running the slowest?

b) Anna did not start the run until 10 minutes after Graham. She completed the whole run at a constant speed of 4 m/s.

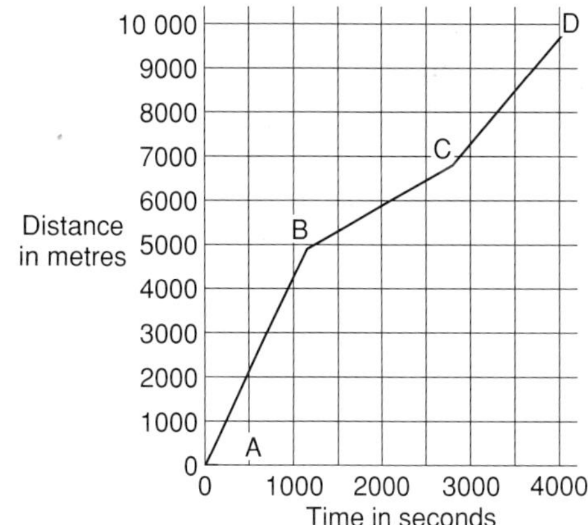

i) Write down the equation that links distance, speed and time.

ii) Calculate, in seconds, how long it took Anna to complete the run. Show clearly how you worked out your answer.

iii) Copy the graph and draw a line to show Anna's run.

iv) How far had Graham run when he was overtaken by Anna?

5 Five forces, A, B, C, D and E act on the van.

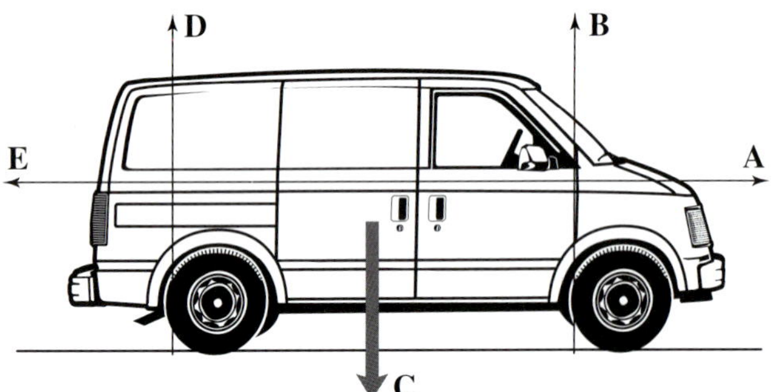

a) Copy and complete the following sentences by choosing the correct forces from A to E.
Force ______ is the forward force from the engine.
Force ______ is the force resisting the van's motion.

b) The size of forces A and E can change. Copy and complete the table to show how big force A is compared to force E for each motion of the van. Do this by placing a tick in the correct box. The first one has been done for you.

Motion of van	Force A smaller than force E	Force A equal to force E	Force A bigger than force E
Not moving		✓	
Speeding up			
Constant speed			
Slowing down			

c) The van was travelling at 30 m/s. It slowed to a stop in 12 seconds. Calculate the van's acceleration.

6 Copy and complete each sentence by choosing the correct word or phrase from the box. Each word or phrase should be used once or not at all.

Milky Way	planet	solar system	star	Universe

The Sun is the nearest ______ to the Earth.
The Sun is in the galaxy called the ______.
Within the ______ there are millions of galaxies.

7 Communications satellites and satellites used to observe the Earth are placed in different orbits.

a) Describe the orbit of a communications satellite.
b) Describe the orbit of a satellite used to observe the Earth.
c) Explain why the satellites are placed in different types of orbit.

Chapter 12

Waves and radiation

Key terms

amplitude • analogue • background radiation • diffraction • digital • electromagnetic spectrum • frequency • half-life • isotope • longitudinal wave • mass number • neutron • proton • radioactive decay • radioactive isotope • refraction • seismic waves • total internal reflection • transverse wave • ultrasonic waves • ultrasound • wavelength

12.1 Looking at waves

Co-ordinated	Modular
DA 12.9	DA 12
SA 12.5	SA 18

Waves are important. They carry energy and information.

Light and sound travel as waves. But we cannot see the waves. So looking at water waves or waves moving along a rope or spring can help us find out more about light waves and sound waves.

Shaking a rope (or spring) up and down makes a wave move along it.

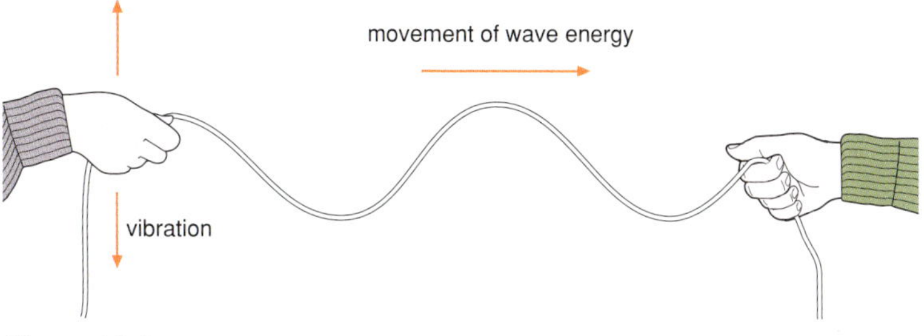

Figure 12.1
A wave moving along a rope

Someone holding the other end of the rope will feel the wave arrive. The wave has carried energy from one person to the other. But the rope itself does not move from one person to the other. So waves carry energy from one place to another without transferring any material (matter).

The vibration making the wave is up and down or side to side. The wave moves at right angles to the vibration. This type of wave is called a **transverse wave**.

Light waves move as transverse waves.

Energy can also be carried through a spring by moving the spring backwards and forwards.

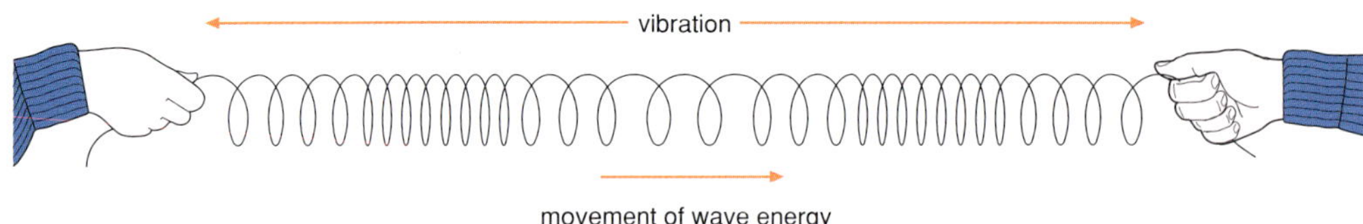

Figure 12.2
A wave moving through a spring

The vibration making the wave is backwards and forwards. The wave moves along in the same direction as the vibration. This type of wave is called a **longitudinal wave**.

Sound waves move as longitudinal waves.

Waves moving along a rope (or spring) can be reflected back down the rope (or spring). Light waves and sound waves can also be reflected.

Describing a transverse wave

Waves moving across a water surface are also transverse. Three quantities, **amplitude**, **wavelength** and **frequency** are used to describe the wave.

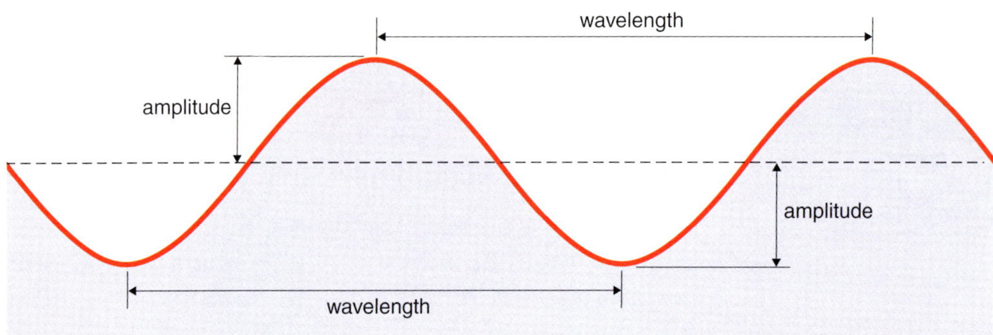

Figure 12.3
A transverse wave moving across a water surface

Amplitude is the maximum disturbance caused by a wave.

Wavelength is the distance from a point on one wave to the equivalent point on the next wave.

Frequency is the number of waves produced each second by a vibrating source. It is also the number of waves that passes a point each second.

Frequency is measured in units called hertz (Hz). One wave sent out each second is one hertz.

The wave equation

If you know the frequency and wavelength of a wave, its speed can be calculated using the equation:

$$\text{wave speed (m/s)} = \text{frequency (Hz)} \times \text{wavelength (m)}$$

Example: Water waves of wavelength 0.6 metres make a moored boat bob up and down 4 times a second. At what speed do the waves travel across the surface of the water?

$$\text{frequency} = 4\text{ Hz; wavelength} = 0.6\text{ m}$$

$$\text{wave speed} = \text{frequency} \times \text{wavelength}$$

$$\text{wave speed} = 4 \times 0.6 = 2.4\text{ m/s}$$

Topic Questions

1 Copy and complete the following sentences.

a) Waves carry ________________ without transferring ____________________.
b) Light waves are _________________ , sound waves are ________________.

2 The diagram shows three water waves, drawn to the same scale.

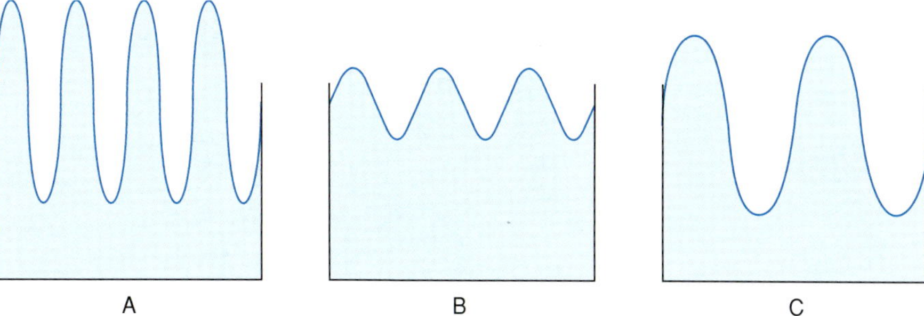

a) Which wave has the smallest amplitude?
b) Which wave has the largest wavelength?

3 A sound wave of frequency 300 Hz has a wavelength in water of 5 m.

a) Write down the equation that links frequency, wavelength and wave speed.
b) Calculate the speed of a sound wave through water.

4 The wave maker at a leisure pool makes 1 wave every 2 seconds. The wavelength of each wave is 4 metres.

a) What is the frequency of the waves?
b) Calculate the speed of the water waves.

12.2 Water waves

Co-ordinated	Modular
DA 12.9	DA 12
SA 12.5	SA 18

Small water waves (ripples) can be made in a tank of water called a ripple tank.

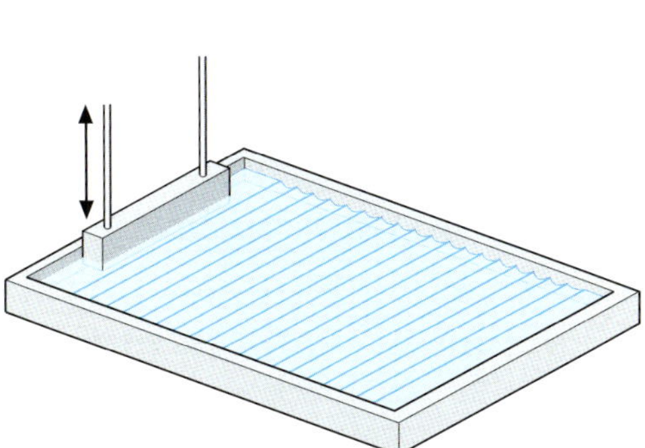

Figure 12.4
Hitting the water surface with a vibrating piece of wood gives straight ripples

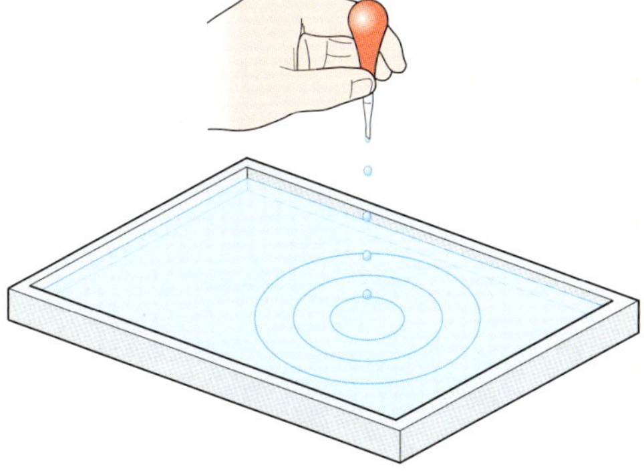

Figure 12.5
Water droplets hitting the surface give circular ripples

The waves move across the water surface away from what made the ripples. But the water itself does not move away from what made the ripples.

The waves carry energy from one part of the tank to another without transferring any of the water.

Looking at reflection

Water waves can be reflected.

Figure 12.6

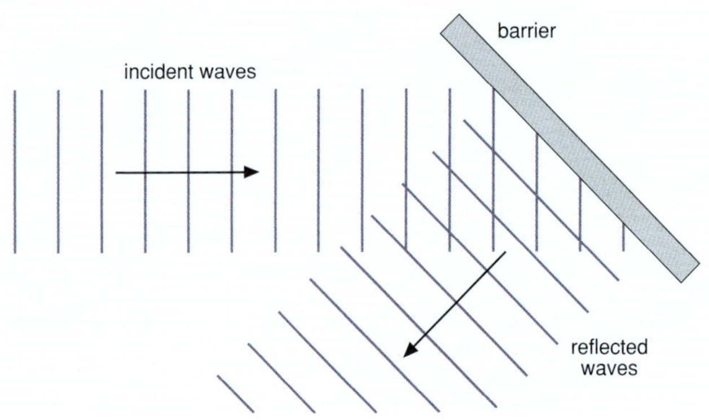

A straight barrier will reflect the waves at the same angle as they hit it.

Looking at refraction

A flat piece of glass in the ripple tank makes an area where the water is less deep.

Figure 12.7
A side view of waves passing across a water surface

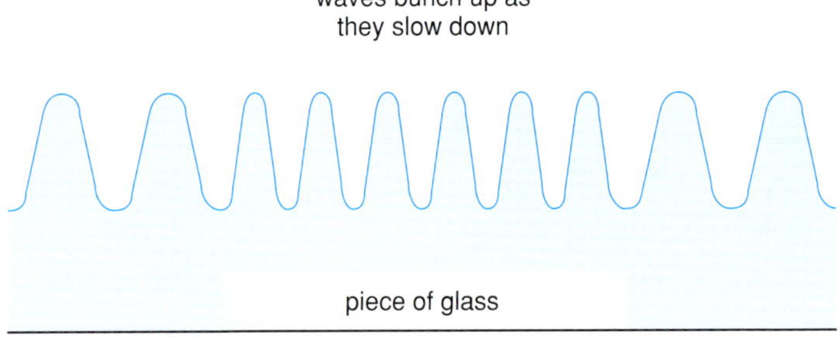

Water waves move faster across deep water than shallow water. So water waves moving across water of different depths change speed. This will usually (but not always) make the waves change direction.

Figure 12.8 shows water waves moving over an area of shallow water. The waves have slowed down but not changed direction. This is because all of the wave hits the edge of the shallow water at the same time.

Figure 12.8

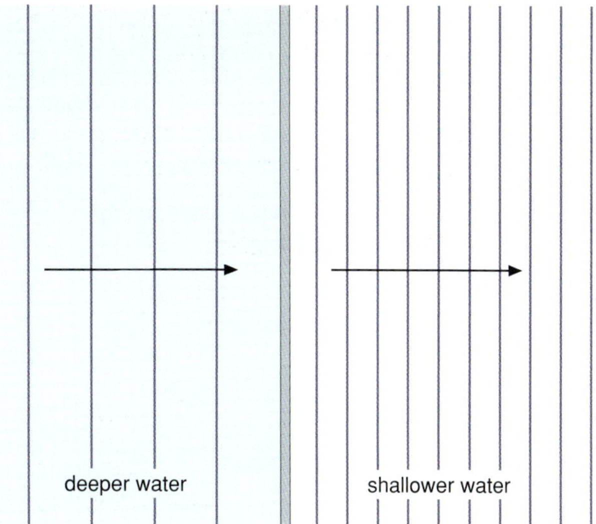

Figure 12.9 shows what happens when the water waves hit the shallow water at an angle. This time the waves slow down and change direction. Then as they leave the shallow water they speed up and again change direction. This is called **refraction**.

So, for water waves to be refracted they must hit shallower or deeper water at an angle.

Figure 12.9

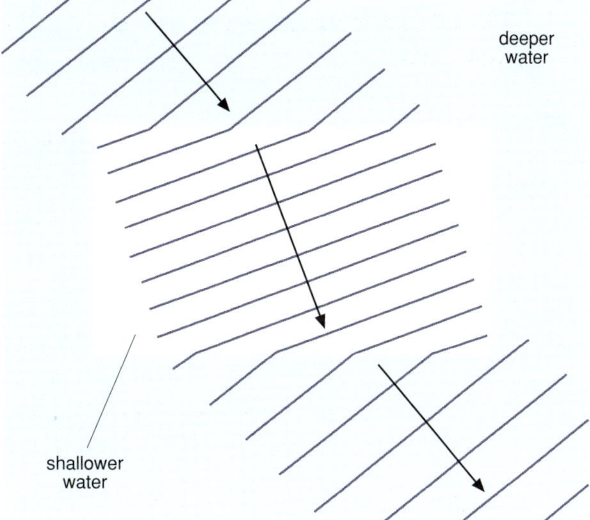

Water waves can also be diffracted, but more of this later (see Section 12.9).

Did you know?

In deep water a tsunami, or tidal wave, is only small, but it is moving at about 200 m/s (over 400 mph). Near land, where the water is shallow, the tsunami slows down and grows to a height of about 10 metres. The energy carried by the wave will cause enormous damage to anything it hits.

Topic Questions

1 Copy and complete the following sentences.

a) When water __________ hit a barrier they are _________________ .
b) When water waves move from deep to shallow water they are usually ___________.

2 Copy and complete the following sentences by choosing the correct phrases.

a) Waves at the seaside *speed up* / *stay at the same speed* / *slow down* when they move from deep water into shallow water.

b) The wavelength of a water wave will *get longer* / *stay the same* / *get shorter* as it moves from deep water into shallow water.

3 A leisure pool has a wave machine. The diagram shows the direction the waves move across the pool

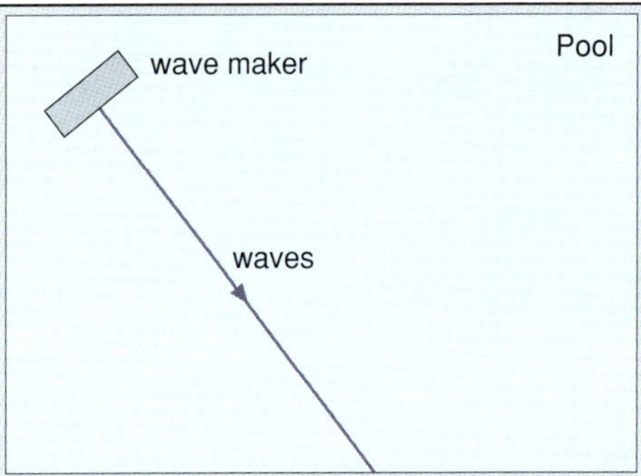

Copy the diagram, then draw a line to show the direction of the waves after they hit the side of the pool.

4 The diagram shows straight water ripples about to move from deep to shallow water.

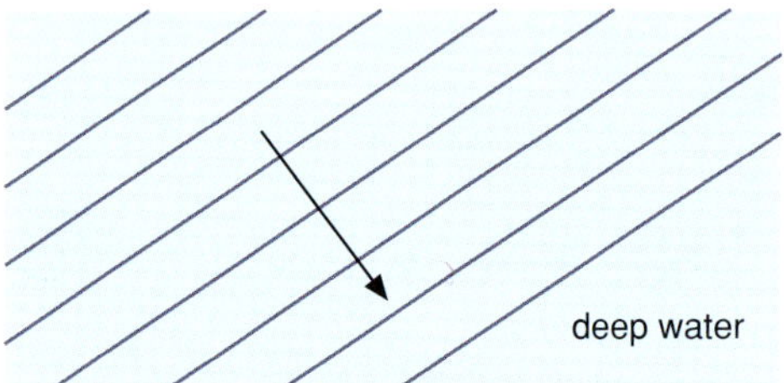

Copy and complete the diagram to show how the ripples travel across the shallow water.

12.3 Light waves

Co-ordinated	Modular
DA 12.9	DA 12
SA 12.5	SA 18

What you know

Mirrors, which can be curved or flat, reflect light. A flat mirror is called a plane mirror. Figure 12.10 shows what happens when a ray of light hits a plane mirror.

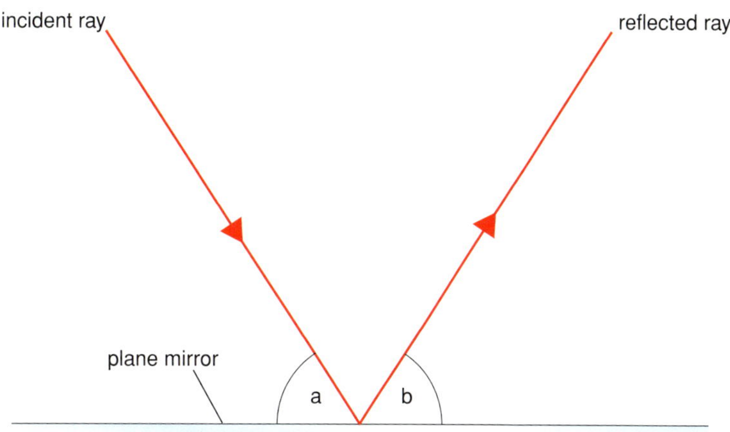

Figure 12.10

The light is reflected off the mirror at the same angle as the light hits the mirror.

angle a = angle b

When a ray of light crosses the boundary between two clear materials it will usually change direction. The ray of light is refracted. The ray of light will not be refracted if it hits the boundary at right angles.

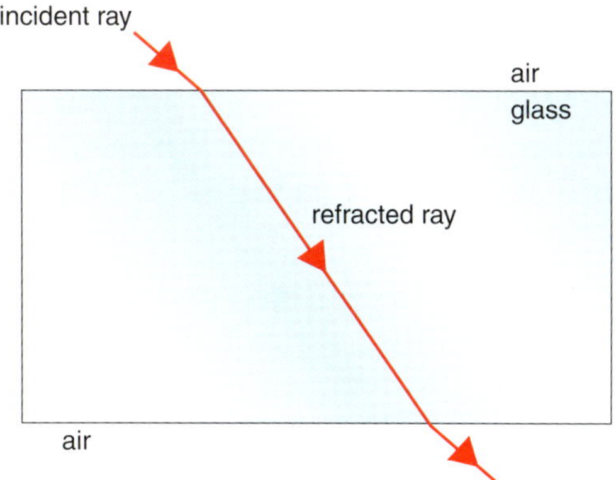

Figure 12.11
Refraction of light through a glass block

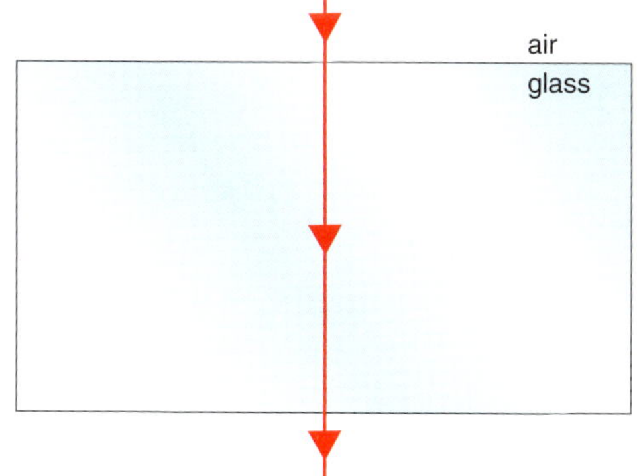

Figure 12.12
The light is not refracted

A ray of white light passing through a prism will be refracted and split into different colours. The light produces a spectrum.

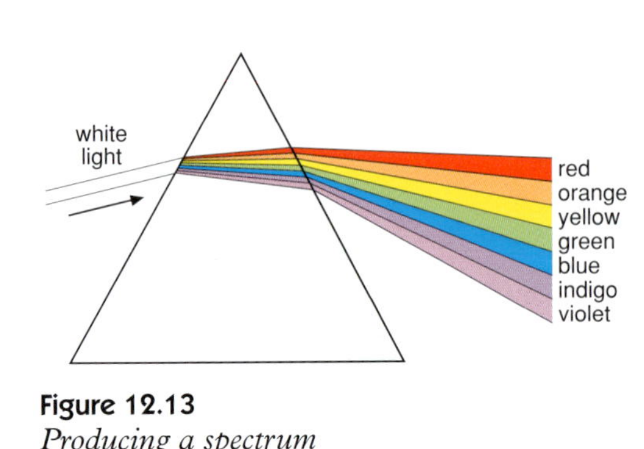

Figure 12.13
Producing a spectrum

This happens because white light is a mixture of different colours. The prism refracts each colour by a slightly different amount. The violet light is refracted the most and the red light the least. So the colours go into the prism together but they come out apart.

Did you know?

When it's sunny and raining you can sometimes see a rainbow. The raindrops have refracted the sunlight and made it split into the colours of the spectrum.

Figure 12.14
A rainbow

Water waves are refracted because they change speed when the depth of water changes (see Section 12.2).

Rays of light are refracted because they change speed as they go from one clear material into another. For example, light moves faster in air than in water or glass. In Figure 12.11 the light slows down as it goes into the glass. As it leaves the glass it speeds up.

Figure 12.15 gives a detailed picture of what happens to a ray of light when it moves from glass into air.

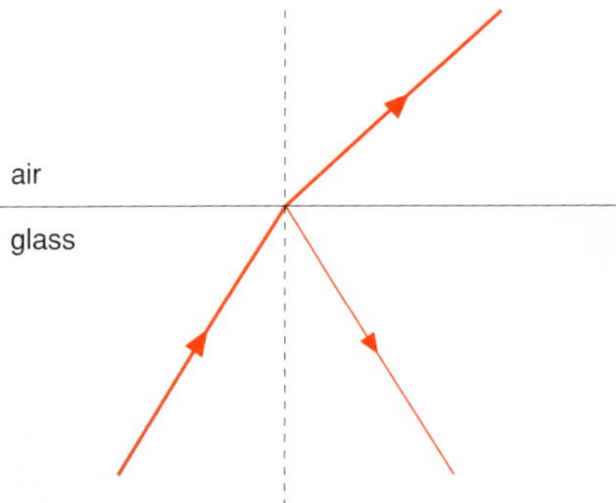

Figure 12.15
Light is reflected and refracted

At the boundary the ray of light splits. Most of the light is refracted but some of the light is reflected. The same thing happens if a ray of light moves from water or Perspex into air. Some of the light is reflected at the boundary.

Sometimes it's possible for all the light to be reflected and none of it to be refracted into the air. This is called total internal reflection (see Section 12.4).

Light waves are different in one big way to water waves and sound waves. Light waves can move through a vacuum (empty space). This is just as well – if light couldn't move through a vacuum it would be night all the time.

Topic Questions

1 Copy and complete the following sentences.

a) Light waves can move through a ______________ but sound ________ cannot.
b) Light moving from glass into air is usually ____________ and ___________.
c) A prism can be used to split white light into a ___________________ .

2 Copy and complete the following sentence by choosing the correct phrase.

When light moves from water into air it *slows down / speeds up / stays at the same speed.*

3 Copy and complete each of the diagrams to show the direction of the ray of light after crossing the boundary.

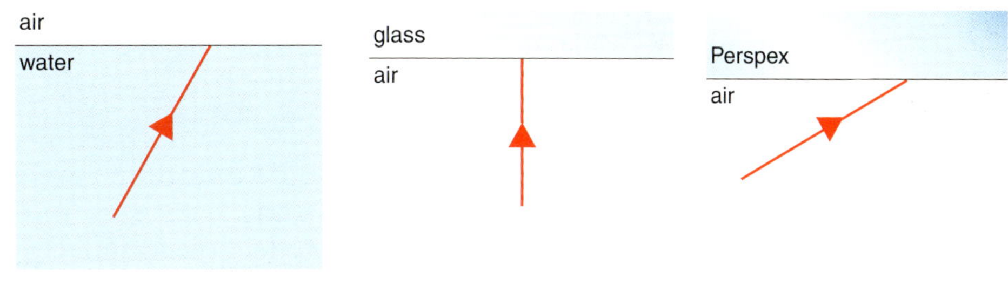

12.4 Total internal reflection

Co-ordinated	Modular
DA 12.9	DA 12
SA 12.5	SA 18

When a ray of light moves from glass into air some of the light may be refracted into the air and some reflected back into the glass. How much of the light is refracted and how much is reflected depends on the angle at which the light hits the glass surface.

Figure 12.16 shows what happens to three rays of light that hit a glass surface at different angles.

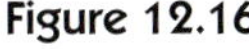
Figure 12.16

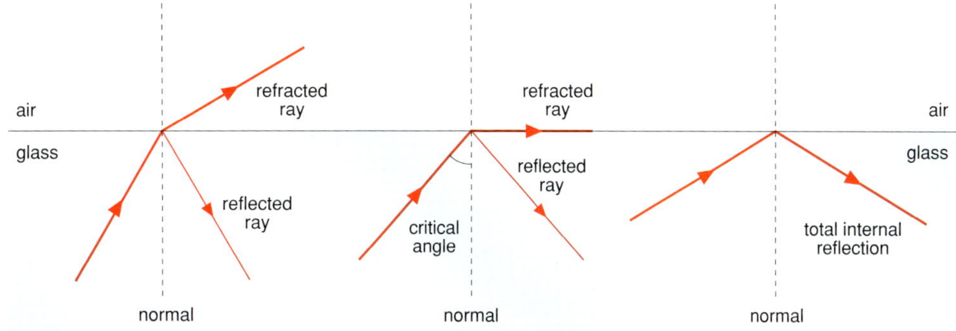

At small angles most of the light is refracted. There is a weak reflected ray.

Some light is refracted along the surface of the glass. When this happens the angle inside the glass, between the ray of light and the normal, is called the critical angle. The critical angle for glass is 42°.

When the ray of light hits the inside of the glass at an angle bigger that the critical angle, all the light is reflected back inside the glass. None of the light is refracted into the air. This is called **total internal reflection**. The glass surface is like a perfect mirror.

The same effect can happen when light moves from water, Perspex or other transparent materials into air. With each different material there is a different critical angle.

Reflecting prisms

A prism can be used to reflect light.

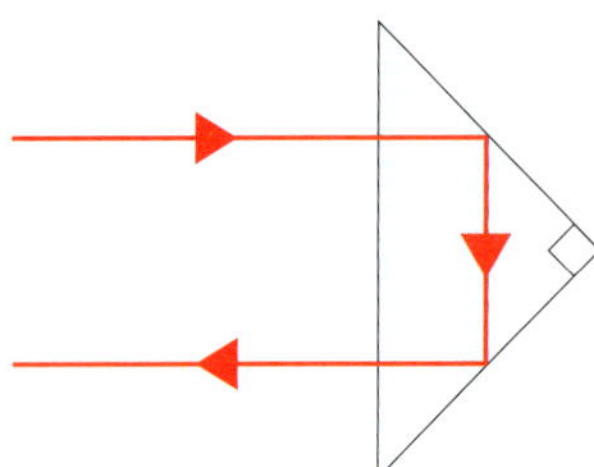

Figure 12.17
Total internal reflection inside a prism

Light hits each inside surface at 45°. This is a bigger angle than the critical angle. So at each inside surface the light is totally internally reflected. The light comes out of the prism parallel to the direction it went in.

The reflectors on cars and bicycles and in some 'cat's eyes' in the road use total internal reflection to reflect light back the way it came.

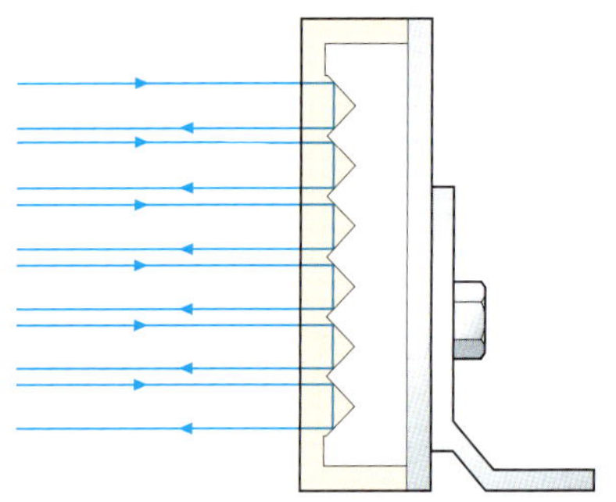

Figure 12.18
A bicycle reflector

Optical fibres

Optical fibres are usually made of glass. They are very thin and flexible. Rays of light going in at one end of the fibre are totally internally reflected many times until they come out at the other end of the fibre.

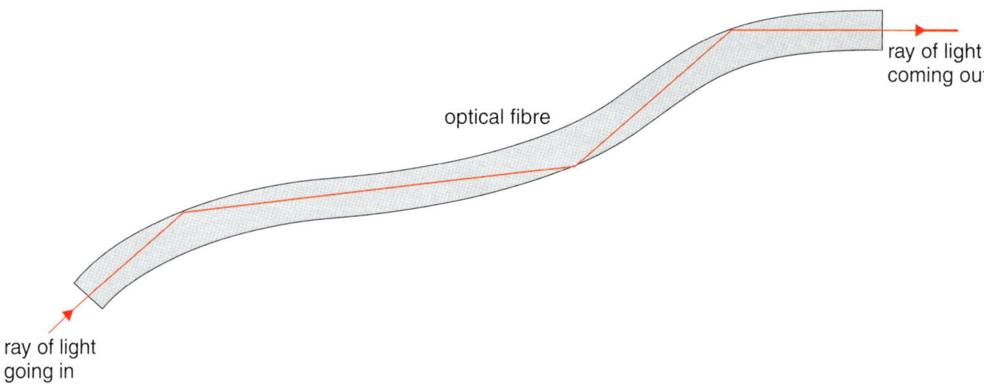

Figure 12.19
An optical fibre

Optical fibres have two very important practical uses.

Medical use

An endoscope can be used by a doctor for looking inside a patient. The endoscope has two bundles of optical fibres. Light is sent down one bundle of fibres. Reflected light passes back through the second bundle of fibres. This lets the doctor see into the patient's body without a major operation.

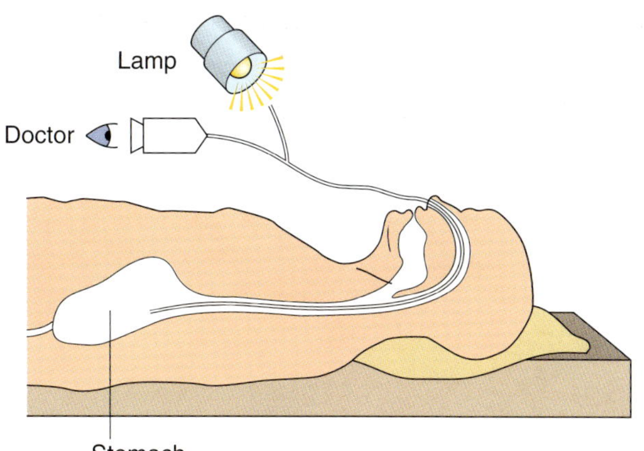

Figure 12.20
An endoscope in use

Communications

Optical fibre cables have replaced many copper telephone cables. Information is coded and sent along optical fibres using pulses of laser light. Just one optical fibre cable can carry thousands of telephone conversations at the same time. This is many times more than can be carried by a copper cable of the same size.

Topic Questions

1 Copy and complete the following sentence by choosing the correct phrase.

Total internal reflection happens when light moving through a transparent material into air hits the surface of the transparent material at an angle *smaller than / equal to / bigger than* the critical angle.

2 The diagram shows two prisms being used as a periscope. Copy the diagram and show the path taken by the ray of light from the ship to the sailor.

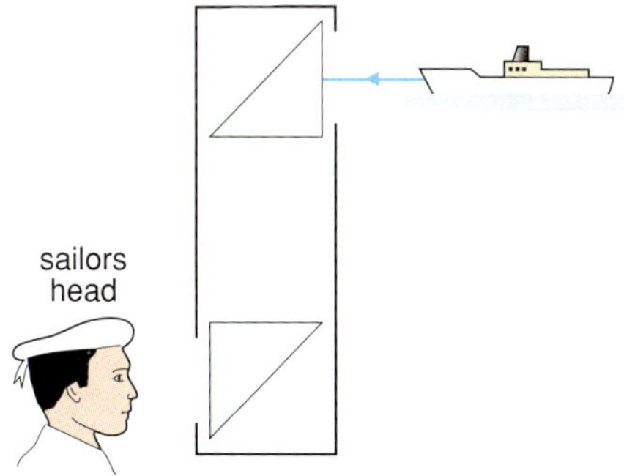

3 What is the value of the critical angle for a ray of light in a glass block?

4 Sunlight enters an underground car park through thick glass blocks. The blocks usually have the shape shown below.

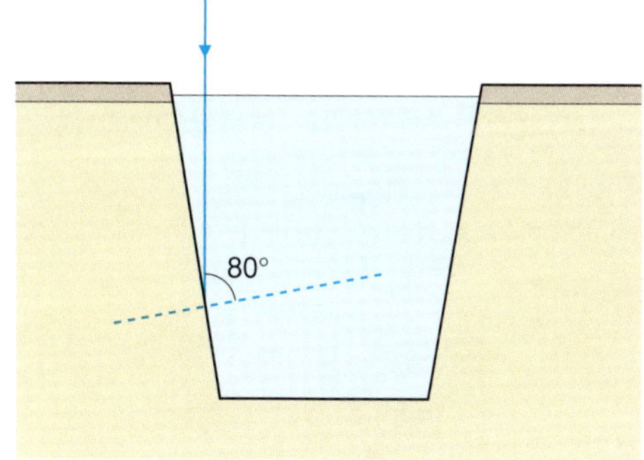

Copy the diagram and show the path of the light ray through the block.

5 Give two practical applications that use the effect of total internal reflection.

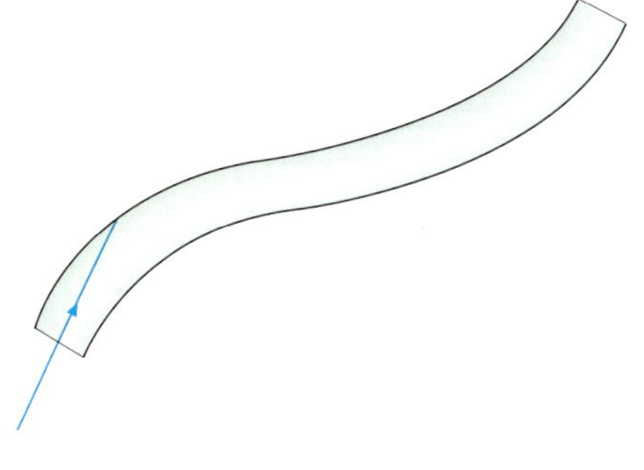

12.5	
Co-ordinated	Modular
DA 12.11	DA 12
SA 12.7	SA 18

Sound waves

What you know

Sound waves are produced when an object vibrates. Figure 12.21 shows some examples of vibrating objects causing sounds.

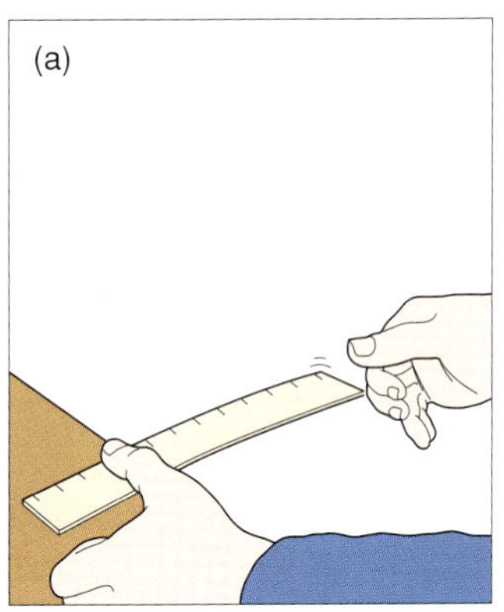

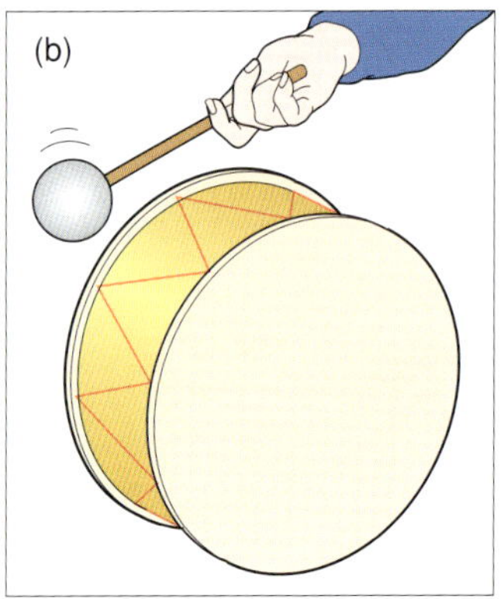

Figure 12.21

When someone shouts, the sound wave has a bigger amplitude than when they whisper. The size or amplitude of a vibration changes the loudness of the sound produced. The bigger the amplitude, the louder the sound.

The number of complete vibrations produced each second gives the frequency of the sound wave. Different sounds have different frequencies. The higher the frequency of a sound, the higher its pitch.

Figure 12.22

A hard surface can reflect sound. An echo is a reflected sound wave.

Sound, like light, can be refracted. This usually happens when sound moves from one substance into another. However, refraction will not happen if the sound wave is moving at right angles to the boundary between the substances.

When a tuning fork vibrates it causes the nearest air particles to vibrate. As these particles vibrate they pass energy to other particles, which also start to vibrate. This is a sound wave. A sound wave is a longitudinal wave.

Sound waves carry energy from one place to another without transferring any material (matter).

Sound can travel through solids and liquids as well as through gases like air. In fact the closer the particles the faster the sound travels. So sound

travels faster through water than through the air. This is because in water the particles are closer together so can pass on the vibrations quicker.

Sound waves cannot go through a vacuum because there are no particles to pass on the vibrations.

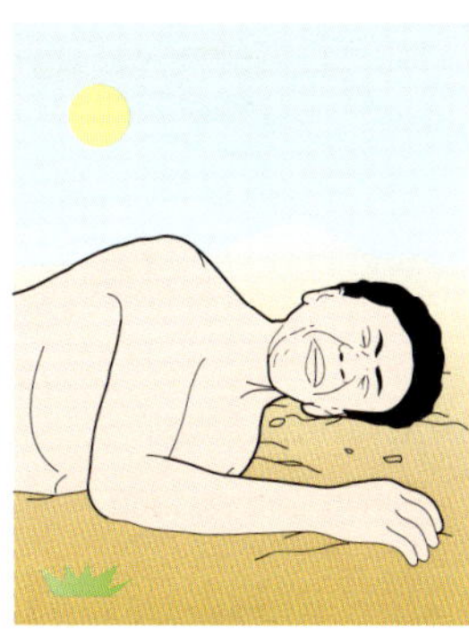

Figure 12.23 *Sound can travel through solids*

Looking at sound waves

A microphone changes sound energy into electrical energy. If the microphone is connected to an oscilloscope (CRO), a wavetrace can be shown on the screen. This represents the sound wave but it is not a picture of a sound wave.

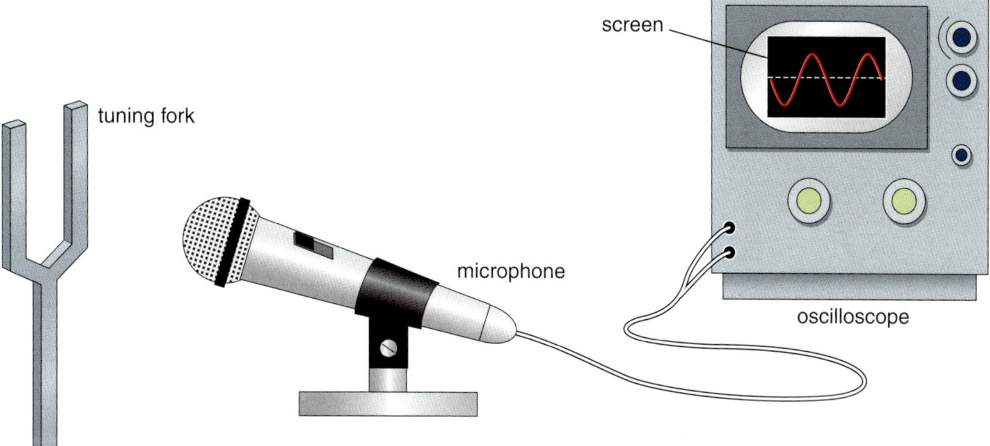

Figure 12.24 *Displaying sound waves on an oscilloscope screen*

The wavetrace on the oscilloscope screen can be used to give information about the amplitude and frequency of the sound wave.

Loud sounds carry more energy than quiet sounds. A loud sound has a bigger amplitude than a quiet sound. On an oscilloscope screen, the louder the sound the taller the wave.

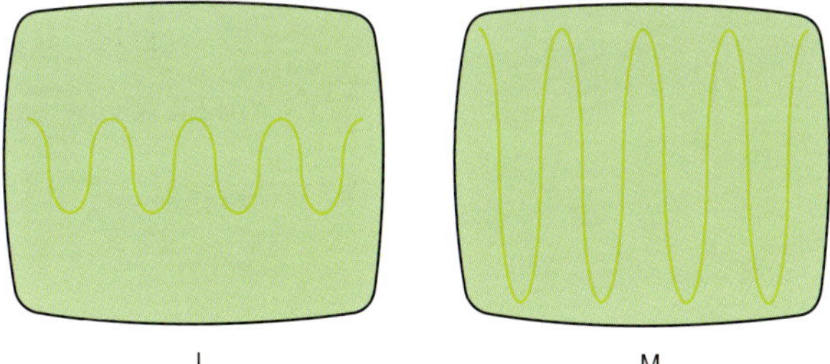
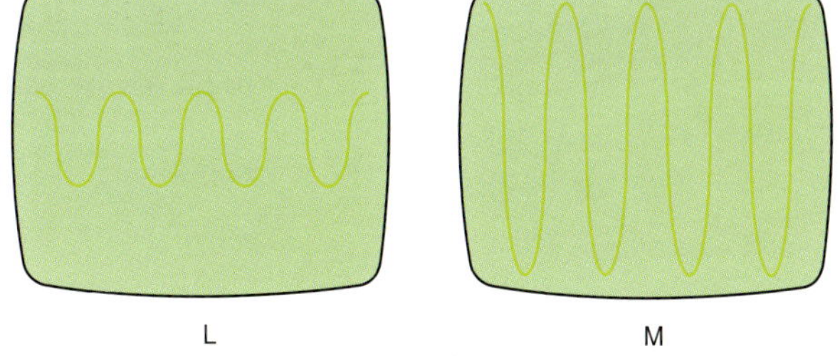

Figure 12.25 *The louder the sound, the taller the wave*

The higher the frequency the closer the sound waves are together. So the higher the frequency the more waves will be seen on the oscilloscope screen.

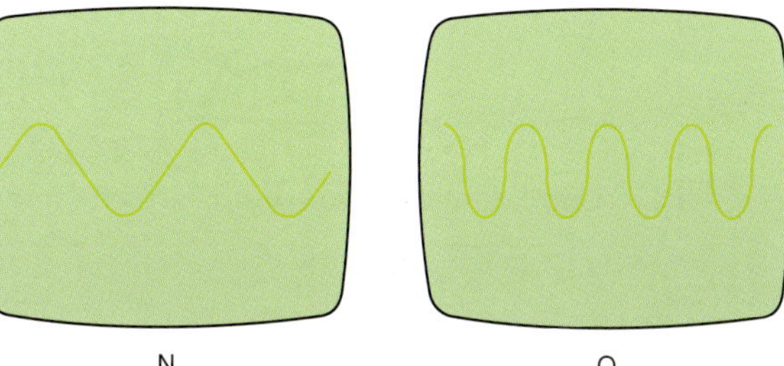

Figure 12.26 *Waves of different frequency*

Did you know?

If you double the frequency of a note, the note is exactly one octave higher in pitch.

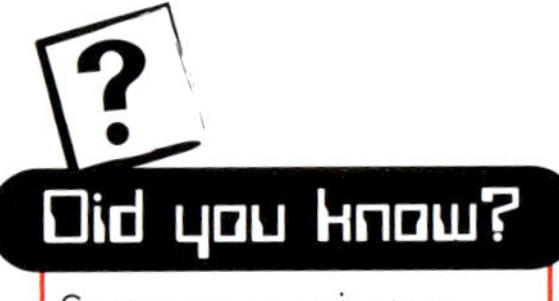

Did you know?

Some opera singers are able to produce high pitched notes with such a loud volume that they can shatter a wineglass!

Topic Questions

1 Copy and complete the following sentences.

a) A reflected sound is called an _______________.
b) A loud sound has a bigger _______________ than a quiet sound.
c) Sound waves are _________________________ waves.

2 The pictures show the oscilloscope wavetraces for three different tuning forks.

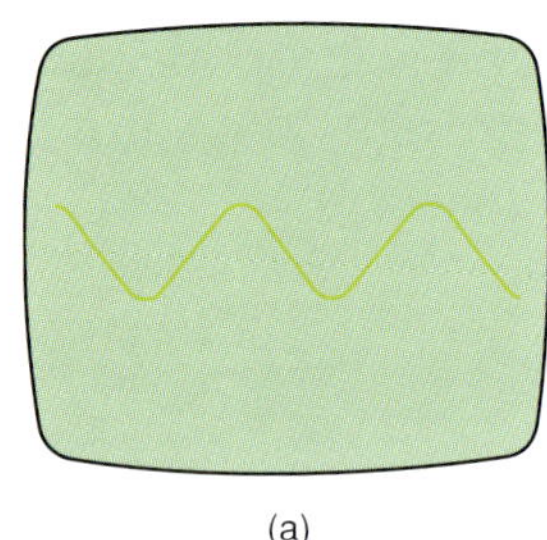

(a)

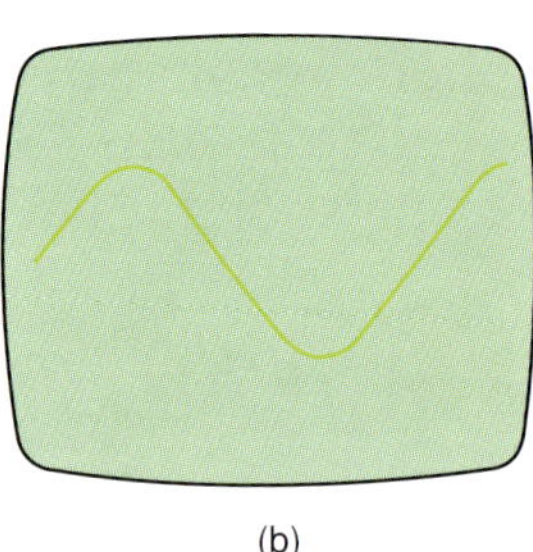

(b)

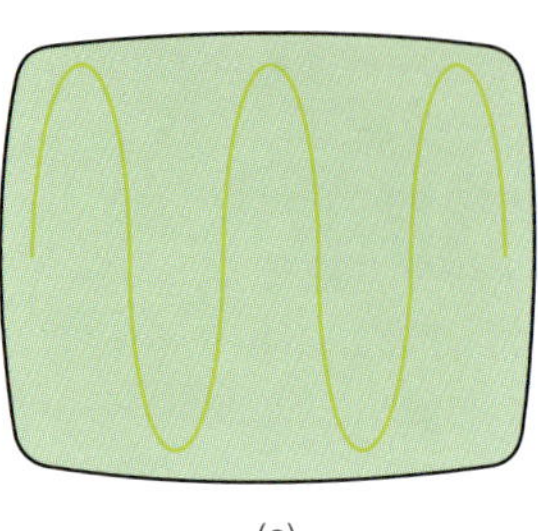

(c)

a) Which of the tuning forks is vibrating with the smallest amplitude?
b) Which of the tuning forks gives the loudest sound?
c) Which of the tuning forks has the lowest frequency?

3 The table gives the frequency of four different sounds.

Sound	Frequency
A	254 Hz
B	512 Hz
C	406 Hz
D	456 Hz

a) Which sound has the highest pitch?
b) Which sound has the lowest pitch?

4 Explain why sound cannot travel through space.

5 In which of the following does sound travel faster – air, water or iron? Give a reason for your choice.

6 The diagrams show two oscilloscope wavetraces produced by different sounds.

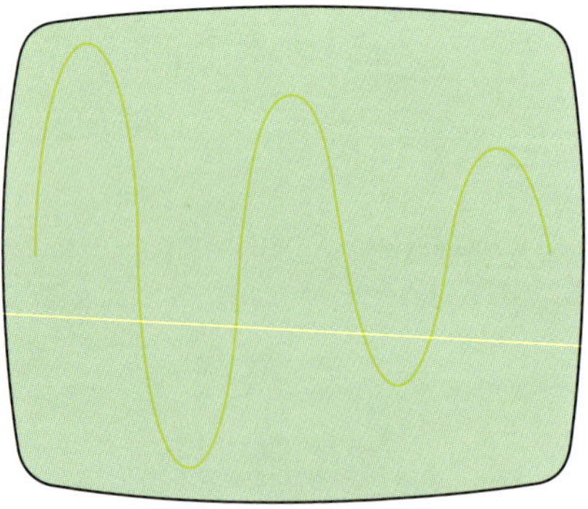

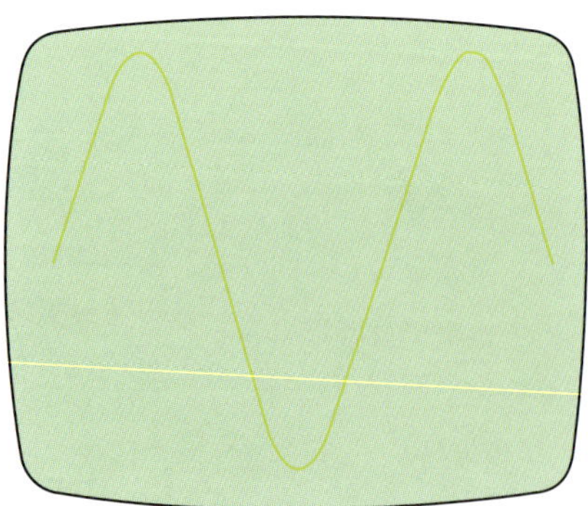

Describe how the two sounds are changing.

12.6	
Co-ordinated	Modular
DA 12.11	DA 12
SA 12.7	SA 18

Ultrasonic waves

Ultrasonic waves produce sound with a frequency too high for humans to hear. These sounds are often called **ultrasound**. Although humans cannot hear ultrasound many animals can.

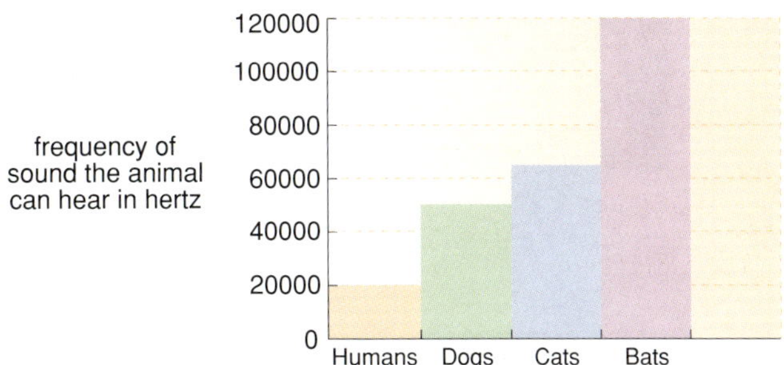

Figure 12.27
The sounds an animal can hear

Ultrasonic waves have many uses in industry and in medicine. Here are some examples.

Cleaning objects

A jeweller can use ultrasonic waves to clean a valuable piece of jewellery. The jewellery is put into a tank of liquid with an ultrasonic wave generator. Any small pieces of dirt are shaken from the jewellery by the rapid vibration of the ultrasonic waves.

Finding cracks

Hidden cracks in metals or concrete can be found by sending ultrasonic waves into the material. If the waves hit a crack some will be reflected as an echo. The echo shows up as a small pulse on an oscilloscope.

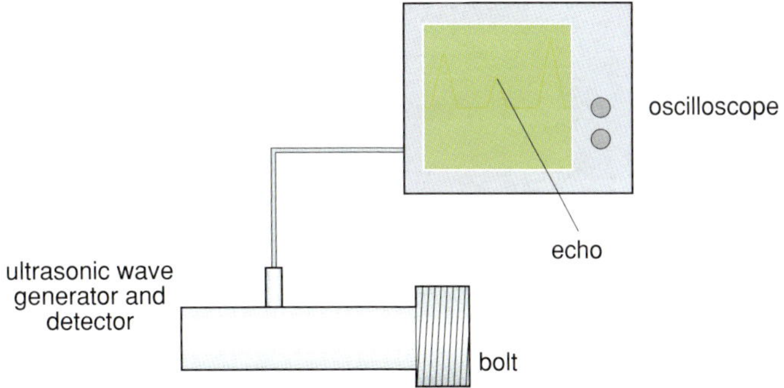

Figure 12.28
Looking for cracks in a metal bolt

Pre-natal scanning

Ultrasonic waves can be used to produce an image of a fetus. Pulses of ultrasound are generated and sent through the mother and fetus. A detector picks up waves reflected back from inside the mother. A computer changes the reflected waves into a picture. This gives the doctor and mother an early image of the growing baby. It also lets the doctor check that the baby is developing without any problems.

Did you know?

A bat uses ultrasonic waves to hunt for insects. The bat sends out ultrasonic waves which are reflected off the insect. The bat picks up the reflected waves and uses them to find exactly where the insect is. Unfortunately for the bat, some insects are also able to hear the ultrasound.

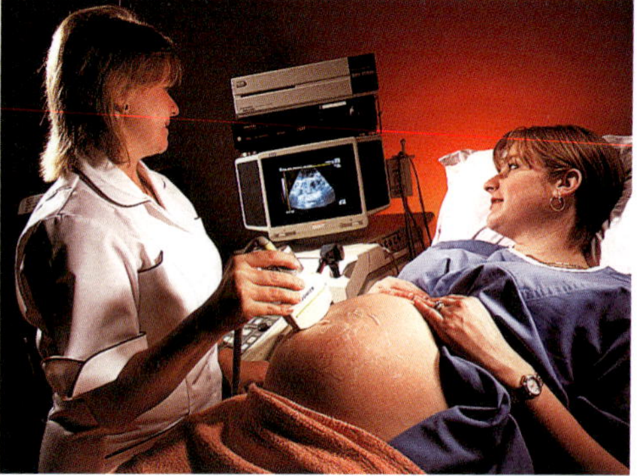

Figure 12.29
Having a pre-natal scan

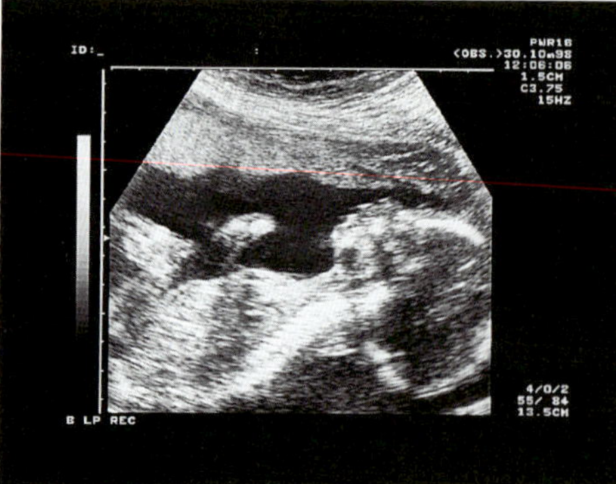

Figure 12.30
Ultrasound fetal image

Topic Questions

1 Copy and complete the following sentences.

a) Ultrasound has a ________________ that is too high for humans to hear.
b) Ultrasound would be _____________________ by a hidden crack in a concrete beam.

2 Name two animals that can hear sounds with a frequency above 60 000 Hz.

3 Give two uses for ultrasonic waves in industry.

4 Name one animal that uses ultrasonic waves to help catch its prey.

5 The diagram shows a trawler using ultrasonic waves to find a shoal of fish.

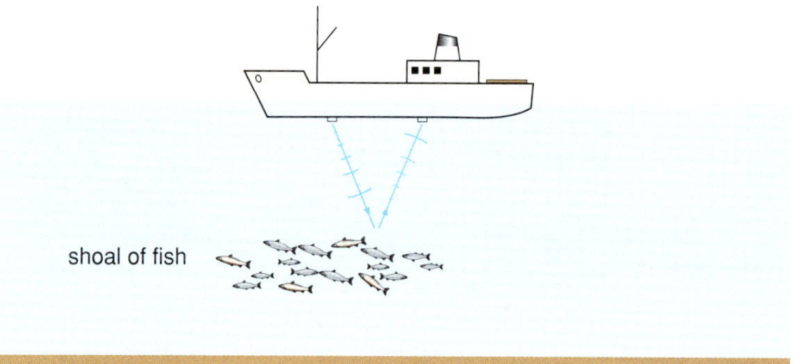

The trawler sends out pulses of ultrasonic waves of frequency of 50 000 Hz and wavelength 0.03 m.

a) What happens to the ultrasonic waves when they hit the shoal of fish?
b) Use the wave equation given in Section 12.1 to calculate the speed of the ultrasonic waves through the water.

12.7	
Co-ordinated	Modular
DA 12.10	DA 12
SA 12.6	SA 18

Electromagnetic waves – 1

Light is just one small part of the family of electromagnetic waves or electromagnetic radiation. Light, together with the other electromagnetic waves, make up the **electromagnetic spectrum.**

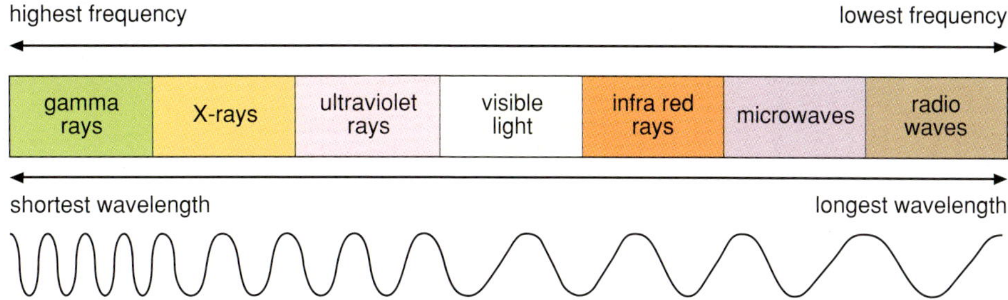

Figure 12.31 *The electromagnetic spectrum*

There are no gaps in the electromagnetic spectrum. It runs smoothly from one wavelength to another. It is a continuous spectrum.

All electromagnetic waves have some things in common:

- They all travel through space (a vacuum) at the same speed (300 000 000 m/s). This means that waves with the lowest frequency must have the longest wavelength.
- They can all be reflected, absorbed or transmitted. But different parts of the spectrum are reflected, absorbed or transmitted by different substances or types of surface. This is because each part of the spectrum has a different range of wavelengths.
- They can all be diffracted (see Section 12.9).

Electromagnetic radiation carries energy. So when electromagnetic radiation is absorbed, the energy it carries is likely to make the absorbing object hotter.

When absorbed, electromagnetic radiation may even produce an alternating current (a.c.) with the same frequency as the radiation.

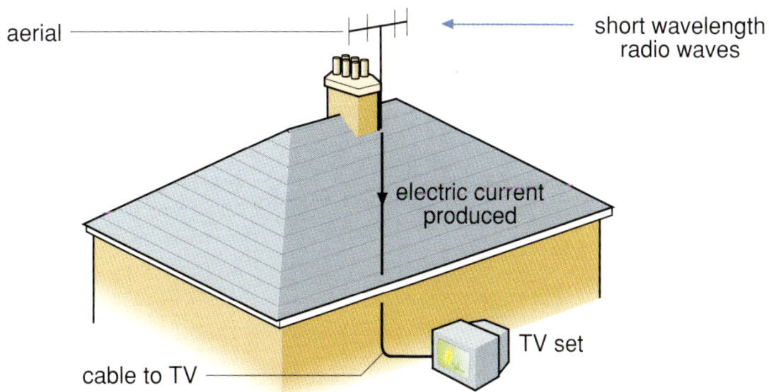

Figure 12.32 *Electromagnetic waves can produce an alternating current*

Radio waves

Radio waves pass through air and non-metallic solids, like brick or glass. But they are absorbed by water.

Figure 12.33
TV programmes are transmitted from telecommunications towers, such as this one

Radio waves are used for communications. Television programmes are transmitted using short wavelength radio waves. These waves go straight from one transmission station to another.

The Earth's atmosphere has an electrically charged layer. This layer reflects long wavelength radio waves. So by using reflection, the waves can be sent around the world, even though the Earth's surface is curved.

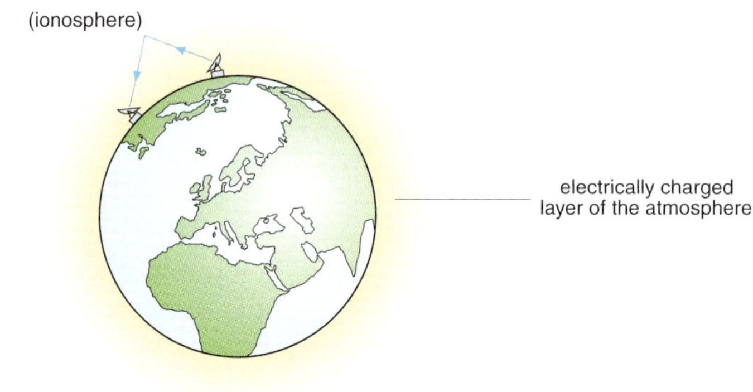

Figure 12.34
Long wavelength radio waves are reflected by the ionosphere

Microwaves

Microwaves are also used for communications.

Microwaves with very short wavelengths (smaller than 1 cm) can pass easily through the Earth's atmosphere. So information carried by microwaves can be sent to and from satellites. Many people now receive some of their television programmes from signals carried by microwaves sent into space then back to Earth.

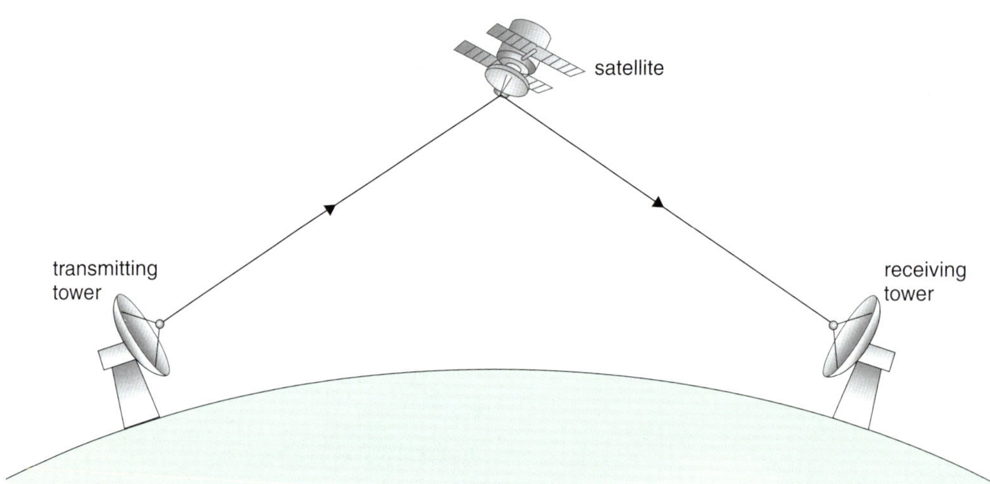

Figure 12.35
Sending a signal using a satellite

Microwaves used by mobile phone networks are sent from place to place via tall aerial masts.

Microwaves can also be used for cooking. Microwaves that pass easily through a container made from plastic, paper or china are strongly absorbed by the water molecules in the food. The energy from the microwaves makes the food hot.

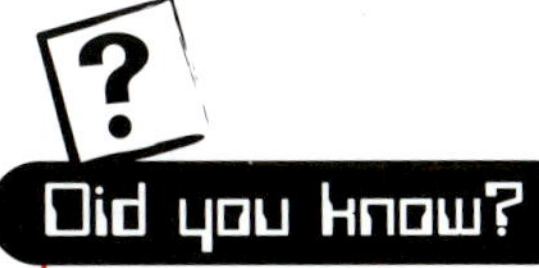

Did you know?

Infra red cameras can detect objects at different temperatures and are used by the Army and police to spot people in the dark. Fire crews use infra red cameras to find people in smoke filled buildings.

Infra red radiation

All objects give out (emit) infra red radiation. Objects that take in (absorb) infra red become hotter. It is the infra red radiation given out by a toaster or grill that is used to cook food.

The remote control for a TV or video works by pulses of infra red radiation being sent from the handset to a detector.

Information can be carried down an optical fibre (see Section 12.4) using pulses of infra red radiation.

Topic Questions

1 Copy and complete the following sentence by choosing the correct phrase.

The speed of microwaves through a vacuum is *faster than / the same as / slower than* the speed of light through a vacuum.

2 Which parts of the electromagnetic spectrum have a shorter wavelength than ultra violet?

3 Which part of the electromagnetic spectrum:

a) is used to cook food under an electric grill?
b) passes through brick but is absorbed by water?
c) carries TV signals from satellites to Earth?
d) is used by the remote control of a video?

4 Draw a diagram to show the path taken by a pulse of infra red radiation passing through an optical fibre.

12.8 Electromagnetic waves – 2

Co-ordinated	Modular
DA 12.10	DA 12
SA 12.6	SA 18

Ultra violet radiation

Ultra violet radiation is given out by the Sun. The ozone in the Earth's atmosphere absorbs most of this. The ultra violet that passes through the atmosphere is what causes sun tanning.

Ultra violet radiation is dangerous. The skin absorbs the energy carried by the ultra violet waves. This will damage the skin cells and may cause skin cancer. So it is important that sun creams are used to protect the skin from this radiation. It is also important that a sun bed is not used for longer than the recommended time. Ultra violet can also damage your eyes. So someone using a sun bed should wear special glasses.

Some substances absorb the energy from ultra violet radiation then give the energy out as light. This is called fluorescence. The inside of a fluorescent lamp is coated with a chemical that absorbs ultra violet and then gives out light.

Some washing powders contain chemicals that cause fluorescence. In a disco, clothes washed with these powders glow brightly. The ultra violet from the disco lights is absorbed by the clothes and light is given out.

Figure 12.36
A fluorescent lamp gives out more light and less heat than an ordinary light bulb

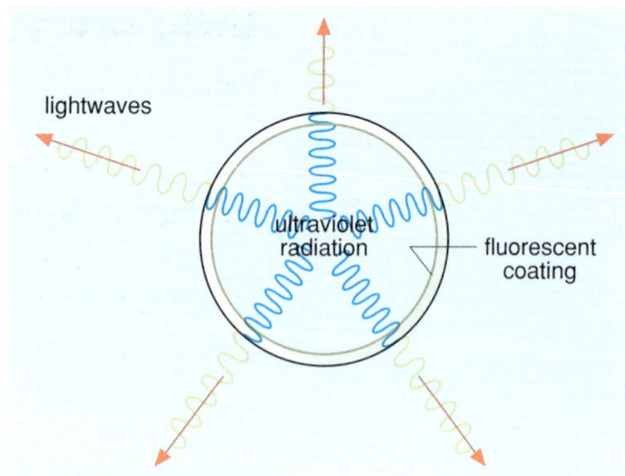

Figure 12.37
Ultra violet is absorbed and given out as light

Ultra violet radiation is also used for security markings. Numbers or names written using special inks are invisible in light but show up in ultra violet radiation.

X-rays

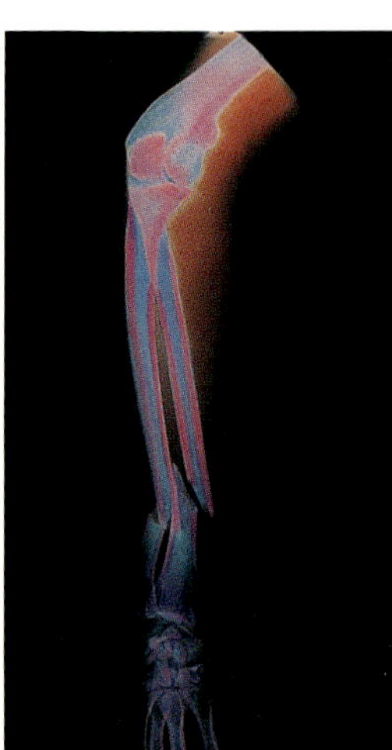

Figure 12.38
An X-ray of a broken arm

X-rays pass through skin and healthy flesh. Bones and diseased tissue absorb them. X-rays also cause photographic film to go dark. So an X-ray photograph can be used to show broken bones or diseased tissue.

X-rays are also used in security checks at airports.

Gamma rays

Gamma radiation can kill or damage living cells. This means that gamma radiation can cause cancer and also be used to treat cancer.

Gamma rays also kill bacteria. This means gamma radiation can be used to:

- sterilise medical instruments
- keep food fresh for longer.

Figure 12.39
The bacteria that causes food to go off is killed by gamma radiation

Radiation and cancers

Low level doses of ultra violet radiation, X-rays and gamma radiation can cause cells to become cancerous.

High level doses of X-rays and gamma radiation can be used to kill cancerous cells.

The radiation is aimed at the cancer cells from different directions. This means healthy cells are less likely to be harmed. But the cancer cells are hit each time the radiation is used. So they receive a high dose of radiation and are killed.

Dangers caused by exposure to electromagnetic radiation

The waves of the electromagnetic spectrum have many good uses, but they can also be dangerous. Many of the dangers are due to the effect that the radiation has on living cells. Figure 12.41 gives some of the dangers caused by exposure to the different radiations and what can be done to lower the risks.

Figure 12.40 *Dangers of exposure to different radiations*

Radiation	Effect
microwave	These are absorbed by water in cells. The heat produced can kill the cells.
infra red	These are absorbed by skin and felt as heat. They cause sunburn.
ultra violet	These can pass through the skin to reach deeper tissue. Exposure to too much ultra violet can cause skin cancer. People with a light skin colour are at greater risk than people with a dark skin colour. The darker the skin the more ultra violet it absorbs, so less reaches the deeper tissues. You should use 'sun block' creams to reduce the amount of radiation going through your skin.
X-rays and gamma radiation	Most of these pass through healthy tissue but some energy is absorbed which could cause cancer. People who use machines that produce X-rays wear special clothing or use lead shields to protect themselves from high doses of the radiation. People who work with sources of gamma or other types of radiation wear special badges that measure the levels of radiation the person has been exposed to (see Section 12.14).

Topic Questions

1 Copy and complete the following sentences.

a) Ultra violet radiation has a shorter ____________ than visible light.
b) A high ________ of gamma radiation can be used to kill __________ cells.
c) Infra red radiation makes skin feel ___________ .

2 What are the properties of X-rays that make them so useful?

3 Why is it important that microwaves cannot escape from a microwave oven?

4 Gamma radiation is used to sterilize medical dressings. What does the gamma radiation do to the bacteria on the dressings?

5 Before sunbathing it's a good idea to apply sun cream to your exposed skin. Why is this?

6 What precautions should a radiographer (person who takes X-ray photographs) take before using an X-ray machine?

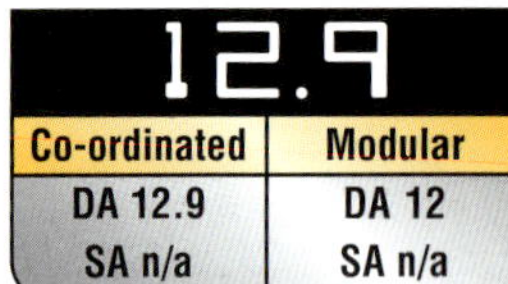

Waves and diffraction

Figure 12.41 shows water waves spreading out after they pass through a gap in a barrier. When the gap is about the same size as the wavelength of the waves, the waves spread out a lot. When the gap is a lot wider than the wavelength of the waves, the waves only spread out a little. This spreading out of a wave as it passes through a gap is called **diffraction**.

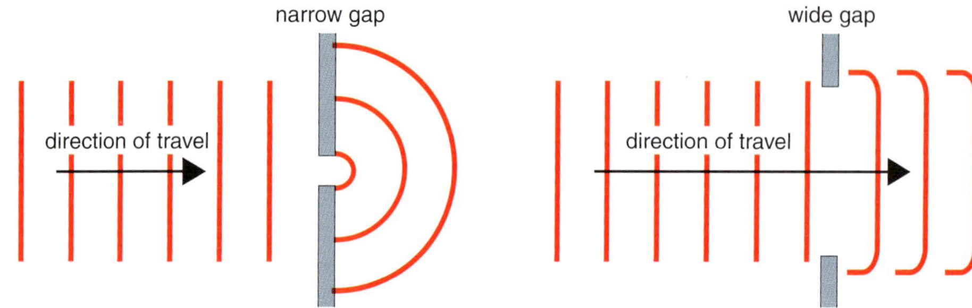

Figure 12.41
Diffraction of waves at a narrow and at a wide gap

Diffraction also happens when waves pass the edge of an object. How much the waves spread out depends on the size of the object and the wavelength of the waves.

Electromagnetic radiation and sound can also be diffracted. This is evidence for the idea that electromagnetic radiation and sound travel as waves. But diffraction effects are not always easy to notice. Light with its very short wavelength (0.000 000 5 metres) needs to pass through a very small gap before diffraction effects are noticed.

Diffraction of radio waves

Large buildings and hills strongly diffract long wavelength radio waves. This means that someone living in the 'shadow' of a hill can still receive long wave radio programmes.

Very high frequency (VHF) radio signals and TV signals have very short wavelengths. This means they do not diffract and spread out strongly. So someone living in the 'shadow' of a hill may not be able to receive short wave radio or TV programmes.

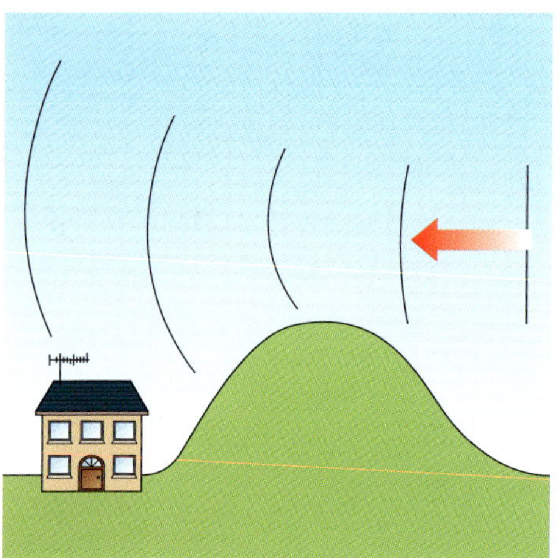

Figure 12.42
Diffraction of long wave radio waves

Diffraction of sound waves

Sound waves are diffracted and spread out as they go through an open doorway or pass the corner of a building. So someone outside a room may hear people speaking inside the room even though they cannot see the people.

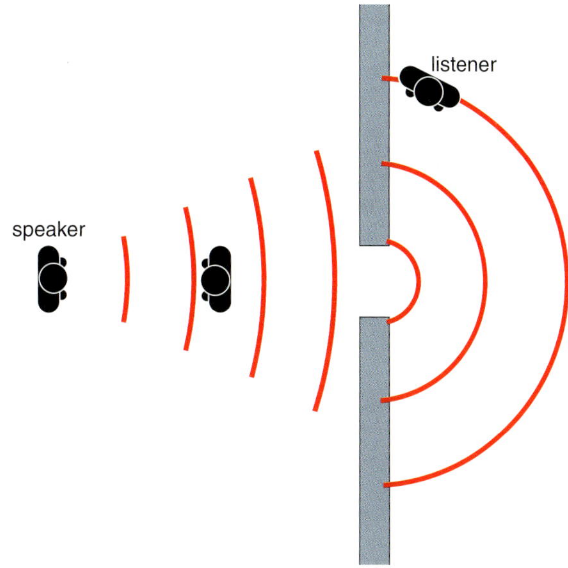

Figure 12.43 *Diffraction of sound waves through a doorway*

Like radio waves, sound waves with longer wavelengths are more strongly diffracted than sound waves with shorter wavelengths.

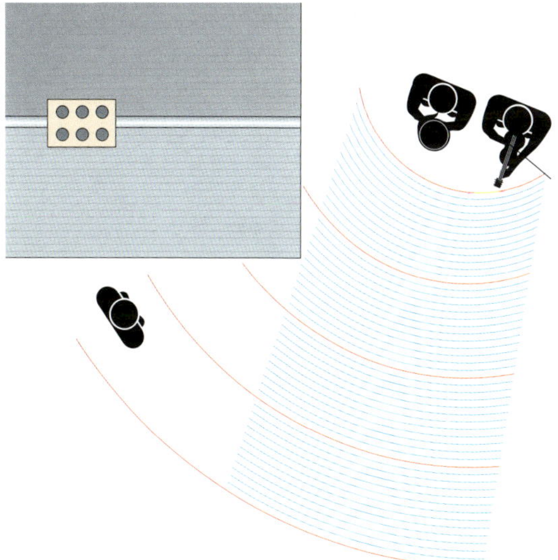

Figure 12.44 *Waves of different wavelength are diffracted different amounts*

Someone standing at the corner of a building can hear some outdoor musicians. The tuba (long wavelength sound) is very clear, but the violin is very faint (short wavelength sound).

Topic Questions

1 Copy and complete the following sentence.

___________ happens when waves pass through a ___________ or pass the edge of an object.

2 Three sounds, A, B and C, each having a different frequency pass through an open window.

A = 1800 Hz B = 256 Hz C = 2500 Hz

Which of the three sounds will spread out the most? Give a reason for your answer.

3 Explain why sounds can easily spread around the corner of a building.

4 The diagram shows water waves moving towards a harbour wall.

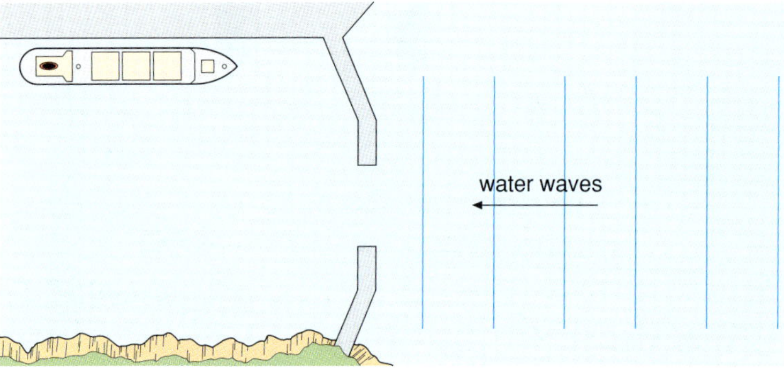

Copy and complete the diagram to show the water waves after they pass through the gap in the harbour wall.

5 The diagram shows what happens to radio waves as they pass a hill.

a) What word describes what happens to the waves as they pass over the hill?
b) Are they long or short wavelength radio waves? Give a reason for your answer.

12.10	
Co-ordinated	Modular
DA 12.10	DA 12
SA 12.6	SA 18

Sending information

The telephone, invented in 1876, is often used to send information over long distances.

When you speak, you produce sound waves. A microphone turns the sound waves into electrical signals. The amplitude and frequency of the electrical signals change as the amplitude and frequency of the sound waves change. These signals, called **analogue signals** are carried through cables to the telephone receiver.

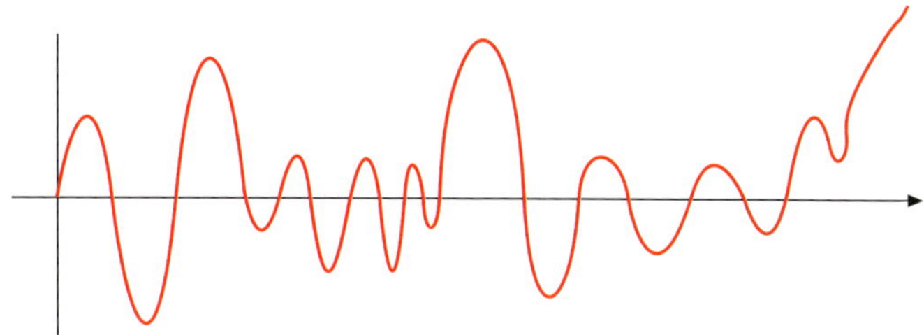

Figure 12.45 *Analogue signal produced by a microphone*

The earphone in the receiver is like a small loudspeaker. It changes the electrical signals back into sound.

Microwaves can also carry information. These signals may be sent to satellites in orbit around the Earth. The signals are then sent back to Earth by the satellite (see Section 11.10).

Information like speech or music, which is carried by sound waves, can also be changed into light or infra red signals. These signals are then sent along optic fibres (see Section 12.4). But before the information can be sent it must be changed from its analogue form into a series of coded pulses. This type of signal is called a **digital signal**.

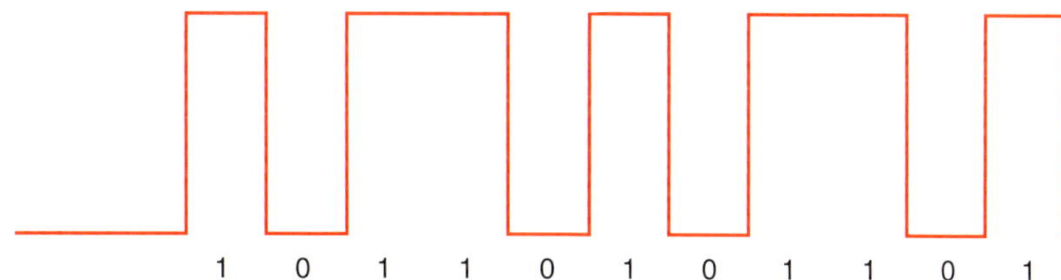

Figure 12.46 *A digital signal is either 'on' or 'off'*

A digital signal can only be 'on' or 'off'. These are called the states of a digital signal. The number 1 is used to represent the 'on' state. The number 0 is used to represent the 'off' state. This means any digital signal can be written as a series of 1s and 0s.

So information can be sent using either digital or analogue signals. But using digital signals has two main advantages over using analogue signals:

- Information carried as a digital signal does not get changed during transmission. The signal that is received is the same as the signal that was sent out. This means the received signals are high quality. When an analogue signal is transmitted it loses quality.
- More information can be sent (in a given time) using digital signals than using analogue signals.

Did you know?

Figure 12.47

The telegraph was the first modern invention that allowed information to be sent using an electric current. The signal was sent using a series of current pulses, according to an agreed code. Often this was the Morse code, where each letter of the alphabet is represented by a series of short and long pulses.

Topic Questions

1 The diagram shows two signals. One is analogue and one is digital.

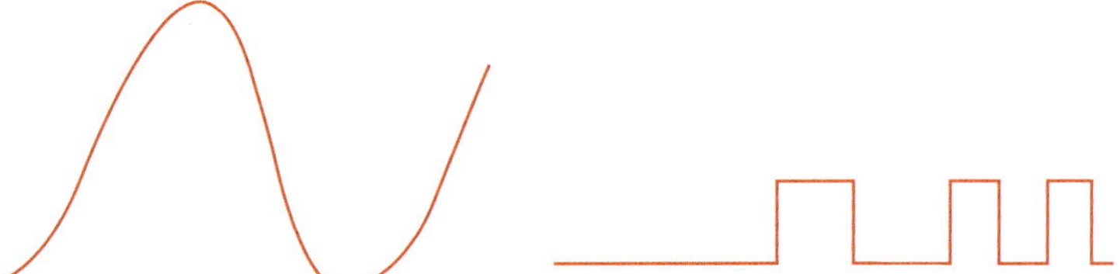

a) Copy the two signals. Label each signal with its correct name.
b) What are the main differences between an analogue and digital signal?
c) Give one advantage of sending a digital signal rather than an analogue signal.

2 Telephone companies are replacing underground copper cables with optic fibre cable.

a) What type of signal is sent along an optic fibre?
b) Which two of the following are used to carry information along an optic fibre – infra red, microwaves, light, or ultraviolet?
c) Give one advantage of changing from copper cables to optic fibre cables.

3 Find out when the first television programmes were broadcast using digital signals.

12.11	
Co-ordinated	Modular
DA 12.12 SA n/a	DA 12 SA n/a

Seismic waves

Earthquakes are caused by underground rock movements. They can release huge amounts of energy, which are carried through the Earth by a series of shockwaves or **seismic waves**. These shockwaves can cause enormous damage to buildings.

Figure 12.48
Damage caused by an earthquake

The vibrations caused by the seismic waves are detected and recorded using a seismograph. In a simple seismograph a pen draws a line on a rotating drum. Seismic waves make the drum vibrate up and down. So the pens draws a series of waveshapes.

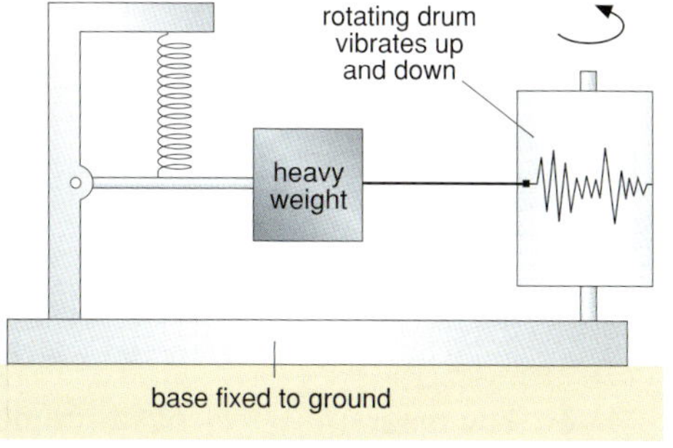

Figure 12.49
A simple seismograph

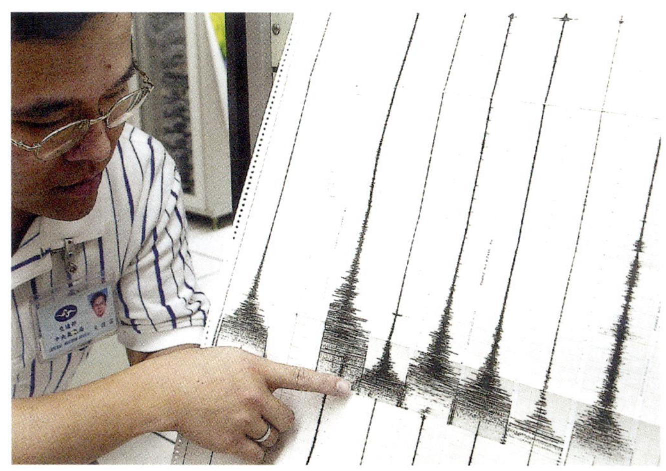

Figure 12.50
Earthquakes recorded on a seismograph

Seismic waves have given scientists evidence that the Earth is made up of three layers – the core, mantle and crust.

Figure 12.51 *The structure of the Earth*

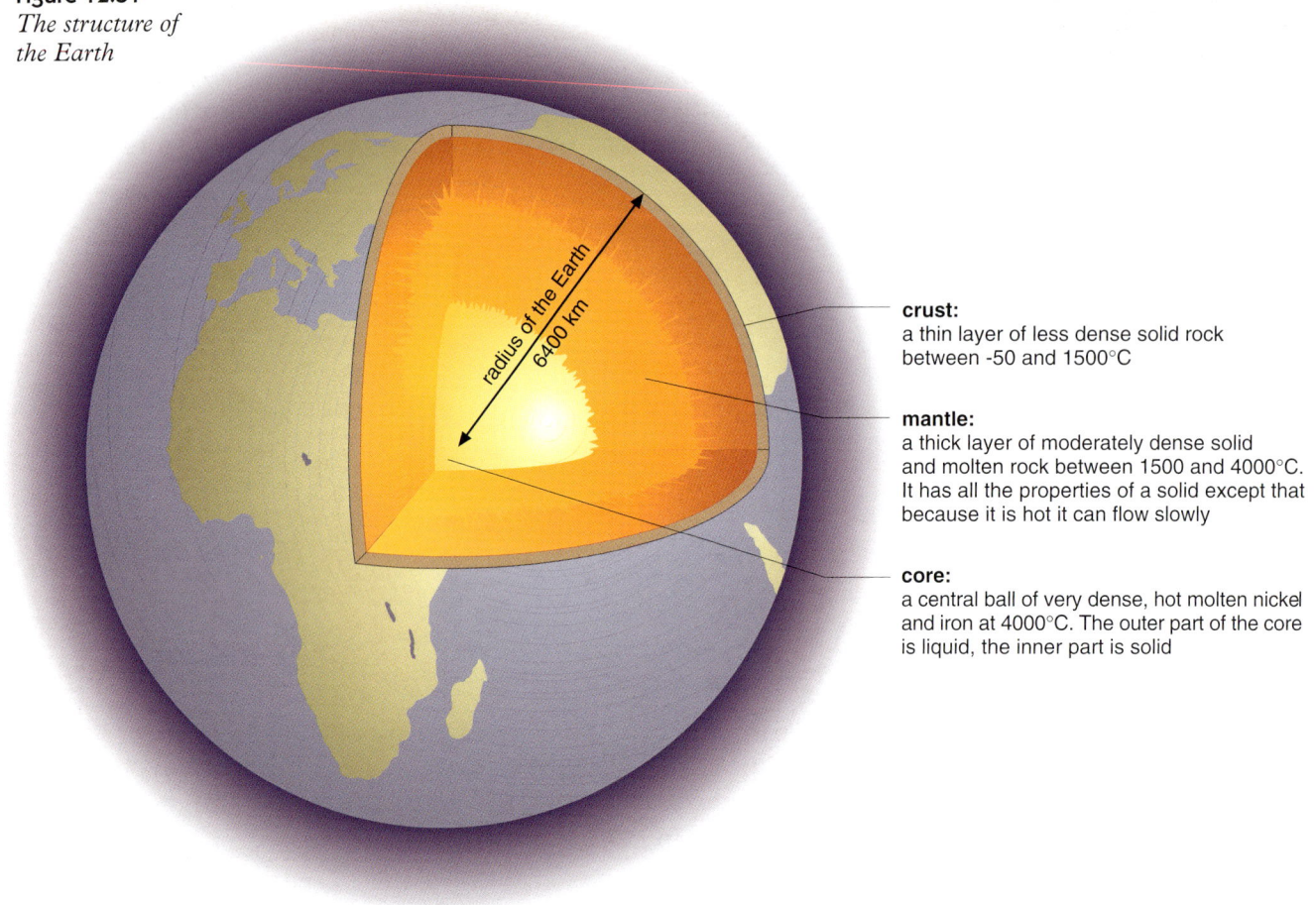

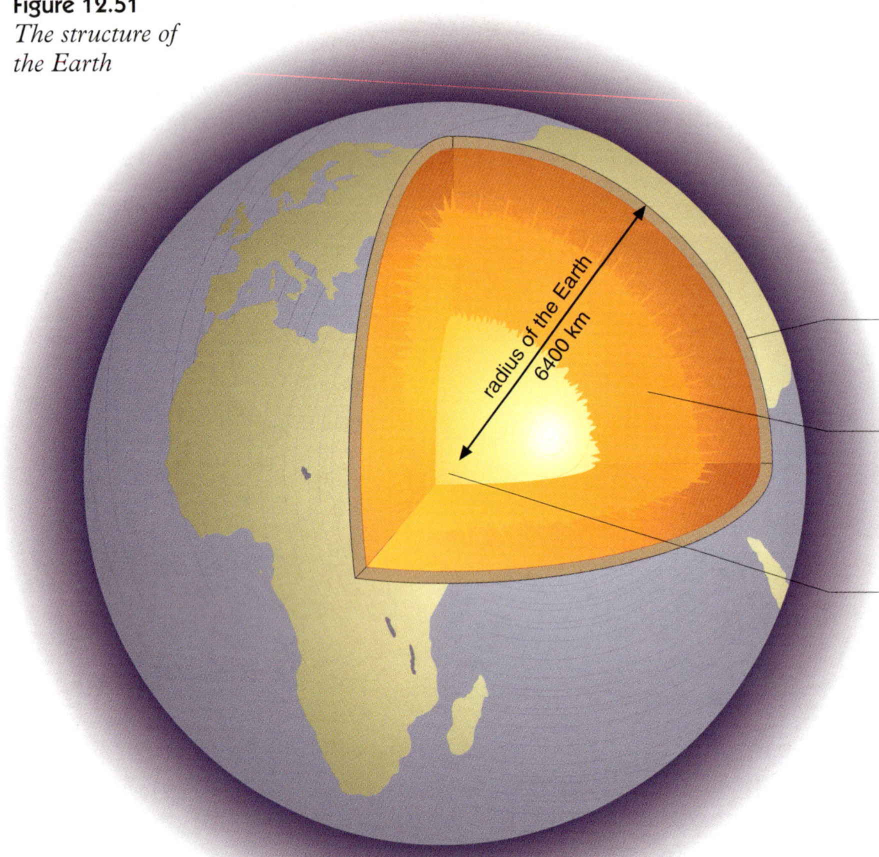

radius of the Earth
6400 km
crust:
a thin layer of less dense solid rock between -50 and 1500°C
mantle:
a thick layer of moderately dense solid and molten rock between 1500 and 4000°C. It has all the properties of a solid except that because it is hot it can flow slowly
core:
a central ball of very dense, hot molten nickel and iron at 4000°C. The outer part of the core is liquid, the inner part is solid

Did you know?

The speed of seismic waves varies from about 4 km/s for waves near the Earth's surface to about 15 km/s for waves deep in the Earth.

Did you know?

Seismic surveys can be used to search for oil. Small explosions send shock waves into the ground. Using the reflected shockwaves scientists can work out the likely position of an oil deposit.

Topic Questions

1 Copy and complete the following sentences.

a) Seismic waves are recorded on a ____________________ .
b) Seismic waves transfer ____________ through the Earth.

2 The diagram represents a simplified section through the Earth. Name the parts labelled A, B and C.

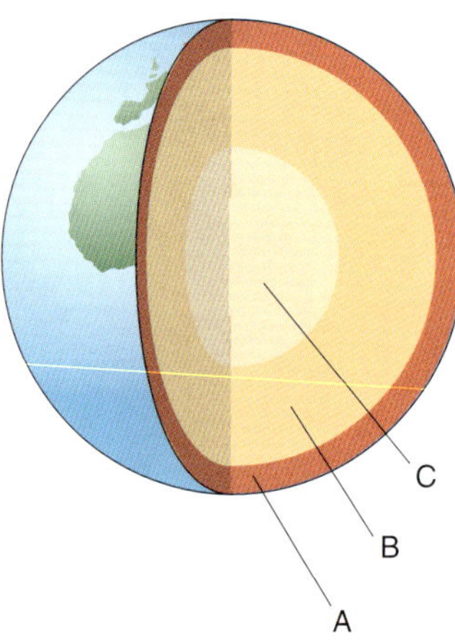

3 What causes an earthquake?

12.12	
Co-ordinated	Modular
DA 12.23	DA 12
SA 12.15	SA 18

Parts of the atom

Everything is made up of one or more elements. All elements are made of atoms. All atoms (except hydrogen atoms) contain three types of particle – electrons, **neutrons** and **protons**.

The protons and neutrons are found at the centre of an atom, in the nucleus. Protons have a positive (+) charge. Neutrons have no charge.

The electrons orbit the nucleus. Each electron has a negative (−) charge equal in size to the positive charge on a proton. The total number of electrons and protons inside an atom is the same. This means that the total charge of a single atom is zero.

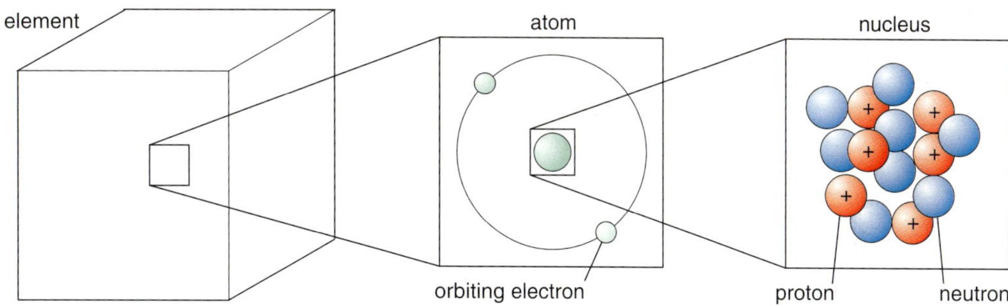

Figure 12.52
Atomic structure

Protons and neutrons are tiny particles. But even so, they both have a mass about 2000 times bigger than an electron. As the masses are so small it is easier to compare them using a simple scale.

Particle	Mass	Charge
proton	1	+1
neutron	1	0
electron	negligible	−1

All atoms of the same element have the same number of protons in the nucleus. A different element will have a different number of protons in the nucleus.

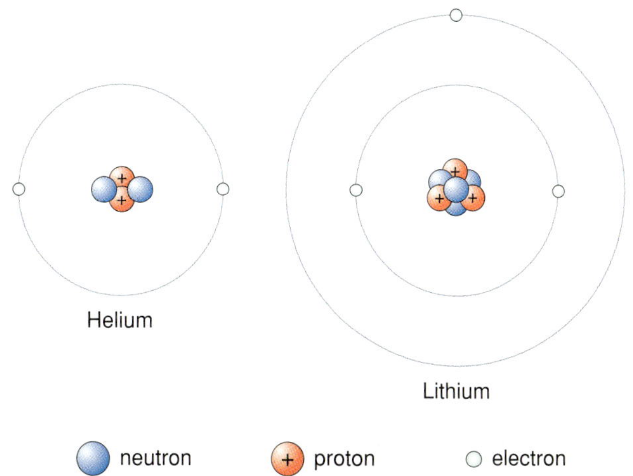

Figure 12.53
Showing the structure of a helium atom and a lithium atom

Figure 12.54
Atoms are difficult to draw to scale. If the nucleus was as big as a football, the electrons would be outside the football ground

Both protons and neutrons can be called nucleons. The total number of nucleons in an atom is called the **mass number** or nucleon number.

Element	Number of protons	Number of neutrons	Mass number	Symbol
helium	2	2	4	He
lithium	3	4	7	Li

Atoms of the same element always have the same number of protons. But they do not always have the same number of neutrons. These different atoms of the same element are called **isotopes**.

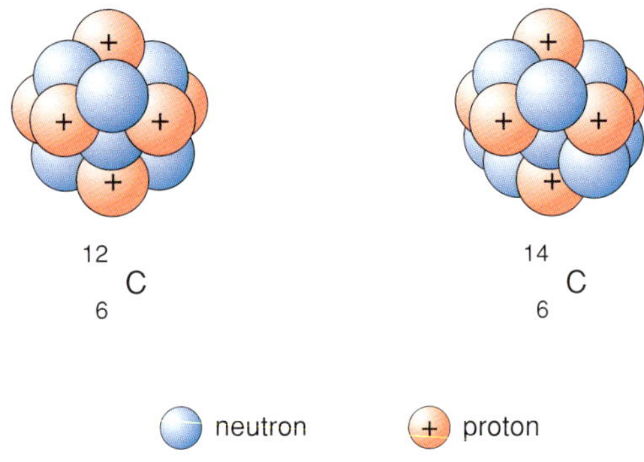

Figure 12.55
The nuclei of two carbon isotopes

Topic Questions

1 Using words from the box, copy and complete the following sentences. Each word can be used once or not at all.

electron	isotope	neutron	nucleon	proton

a) Atoms of different elements have different numbers of ____________ in the nucleus.
b) Atoms of the same element with different numbers of neutrons are called _______.
c) A proton or a neutron can be called a _______________ .

2 Which part of an atom is described by each of the following statements?

a) It contains both protons and neutrons.
b) It orbits the nucleus.
c) It is uncharged.

3 Two isotopes of chlorine have mass numbers 35 and 37. How are the two isotopes different?

4 An isotope of calcium (Ca) has 20 protons and 24 neutrons. Write this information in symbol form.

5 A sodium atom has 11 electrons, 11 protons and 12 neutrons.

a) What is the mass number of the sodium atom?
b) Why is the total charge of a single sodium atom zero?

6 How is a hydrogen atom different to the atoms of all other elements.

12.13 Why do some atoms produce radiation?

Co-ordinated	Modular
DA 12.22	DA 12
12.23	
SA 12.14	SA 18
12.15	

Most atoms have a stable nucleus. This means that the numbers of protons and neutrons in the nucleus do not change.

Some atoms have an unstable nucleus. Atoms with an unstable nucleus are called **radioactive isotopes** (or radioisotopes or radionuclides). Radiation is produced when an unstable nucleus splits, giving out a particle or ray (or both). This process is called radioactivity.

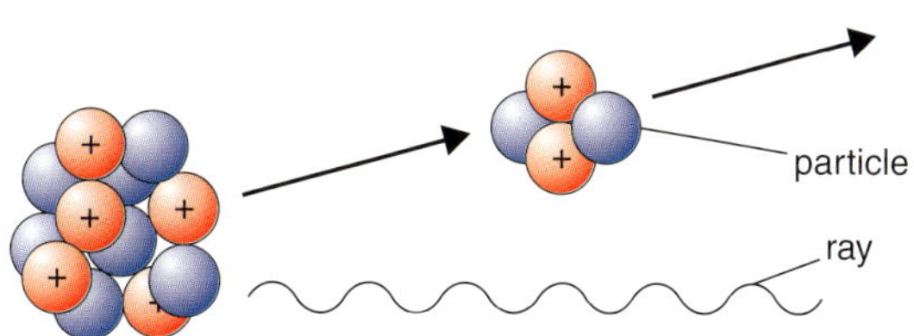

Figure 12.56 *Unstable nucleus emitting a particle and a ray*

When a particle is given out the unstable nucleus changes. The number of protons in the new nucleus is different. This means that an atom of another element has been made. This process is called **radioactive decay**.

Figure 12.58 shows how the nucleus of a carbon-14 atom changes when it gives out a radioactive particle.

Figure 12.57

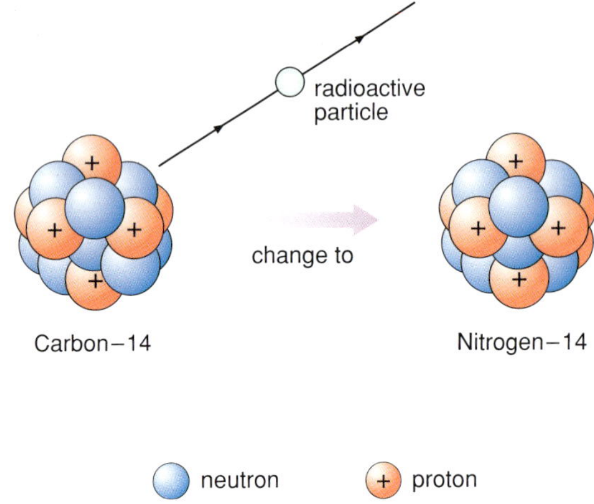

Substances that give out radiation are called radioactive. Nothing can change the radiaoactivity of the substance.

Types of radiation

There are three types of radiation that can be given out (emitted) by a radioactive substance. They are called alpha (α), beta (β) and gamma (γ).

Alpha (α) radiation is easily absorbed by a thin sheet of paper or by a few centimetres of air.

Beta (β) radiation will pass through air and paper, but it is absorbed by a few millimetres of metal.

Gamma (γ) radiation can go through anything, it is never completely absorbed. However, several centimetres of lead or metres of concrete will absorb most of it.

Figure 12.58
Absorption of radiation

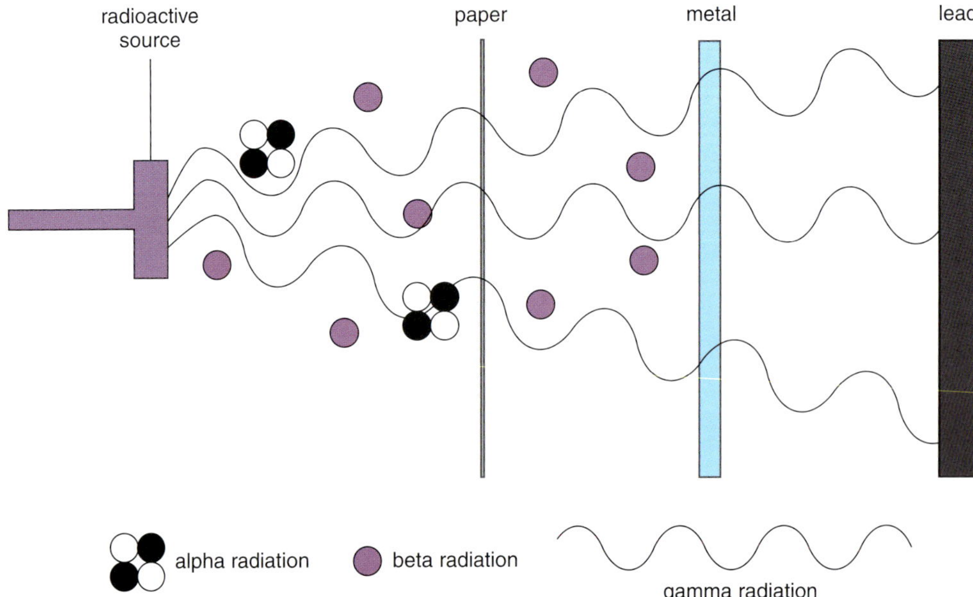

Figure 12.59
Radioactive material is transported in this thick metal container

Background radiation

Radioactive substances are all around us. The air we breathe, the food we eat, the rocks around us are all sources of radiation. These radiations are called **background radiation**. The level of background radiation in most places is so low that it is not a health risk.

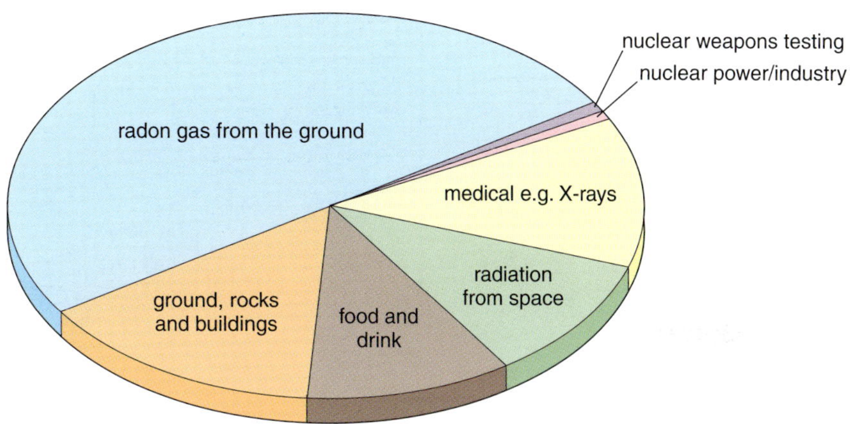

Figure 12.60
Sources of background radiation in the UK

Half-life

The average number of particles or rays given out by a radioactive material in one second is called the count rate. As the unstable nuclei decay the count rate goes down. Eventually, the count rate will have gone down to half what it was to start with. This happens when only half the unstable atoms in the original material are left. How long it takes for this to happen is called the half-life.

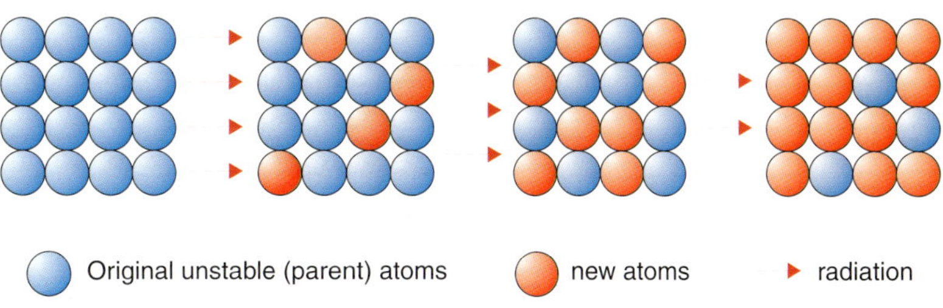

Figure 12.61
Unstable atoms decay to form new atoms

The graph in Figure 12.62 shows how the count rate for one type of radioactive material changes as the material gets older. Every half-life, the count rate goes down by half. All radioactive materials give the same shape graph. But all of them have a different half-life.

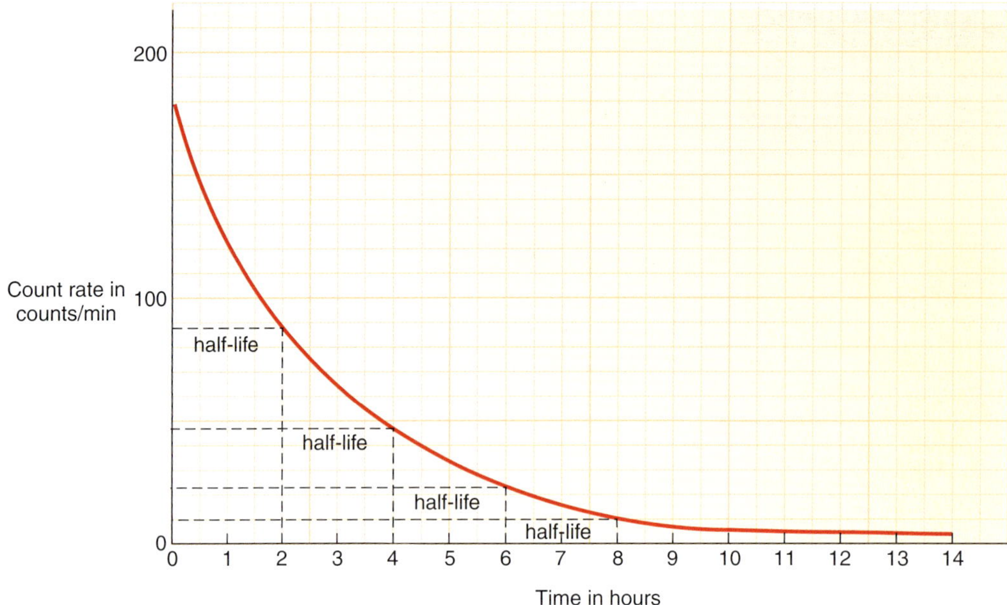

Figure 12.62
This material has a half-life of 2 hours

Topic Questions

1 Copy and complete the sentences, using some of the words from the box.

different	nucleus	particle	stable	unstable

a) Radiation comes from an _______________ nucleus.
b) When a ________________ leaves a nucleus, the ________________ changes to that of a __________________ atom.

2 Which type of radiation, alpha (α), beta (β) or gamma (γ):
a) is only stopped by several centimetres of lead or several metres of concrete?
b) can only travel through a few centimetres of air?
c) is absorbed by a thin sheet of paper?

3 a) What is the radiation that is around us all of the time called?
b) What is the biggest source of this radiation?

4 The table shows the count rate for a radioisotope.

Count rate (in counts /sec)	Time (in days)
8000	0
4000	2
2000	4

a) What is the half-life of this radioisotope?
b) What will be the count rate after 8 days?

12.14	
Co-ordinated	Modular
DA 12.22	DA 12
SA 12.14	SA 18

Dangers of radioactivity

Radioactive particles and rays can pass into and sometimes pass through substances. The radiation has enough energy that it can remove electrons from atoms of the substance. By removing one or more electrons, the radiation has changed the atom from being neutral to having an overall charge. The radiation has ionised the atom (which should now be called an ion), causing the substance to change.

It is because radiation can ionise atoms that it is dangerous to living cells. Radiation absorbed by our bodies can change the chemical structure of cells so that they do not work properly. In some cases, the individual cells are so changed they affect normal cells and cause cancers. The higher the level (or dose) of radiation absorbed, the higher the risk of cell damage and cancer.

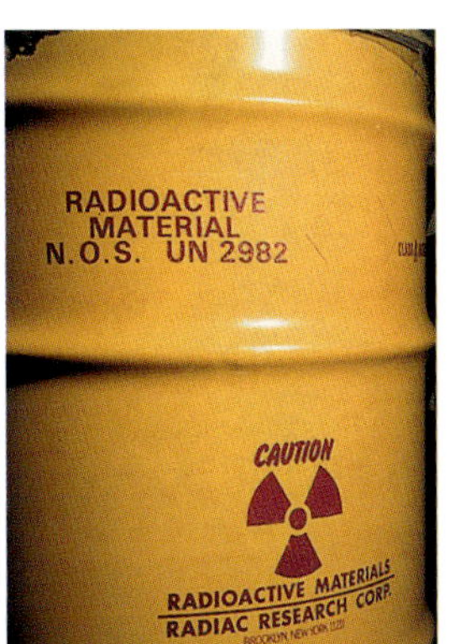

Figure 12.63
Radioactive material must be properly stored. Any radiation passing through the container would be a danger to health

Did you know?

Henri Becquerel discovered the radioactivity of uranium in 1896. He was awarded the Nobel Prize for Physics in 1903 jointly with Pierre and Marie Curie. The Curies isolated the radioisotopes radium and polonium from uranium ore. Marie Curie died from leukaemia, a disease caused by the radiaoactive substances she worked with.

Figure 12.64
Henri Becquerel; Pierre and Marie Curie

When sources of radiation are outside the body

Alpha radiation is the least dangerous to the body. Clothing or the first few layers of exposed skin will easily stop it. Beta and gamma radiations are more dangerous.
Both radiations can go through the skin. This means they can reach and damage the cells of internal organs.

People who work with radioactive substances wear a special photographic film badge. Every few weeks the film is developed. This is done to check that the badge wearer has not been exposed to too much radiation. The more the wearer has been exposed to radiation, the darker the film will be after being developed.

Figure 12.65
A radiation-detecting badge

Did you know?

The Chernobyl nuclear reactor went out of control in April 1986. Eventually it caught fire. The remains are still very radioactive and will be for a long time yet. A thick concrete shield over and under the reactor tries to stop radiation getting into the air. Despite the dangers, engineers continue to check the levels of radioactivity and the safety of this concrete 'tomb'.

Figure 12.66

When sources of radiation are inside the body

Once inside our body, alpha radiation is the most dangerous. Because it is so strongly absorbed, it will ionise and damage cells easily. Beta and gamma radiations are less dangerous since cells are less likely to absorb either type of radiation.

Did you know?

The dust in uranium mines has a hidden danger. The dust contains atoms that give out alpha radiation. If the dust is breathed into the lungs, the alpha radiation is absorbed and will destroy living tissue.

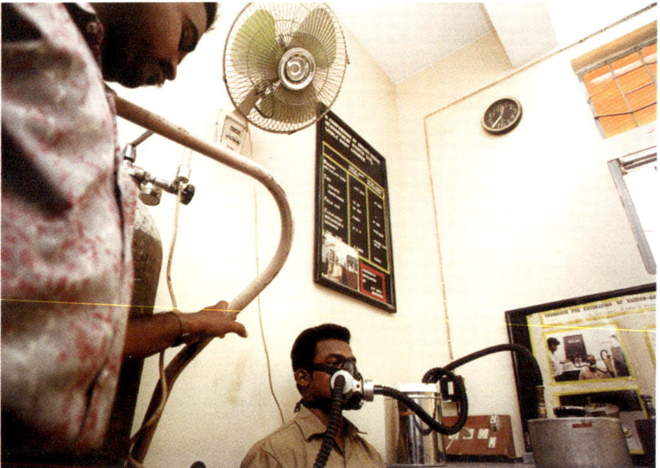

Figure 12.67
Uranium miner being tested for uranium contamination

Topic Questions

1 Which type or types of radiation, alpha, beta or gamma:

   a) is very dangerous when the source of radiation is outside the body?
   b) is most dangerous when breathed into the lungs?

2 Which of the following is the correct name for an atom that has lost an electron – electrode, ion, or molecule?

3 Why is radiation a danger to living cells?

4 A radiation worker keeps their film badge in a pocket. Why is this not a good idea?

5 Many years ago, the workers who painted the dials of watches with luminous paint did not often live into old age. Find out why.

12.15

Co-ordinated	Modular
DA 12.22 12.23 SA n/a	DA 12 SA n/a

Uses of radioactivity

Killing harmful bacteria

Harmful bacteria in fresh food can be killed by gamma radiation. This helps to keep the food fresh for longer. Bacteria on hospital instruments and dressings can also be killed using gamma radiation (see Section 12.8).

Treating cancer

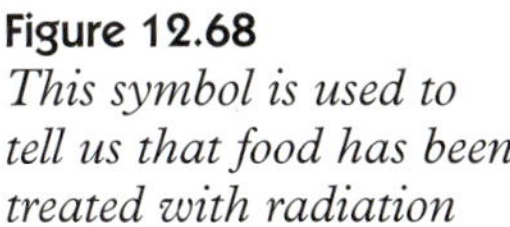

Figure 12.68
This symbol is used to tell us that food has been treated with radiation

Radiation can be used to kill cancer cells. Large doses of gamma radiation are carefully given from a source outside the patient's body. The gamma radiation is aimed so that the cancer cells receive the most radiation and the nearby healthy cells very little (see Section 12.8).

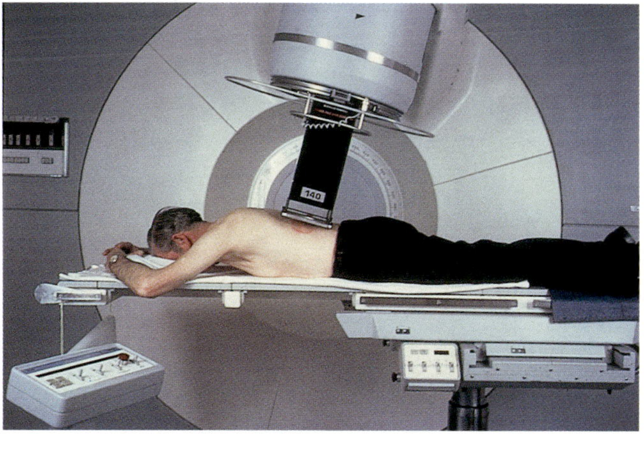

Figure 12.69
A patient having radiation treatment

For some cancers, the radiation must come from a source inside the patient's body. In this case the radiation used is usually beta. This is so that the radiation does not penetrate too far into the patient's body.

Thickness control

A thick sheet of material will absorb more radiation than a thin sheet of material. So radiation can be used to control the thickness of a material as it is being made.

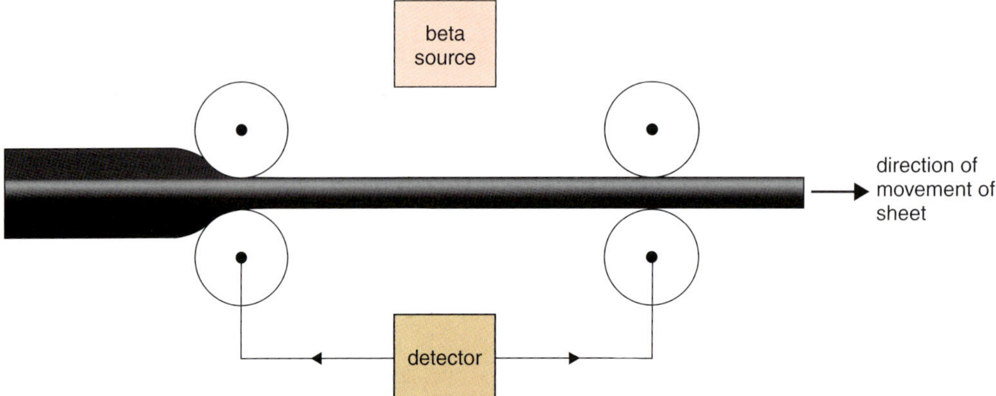

Figure 12.70 *Using a beta source to control the thickness of aluminium sheet*

If the aluminium sheet in Figure 12.70 is too thick, not enough beta radiation will get to the detector. A signal from the control box will make the rollers squeeze harder so the aluminium will be rolled out thinner. By constantly measuring the amount of radiation getting through to the detector the aluminium can be kept at the right thickness.

Dating materials

Living organisms are constantly taking in carbon from the atmosphere. A tiny part of this carbon is the radioactive isotope carbon-14. But the carbon-14 already in the organism also decays. So while the organism is alive the proportion of carbon-14 in the organism stays the same. When the organism dies, the amount of carbon-14 it has in its body starts to go down. So the longer the organism has been dead, the smaller the proportion of carbon-14 left. By measuring the count-rate from the carbon-14 for a sample from a dead organism, the time since the death of the organism can be worked out. This is called radiocarbon dating.

Figure 12.71
The age of Lindow Man was found by radiocarbon dating

The age of some types of rock can be found by comparing the amounts of certain isotopes in the rock.

Example: An atom of potassium-40 changes by radioactive decay into an argon atom. Potassium-40 has a half-life of 1.3 billion years. How old is a rock that initially contained no argon atoms but now contains equal numbers of potassium-40 and argon atoms?

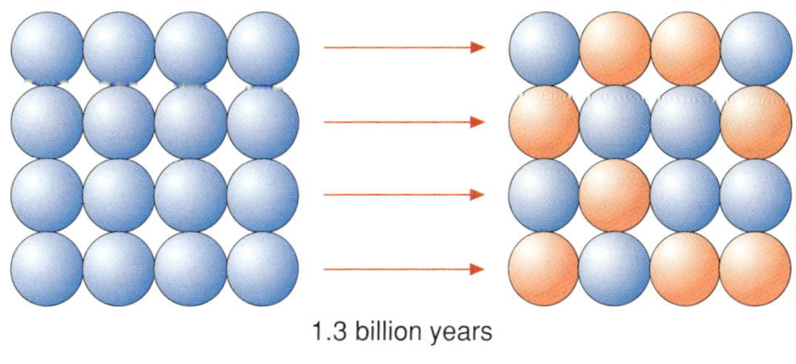

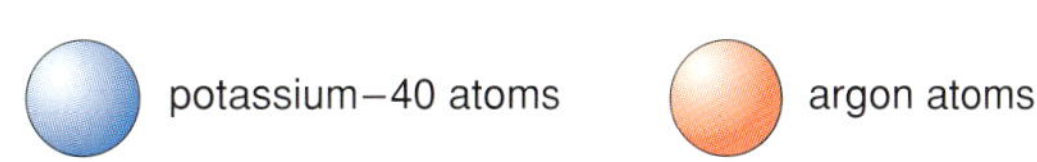

Figure 12.72
Decay of potassium-40 into argon

The rock must be about 1.3 billion years old.

Most smoke alarms contain a weak radioactive source that gives out alpha radiation.

Without the radioactive source the alarm would not work.

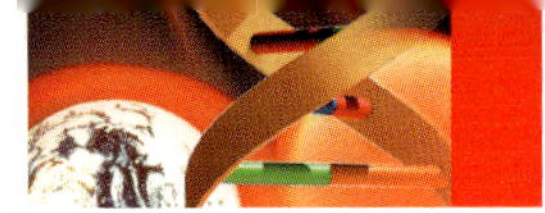

Topic Questions

1 Copy and complete the following sentences.

a) When radiation is used to treat cancer, doctors can use a ____________ radiation source outside the patient's body or a ____________ radiation source inside the patient's body.
b) Using the decay of carbon-14 to find the age of an ancient __________ axe handle is called ________________ dating.

2 Use the information in Section 12.13 to answer this question. Which type of radiation, alpha (α), beta (β) or gamma (γ), should be used to monitor, as it is being produced, the thickness of:

a) plastic clingfilm?
b) aluminium cooking foil?
c) lead roofing sheets?

Give a reason for each of your choices.

3 A firm makes plastic syringes for a hospital. The hospital says that the syringes must have no bacteria on them (they must be sterile). Describe what the firm should do to make sure the syringes are sterile.

4 The half-life of carbon-14 is about 6000 years. The count-rate from the carbon in a freshly cut piece of wood is 16 counts per minute. The count-rate from an ancient piece of wood is 4 counts per minute. How old is the ancient piece of wood?

12.16 Models of the atom

Co-ordinated	Modular
DA 12.23	DA 12
SA 12.15	SA 18

Modern ideas about the existence of atoms and descriptions of their structure (called a model) started with John Dalton. In 1803, he suggested that all elements were made up of tiny particles called atoms. These particles, he said, could not be divided into anything smaller.

The 'plum pudding' model

In 1897, J J Thomson discovered a particle much smaller than an atom. This particle was called an electron. Thomson knew that the electrons had a negative charge so the rest of the atom should have a positive charge. If it didn't the atom would fly apart.

In 1904, he described what is called the 'plum pudding' model of the atom. The pudding was the sphere of positively charged material. The negatively charged electrons were the bits of plum, all mixed into the pudding.

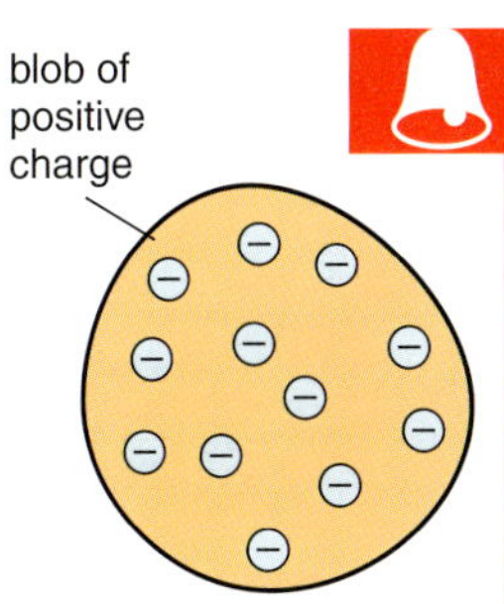

Figure 12.73
J J Thomson's 'plum pudding' model of the atom

The 'nuclear' model

By 1910, Ernest Rutherford had begun to think that the 'plum pudding' model was too simple. In 1911, Hans Geiger and Ernest Marsden carried out an experiment to investigate Rutherford's ideas about the structure of an atom. They fired a beam of alpha radiation, which they knew was made up of positively charged particles, at a very thin gold foil. If Rutherford was right, most of the alpha radiation, but not all, would go straight through the gold foil.

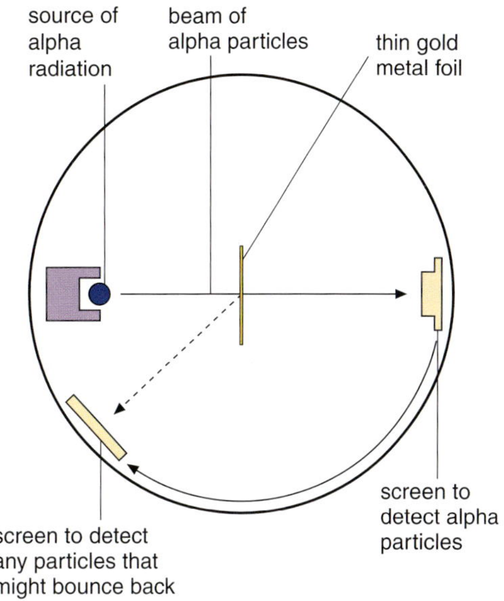

Figure 12.74

After taking over 100 000 measurements they found that most alpha particles went straight through the foil as they expected. But some were scattered through quite wide angles or even bounced back towards the source. This was what Rutherford had predicted would happen.

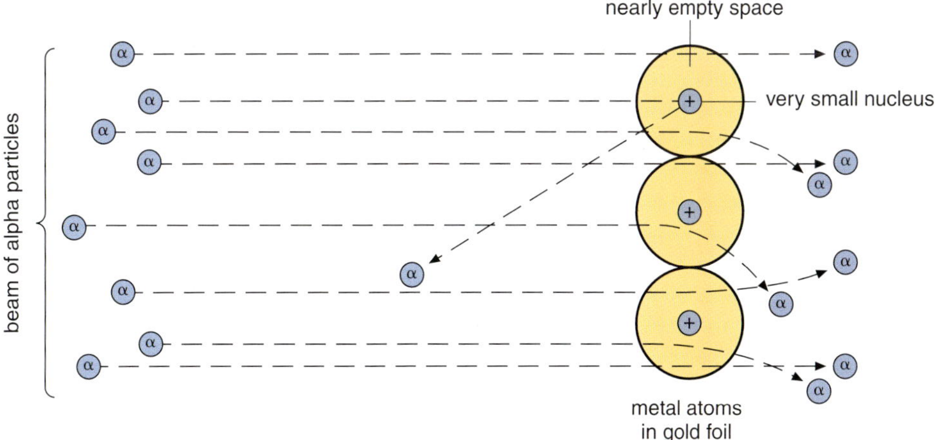

Figure 12.75

To explain the results of the experiment Rutherford came up with a new model for the atom, the 'nuclear' model.

- Most of an atom is empty space.
- The positive charge is concentrated at the centre of the atom, in the nucleus.

- The nucleus is very small.
- The electrons orbit around the nucleus.
- The positive charge in the nucleus equals the negative charge on the electrons.

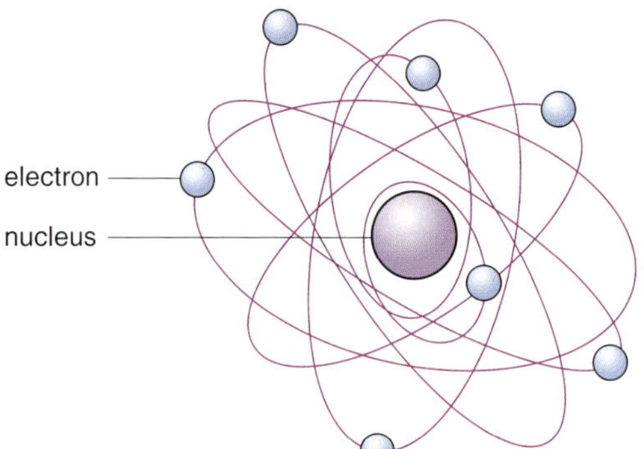
electron
nucleus
The Rutherford Model of an atom
A miniature solar system with the electrons moving like planets around the nucleus.

Figure 12.76 *The Rutherford model of the atom*

The 'nuclear' atom was quickly accepted as the correct structure for the atom.

In 1913, Niels Bohr took Rutherford's model one step further. He suggested that electrons could only be in certain orbits around the nucleus. Using his ideas he could explain why electrons do not fall into the nucleus.

Although ideas and models of the atom have changed since 1913, the 'nuclear' model is still a useful way of thinking of the atom.

Did you know?

The neutron was not discovered until 1932.

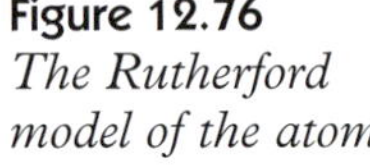

Did you know?

Ernest Rutherford (1871–1937) was born in New Zealand. He was considered the greatest experimental physicist of his generation. At the age of 28 he was a professor. He was knighted in 1914 and made a lord in 1931.

Figure 12.77 *Ernest Rutherford*

Neils Bohr (1885–1962), a Danish theoretical physicist, often called the 'father of atomic theory' won the Nobel Prize for Physics in 1922.

r og Nuttall's Apparat

Figure 12.78 *Neils Bohr*

Topic Questions

1 Copy and complete the following sentences.

a) Rutherford called the ___________ charged centre of an atom the ____________.

b) In the 'nuclear' model of the _________, electrons __________ the nucleus.

2 Draw a labelled diagram to show what an atom is like in the 'plum pudding' model.

3 What are the differences between the 'plum pudding' model of the atom and the 'nuclear' model of the atom?

4 Why in the alpha scattering experiment does most of the alpha radiation pass straight through the gold foil?

5 Why did Rutherford conclude that the nucleus of an atom is very small?

Summary

- Waves carry energy and information.
- Waves transfer energy without transferring any material (matter).
- A transverse wave moves at right angles to the vibration causing it.
- A longitudinal wave moves in the same direction as the vibration causing it.
- The amplitude of a wave is the maximum disturbance caused by the wave.
- Wavelength is the distance from one point on a wave to the same point on the next wave.
- Frequency is the number of waves sent out each second. It is measured in hertz.
- Refraction is caused by changes in the speed of a wave as it passes from one material into another.
- When waves move through gaps or pass the edges of objects they can spread out. This is called diffraction.
- Under certain conditions light can be totally internally reflected.
- Ultrasonic waves (ultrasound) have a frequency too high for humans to hear.
- The electromagnetic spectrum is a range of waves that travel at the same speed through a vacuum.
- Electromagnetic waves can be reflected, absorbed, transmitted and diffracted.
- Information can be coded and sent either as digital signals or as analogue signals.
- The Earth has a layered structure made up of the crust, mantle and core.
- Radiation occurs because of changes in atoms with unstable nuclei.
- The radiation from space, food, rocks air and buildings is called background radiation.
- Radiation can damage living cells.
- Radiation has a number of medical and industrial uses.
- The older a particular radioactive material, the less radiation it gives out.
- The half-life of a radioactive material is the time it takes for the count-rate to halve. It is also the time for half the radioactive nuclei in any size sample to decay.
- Radioactive decay can be used to estimate the age of rocks and organic material.
- The work of Rutherford helped to develop the modern ideas about the structure of the atom.

Examination Questions

1 a) A Geiger-Müller tube and counter is used to measure the small amount of radiation that is around us all of the time.

i) What name is given to the radiation that is around us all the time?

ii) Name two different sources of this radiation.

1 ___________________________

2 ___________________________

b) Gamma radiation is used to sterilise dressings. What does the gamma radiation do to the bacteria on the dressings?

2 a) The diagram shows a cross-section through the Earth.

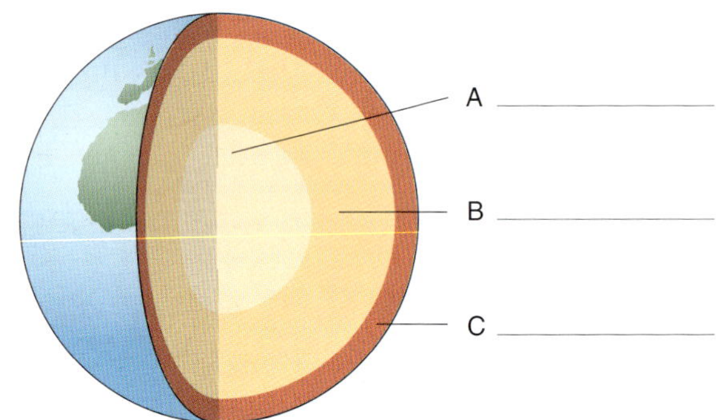

Use the words in the box to name the parts labelled A, B and C. Each word should be used once only.

core	crust	mantle

c) (i) Earthquakes create both longitudinal and transverse waves. Which one of the following waves is also longitudinal? Underline your answer.

light wave	sound wave	water wave

(ii) Complete the following sentence by choosing the correct word from the box.

Energy	mass	pressure

All waves, not just earthquake waves, transfer ___________________________.

3 The boxes on the left show some types of electomagnetic radiation. The boxes on the right show some uses of electromagnetic radiation.

Copy the diagram and draw a straight line from each type of radiation to its use. The first one has been done for you.

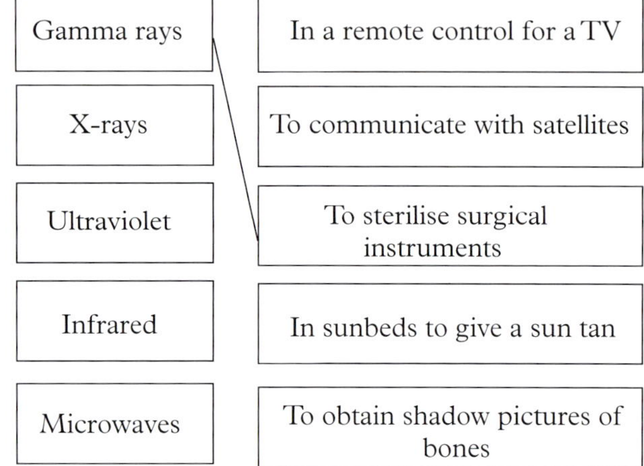

4 a) The diagram represents the electromagnetic spectrum. Four of the waves have not been named. Copy the diagram and draw lines to join each of the waves to its correct position in the electromagnetic spectrum. One has been done for you.

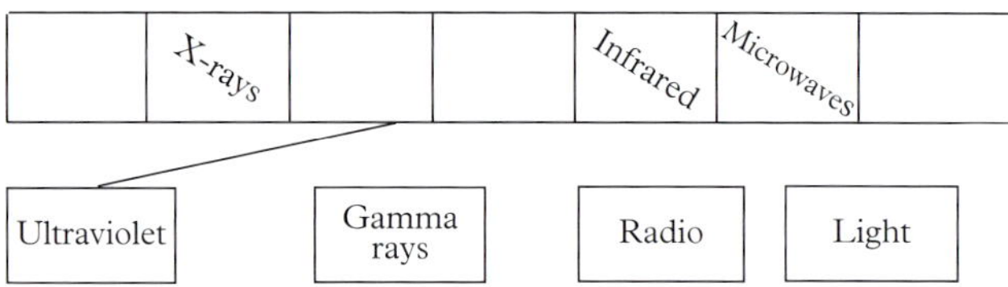

b) Copy and complete the following sentence by choosing the correct answer from the three lines in the box.

The speed of radio waves through a vacuum is

faster than the same as slower than	the speed of light through a vacuum.

c) i) Before sunbathing it's a good idea to apply a sun cream to your exposed skin. Why?

ii) From which type of electromagnetic wave is sun cream designed to protect the skin?

d) The diagram shows an X-ray photograph of a broken leg.
Bones show up white on the photographic film. Explain why.

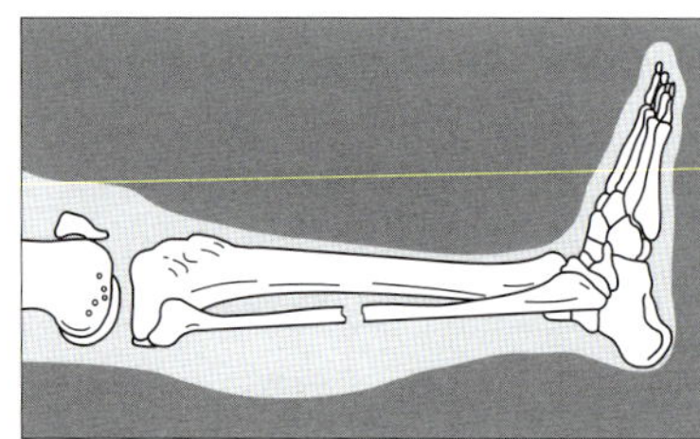

5 a) The diagram shows an electric bell inside a glass jar. The bell can be heard ringing.

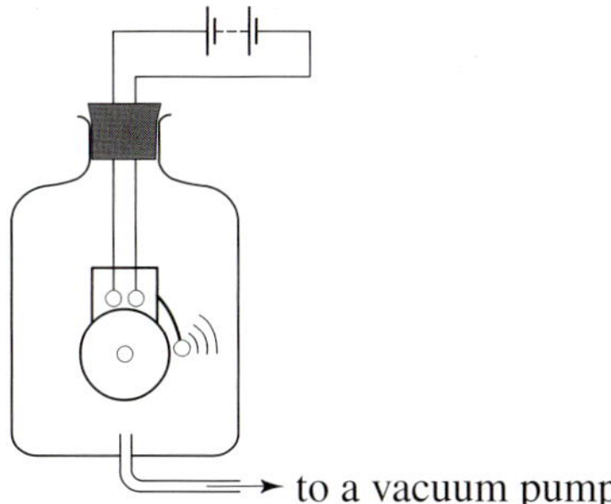

Copy and complete the following sentences, by choosing the correct line in each box.

When all the air has been taken out of the glass jar,

the ringing sound will | stop. / get louder. / get quieter. |

This is because sound | travels faster / travels slower / cannot travel | through a vacuum.

b) The microphone and cathode ray oscilloscope are used to show the sound wave pattern of a musical instrument.

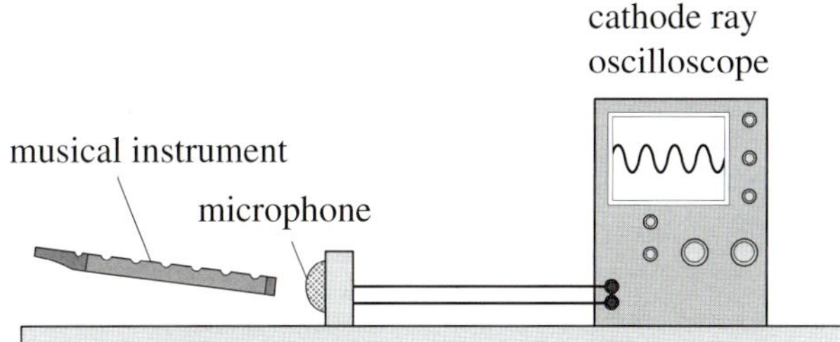

One of the following statements describes what a microphone does. Identify the correct statement.

- A microphone transfers sound energy to light energy.
- A microphone transfers sound energy to electrical energy.
- A microphone transfers electrical energy to sound energy.

c) Four different sound wave patterns are shown. They are all drawn to the same scale.

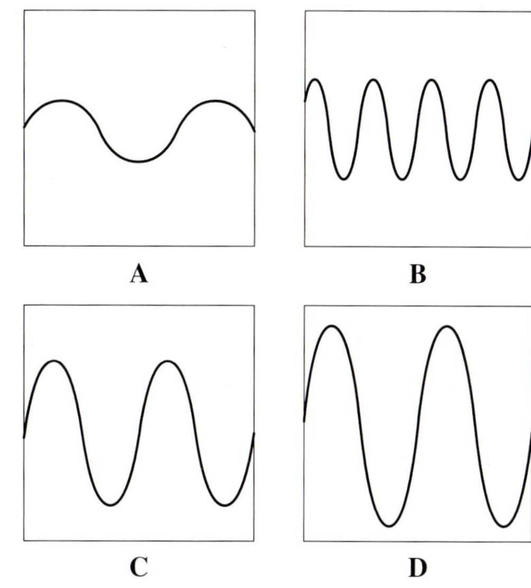

i) Which sound wave pattern has the highest pitch? Give a reason for your answer.

ii) Which sound wave pattern is the loudest? Give a reason for your answer.

d) i) The frequency of some sounds is too high for humans to hear. Which of the following words describes this sound.

| microwave | ultrasound | ultraviolet |

ii) Give one use for this type of sound wave.

6 a) The different sources of radiation to which we are exposed are shown in the pie chart. Some sources are natural and some artificial.

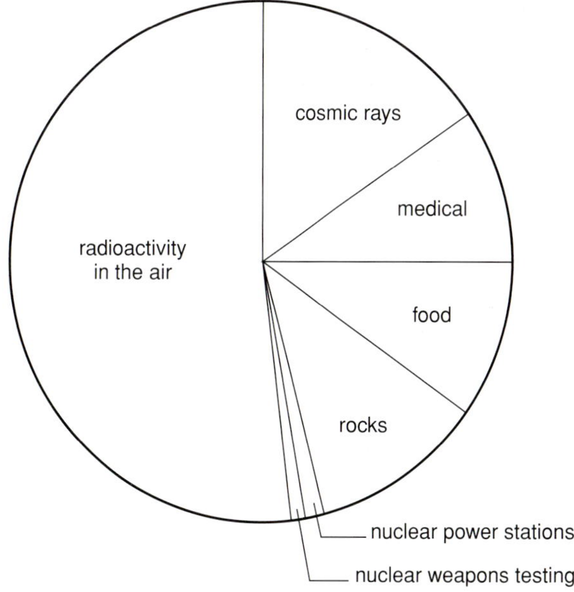

i) Name *one* natural source of radiation shown in the pie chart.

ii) Name *one* artificial source of radiation shown in the pie chart.

b) A radioactive source can give out three types of emission: alpha particles, beta particles, gamma radiation.
The diagram shows the paths taken by the radiation emitted by two sources, X and Y.
What types of radiation are emitted by each of the sources?

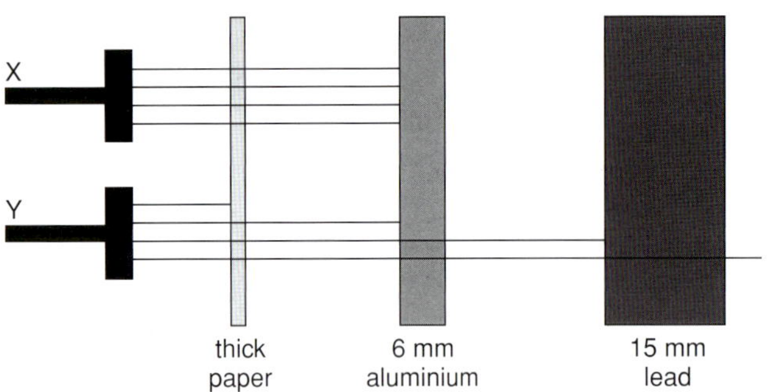

c) The diagram shows a disposable syringe sealed inside a plastic bag. After the bag has been sealed the syringe is sterilised using radiation. Explain why radiation can be used to sterilise the syringe.

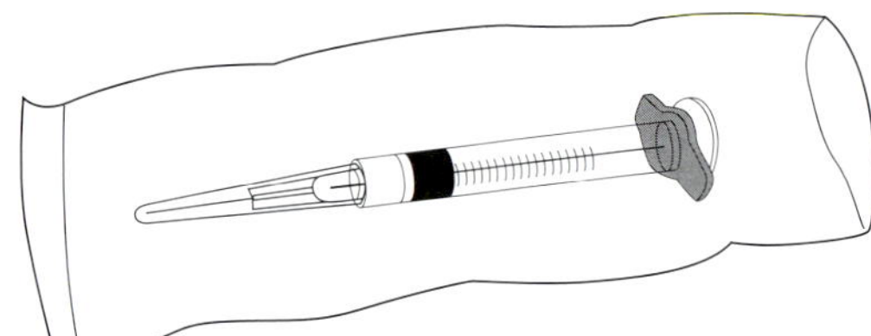

7 The isotope of sodium with a mass number of 24 is radioactive. The following data were obtained in an experiment to find the half-life of sodium-24.

Time in hours	Count rate in counts per minute
0	1600
10	1000
20	600
30	400
40	300
50	150
60	100

a) Draw a graph of the results and find the half-life for the isotope. On the graph show how you obtain the half-life.

Glossary

Absorption (digestion) The process in digestion whereby water or small soluble particles pass through the lining of the intestine into the bloodstream.

Absorption (plants) The process whereby plants take in water or dissolved mineral ions through their root hairs.

Acceleration The rate at which velocity changes.

Acid A substance that dissolves in water to give a solution with a pH of less than 7.

Active transport The process by which cells take up substances against a concentration gradient. The process requires energy from respiration.

Adaptation A feature of an organism that helps it survive in a particular environment.

Addiction Being unable to do without a drug such as nicotine, alcohol or heroin.

ADH (anti-diuretic hormone) A hormone released from the pituitary gland which targets the kidney tubules to ensure water is reabsorbed into the blood so reducing the amount of water in urine and controlling water balance.

Aerobic respiration Respiration that takes place in the presence of oxygen.

Air A mixture of gases made up of approximately 4/5 nitrogen and 1/5 oxygen.

Air resistance (drag) A force of friction.

Alkali A base (metal oxide or hydroxide) that dissolves in water to form a solution with a pH greater than 7.

Alkali metals The name given to the metals in Group 1 of the Periodic Table.

Alleles One of the different forms of a particular gene. For example the allele for eye colour can be for blue or brown eyes.

Alloys A mixture of metals (or of carbon with a metal). Alloys can have properties different from the parent metal(s).

Alpha (radiation/particles) A radioactive emission with low penetrating power blocked by paper.

Alternating current (a.c.) An electric current that reverses direction regularly.

Alveoli (singular: alveolus) A microscopic air sac in the lungs which acts as the surface for gaseous exchange.

Amino acids The breakdown products of the digestion of proteins, and the building blocks for making new proteins.

Ampere (amp) The unit of electric current.

Amplitude The maximum disturbance caused by a wave.

Amylase An enzyme that digests starch into sugar.

Anaerobic respiration Respiration which takes place in the absence of oxygen.

Analogue signals These are continuous waves that vary in amplitude and/or frequency.

Anhydrous Crystals from which water has been removed.

Anode The positively charged electrode.

Antibodies A protein produced by some white cells which neutralises the effects of foreign cells.

Antitoxins Substances which neutralise a toxin (poison) produced by bacteria by combining with it.

Anus The opening at the end of the digestive system.

Aorta The main artery of the body which carries oxygenated blood away from the heart.

Artery A blood vessel that carries blood away from the heart.

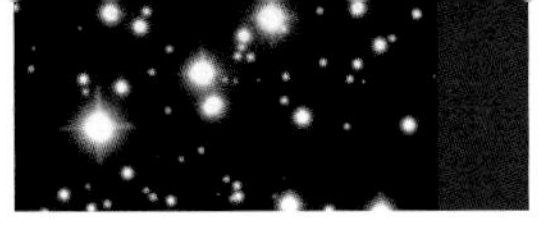

Artificial satellite A satellite put into orbit around the Earth.

Artificial selection The process of deliberately selecting and breeding organisms with desired characteristics.

Asexual reproduction Reproduction that does not involve the formation and fusion of gametes. The offspring have identical genetic information to the parent.

Atmosphere The layer of gases around the Earth.

Atom The smallest part of an element that can exist. Atoms have a nucleus consisting of protons and neutrons around which are shells of electrons.

Atomic number The number of protons present in an atomic nucleus (and the number of electrons present in the neutral atom).

Atria (singular: atrium) The upper chambers of the heart which receive blood from the veins.

Background radiation The radiation that is around us all of the time.

Bacteria (singular: bacterium) A single-celled organism consisting of cytoplasm and a membrane surrounded by a cell wall, with genes not organised to form a distinct nucleus.

Base An oxide or hydroxide of a metal.

Battery Two or more electrical cells joined together.

Bauxite The main ore of aluminium containing aluminium oxide (Al_2O_3).

Beta (radiation/particles) A radioactive emission with moderate penetrating power blocked by thin sheets of metal.

Bile A liquid produced by the gall bladder that breaks up fats into droplets.

Biodegradable Materials which can be broken down (decomposed) by bacteria.

Biomass The dry mass of living material in an ecosystem.

Bladder The sac which fills with urine from the kidney.

Blast furnace The industrial method used for extracting metals such as iron, from their ores.

Bonding The forces that hold atoms together.

Brain The part of the central nervous system which controls and co-ordinates most of the body's activities.

Braking distance How far a car travels before stopping, after the brakes have been applied.

Breathing The action of passing air into and out of the lungs.

Cancer A group of cells that are dividing very much more rapidly than is normal.

Capillaries Fine, thin-walled blood vessels which form a network for the exchange of substances with the tissue cells.

Carbon dioxide A gas formed during respiration and in the combustion of hydrocarbons. It turns clear limewater milky.

Catalyst A substance that changes the speed of a reaction but remains unchanged after the reaction.

Cathode The negatively charged electrode.

Cell (biological) The smallest part of an animal or plant.

Cell (electrical) The single unit from which batteries are made.

Cellulose The material that makes up the cell wall of plant cells.

Cell division The processes of mitosis and meiosis.

Cell membrane The very thin membrane on the outside of a cell that controls the movement of substances in and out of the cell.

Cell wall The outer part of a plant cell that gives strength and shape to the plant cell.

Chain reaction A reaction in which a nucleus is split and neutrons released that can split other nuclei to produce a continuous chain of events.

Characteristic A feature of an organism that can be observed.

Charge A feature of atomic particles. Protons and electrons have a charge. Electrons have a negative charge and protons have a positive charge. Opposite charges attract; like charges repel.

Chlorophyll A green pigment found in the leaves and stems of plants. It traps light energy for use in photosynthesis.

Chloroplast That part of the plant cell containing chlorophyll.

Chromosome One of the thread-like structures found in the nucleus which contains genetic material. They are made of a single very long molecule of DNA. In humans there are 23 pairs of chromosomes in each body cell, but 23 single chromosomes in each gamete.

Ciliary muscles Muscles in the eye which control the thickness of the lens when focussing on near and distant objects.

Circuit breaker A device that uses the action of an electromagnet to break a circuit very quickly.

Clone An organism produced asexually from one parent. The clone will be genetically identical to the parent.

Coke The substance used as the reducing agent in the blast furnace. It is almost pure carbon and is made by heating coal.

Combustion The burning of a substance in oxygen to release heat energy.

Comet An object made of ice and rock which orbits the Sun in a different plane to the planets.

Competition The interaction of organisms which are trying to obtain the same food or occupy the same space.

Compound A substance which contains two or more elements chemically joined together.

Connector neurone (relay neurone) Neurones found in the spinal cord which link sensory and motor neurones.

Consumer An organism which has to rely on food made by green plants (producers) or other animals.

Core (Earth) The innermost part of the Earth.

Cornea The transparent front part of the eye. Plays an important part in focussing light onto the retina.

Corrosion A reaction between a metal and substances in the atmosphere. The process that causes iron to rust.

Cosmic rays Rays and particles from space reaching Earth.

Count rate The number of radioactive emissions in a certain time.

Covalent bond The bonding of atoms caused by the sharing of pairs of electrons in their outer electron shells.

Cracking A form of thermal decomposition in which large hydrocarbon molecules are broken down into smaller ones.

Crude oil A mixture of substances, most of which are hydrocarbons, formed by the decomposition of marine organisms over a long period of time.

Crust The surface layer of the Earth.

Current A flow of electrons, ions or electric charge.

Cytoplasm All the material in a cell inside the membrane (apart from the nucleus), where chemical reactions take place under the control of enzymes.

Decay (atomic) The break up of unstable nuclei resulting in the production of radioactive emissions.

Decay (biological) The breaking down by decomposers of complex organic materials into simple ones.

Decelerate To slow down.

Decomposers Organisms which break down complex organic materials into simple ones during decay. Most are bacteria and fungi.

Deforestation The removal of trees from woodland and mountain sides. It often leads to flooding of rivers as the trees can

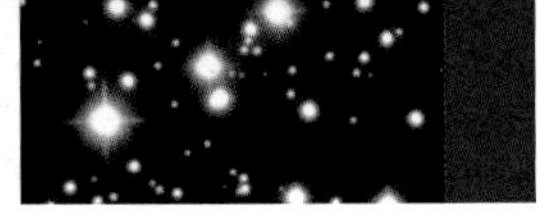

no longer take up the rain that falls on the land.

Denaturing The process by which enzymes are destroyed when heated above a temperature of about 40°C.

Denitrifying bacteria Bacteria which convert nitrates in the soil into nitrogen gas.

Deoxygenated blood Blood that is not rich in oxygen.

Deposition The laying down in water of a layer of rock fragments.

Detrivores Organisms, such as termites, earthworms, fungi and bacteria that obtain energy and nutrients by feeding on dead organic matter. The decomposers are a class of detrivores.

Diabetes A disease caused by lack of insulin production from the pancreas. Blood sugar levels cannot be controlled properly.

Dialysis The treatment of kidney failure by taking blood from the patient and removing the urea and other waste products by diffusion.

Diaphragm A sheet of muscle which separates the thorax from the abdomen. Flattening of the diaphragm results in air entering the lungs.

Diffraction The spreading out of a wave as it passes through a gap or moves past an object.

Diffusion The movement of particles resulting in a net movement from a region where they are at a high concentration to a region where they are at a lower concentration.

Digestion The process by which food is broken down into particles small enough to be absorbed into the blood.

Digital signals These are made up of a series of 'on' and 'off' pulses. The pulses are represented by the numbers of 1 and 0.

Dilate To get wider.

Diode An electrical component that only conducts electricity flowing in one direction.

Direct current (d.c.) An electric current that always flows in the same direction.

Displacement reaction A reaction in which one metal displaces another metal.

DNA The chemical that carries the genetic information on the genes.

Dominant The allele that controls an observable characteristic (phenotype) in the offspring even when it is present on only one chromosome (heterozygous).

Drag The force from the air that opposes movement.

Dynamo (generator) Device supplying a voltage from the relative motion of a conductor with a magnetic field.

Earthing A metal object is earthed when a low resistance conductor joins the object to the Earth's surface.

Ecosystem The organisms living and surviving in a particular place.

Effector neurone Nerve cell which carries impulses away from the spinal cord towards effectors.

Effectors Structures such as muscles or glands which carry out responses to stimuli.

Efficiency The ratio of useful energy transferred by a device to the total energy transferred to the device.

Elastic potential energy The energy stored in an elastic object when work is done to change its shape.

Electric current A flow of electrons or ions.

Electrode A negatively or positively charged conductor.

Electrolysis The process of splitting up a compound using electricity.

Electromagnet A device made by passing an electric current through a coil of wire.

Electromagnetic spectrum The family of electromagnetic waves arranged in order of wavelength and frequency.

Electromagnetic waves Transverse waves that have a common speed in air or a vacuum; can travel through a vacuum.

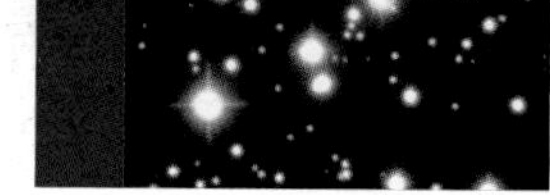

Electrons Negatively charged particles orbiting the nucleus of an atom in energy levels.

Element A substance made up of atoms which contain the same number of protons, and which cannot be broken down into anything simpler by chemical means.

Emulsifying Breaking down of a liquid into very fine droplets.

Endocrine gland A gland which discharges its products, called hormones, straight into the blood.

Endothermic reaction A reaction that takes in heat energy.

Environment The surroundings and conditions that affect the growth and behaviour of plants and animals.

Enzyme A protein that can act as a catalyst for a reaction. It can be easily destroyed (denatured) by heating.

Erosion The wearing away of the Earth's surface.

Eutrophication A process caused when large amounts of nitrates and phosphates are discharged into rivers and streams. The nutrients cause the rapid growth of algae and water plants. The eventual death of the algae and plants soon leads to the rapid growth of aerobic bacteria. These decomposers soon use up all the available oxygen in the water. This in turn causes other animal life in the water to suffocate and die.

Evaporation The loss of the more energetic particles from the surface of a liquid.

Excretion The removal of chemical waste material made in the body or a cell.

Exhale To breathe out.

Exothermic reaction A reaction that gives out heat energy.

Extinct A description of an organism no longer living today but which according to the fossil record has lived in the past.

Extrusive rocks Igneous rocks formed when lava cools on the surface of the Earth. Because the rock cools rapidly, the crystals are small.

Eye A sense organ that contains the receptors sensitive to light.

Faeces The indigestible food remaining once digestion has taken place.

Fatty acids The breakdown products of fats.

Fermentation The changing of glucose into ethanol (alcohol) and carbon dioxide by the action of enzymes in yeast.

Fertile The ability of a male or female to produce sex cells which are capable of producing viable offspring.

Fertilisation The fusion of an egg and a sperm.

Fertiliser A substance which can be natural or artificial applied to soil to improve the growth of plants.

Fertility drugs Drugs that stimulate the release of eggs from the ovaries.

Fetal imaging Using ultrasound waves to view a fetus.

Fetus The name given to an unborn child more than 8 weeks after conception.

Filtration (kidney) Filtration of water and ions from the bloodstream, to form urine.

Flammable Easily set on fire.

Focus The formation of a sharp image of near and distant objects by altering the shape of the lens.

Food chain A diagram which shows feeding relationships of some organisms in an ecosystem. All food chains start with producers which trap light energy.

Formula mass The relative mass of a compound found by adding together the relative atomic masses of every atom in the molecule.

Fossil The remains or imprints of dead plants or animals trapped in sedimentary rocks when the rocks were formed. The remains or imprints may have been mineralised and turned into stone.

Fossil fuels The non-renewable energy resources: coal, oil and natural gas.

Fossilisation The process that produces fossils.

Fractional distillation A method of separating liquids whose boiling points are close

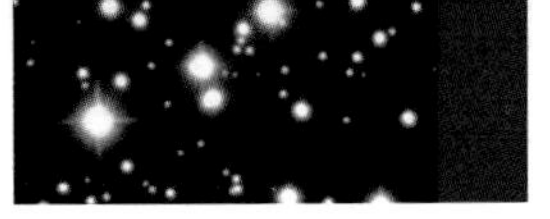

together. The process used to separate the different substances in crude oil.

Free electrons The electrons in metals that move around inside the metal and do not remain in orbit around a nucleus. The presence of these free electrons allows the metal to conduct electricity and heat.

Frequency The number of waves produced each second by a vibrating source.

Friction A force that tries to stop or does stop movement.

FSH (follicle-stimulating hormone) The hormone secreted by the pituitary gland that causes eggs to mature and stimulates the ovaries to produce oestrogen.

Fuse A wire designed to melt if too large a current flows through it.

Fusion (biological) The process that occurs when the nucleus of a male gamete combines with the nucleus of a female gamete.

Galaxy A vast number of stars held together by gravity.

Gall bladder A small organ joined to the liver that stores bile.

Gamete A sex cell.

Gamma (radiation) A radioactive emission with high penetrating power blocked by thick concrete or lead.

Gaseous exchange Occurs in the alveoli in the lungs when oxygen diffuses across the alveolar membrane from the lungs to the capillaries and carbon dioxide diffuses from the blood capillaries into the alveoli.

Gene A unit of inheritance controlling one particular characteristic and made up of a length of chromosomal DNA.

Genetic Related to inheritance.

Genetic engineering The deliberate changing of the characteristics of an organism by manipulating chromosomal DNA.

Genotype The genetic make-up of an individual. The sum total of all the genes even if they are not shown in the individual.

Geostationary satellite A satellite that takes 24 hours to orbit the Earth.

Geothermal energy The heat energy produced in the Earth by the decay of radioactive isotopes.

Giant structure Ionic compounds that have high melting points, usually dissolve in water and are good conductors of electricity when molten or in aqueous solution.

Gland A structure that releases hormones into the bloodstream.

Global warming The name given to the increase in the average temperature of the Earth.

Glucagon A hormone released by the pancreas that causes the liver to convert glycogen into glucose.

Glycerol A breakdown product from the digestion of fats and oils.

Glycogen The form in which excess glucose in the blood is stored in the liver and muscles.

Gravitational potential energy The energy stored in an object due to the vertical height it has been lifted.

Gravity A force of attraction that acts between all bodies.

Greenhouse effect The effect in the atmosphere of heat energy being absorbed by gases such as carbon dioxide and methane.

Group A vertical column of elements in the Periodic Table having similar chemical properties due to the atoms of the elements having the same number of electrons in their outer shells.

Guard cells Pairs of cells which surround the stomata on the surface of leaves which by means of osmosis open and close the stomata thus regulating the flow of gases into and out of the leaf.

Gullet *see oesophagus.*

Haematite A common ore of iron.

Haemoglobin The red pigment in the red blood cells which combines with and transports oxygen.

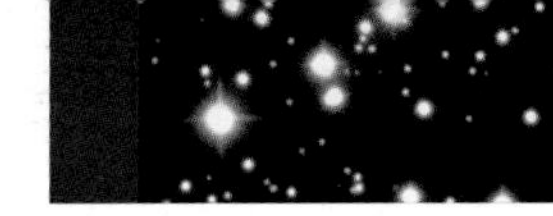

Half-life The time it takes for the count rate from a radioactive material to halve.

Halide The compound formed when a halogen reacts with another element.

Halogens The name given to the elements in Group 7 in the Periodic Table.

Heart A double pump with the right side pumping blood at low pressure to the lungs to release carbon dioxide and collect oxygen and the left side pumping oxygenated blood at higher pressure around the body.

Herbicide A chemical used to destroy unwanted plants.

Herbivore An organism that feeds only on plants.

Hertz The unit of frequency (symbol Hz).

Heterozygous The inheriting of one dominant allele and one recessive allele for a particular characteristic.

Homeostasis The automatic control system by which the internal conditions of an organism are kept steady.

Homozygous The inheriting of two dominant or two recessive alleles for a particular characteristic.

Hormone A substance secreted by endocrine glands directly into the blood in one part of the body and carried in the blood plasma to a target organ. Plant hormones are called auxins.

Hydrocarbons Compounds containing only hydrogen and carbon.

Hydroelectric power Electrical energy generated by the flow of falling water.

Hydrogen The chemical element with the lowest density. Small amounts of it burn with a squeaky pop.

Igneous rocks Rocks formed by magma rising upwards from the mantle, cooling and solidifying into a hard crystalline rock.

Image The picture formed on the retina of the eye.

Indicator A dye which changes colour when mixed with acidic, alkaline or neutral solutions.

Induced current The current produced when a wire that is part of a complete circuit cuts through a magnetic field.

Inert Unreactive.

Inhale To breathe in.

Insoluble A substance that will not dissolve in a liquid, usually water.

Insulin A hormone released by the pancreas which helps to control sugar level in the blood.

Insulin controls the conversion of excess glucose into glycogen which is then stored in the liver and muscle cells.

Intrusive rock Igneous rocks formed when lava cools beneath the surface of the Earth. Because the rock cools slowly, the crystals are large.

Ion An atom that has lost or gained one or more electrons.

Ionic bond The electrostatic attraction between opposite charges responsible for holding metal and non-metal elements together in a compound. The ions are formed when the metal atoms transfer electrons to the non-metal atoms in order to achieve full outer electron shells.

Ionise To remove or add electrons to atoms or groups of atoms so giving them positive or negative charges.

Iris A ring of muscle which controls the amount of light entering the eye.

Isotopes Atoms which contain the same number of protons but different numbers of neutrons.

Joule (J) The unit of energy and work.

Kidney An organ which removes excess water from the blood and excretes urine made from the urea produced in the liver.

Kilowatt (kW) 1000 watts.

Kilowatt hour (kWh) A unit of electrical energy used by companies that supply electricity.

Kinetic energy The energy possessed by an object due to its motion.

Lactic acid A product of anaerobic respiration in very active human muscles which is a mild tissue poison (causes the muscles to hurt).

Large intestine The part of the digestive system where water is removed from indigestible food.

Lava Magma that has erupted through the Earth's crust.

Law of Conservation of Mass This states that during any chemical reaction matter (material) is neither created nor destroyed.

Lens A transparent structure within the eye that is flexible and helps light to form a sharp image on the retina during focussing.

LH (luteinising hormone) The hormone secreted by the pituitary gland that stimulates the release of an egg.

Light dependent resistor An electrical component whose resistance goes down when a bright light shines on it.

Light year The distance a light ray travels in one year.

Limewater A solution of calcium hydroxide that turns milky when carbon dioxide is passed through it.

Limiting factor The factor such as light intensity (brightness), light wavelength, water and carbon dioxide that limits the rate of photosynthesis at a given time.

Lipids Foods made up of fats and oils.

Lipase An enzyme that digests fats to fatty acids.

Lithosphere The outer shell of the Earth made from the crust and the upper part of the mantle.

Liver An organ where excess glucose in the blood is stored as glycogen, where bile is produced and where poisons such as alcohol are removed from the blood.

Longitudinal wave A wave in which the energy travels in the same direction as the vibrations causing the wave.

Magma Molten rock below the Earth's crust.

Magnetic field The space around a magnet where its magnetic force acts on a magnetic object.

Mantle The layer of the Earth between the crust and the core.

Mass The amount of matter an object contains. Measured in kilograms (kg).

Mass number The total number of protons and neutrons in the nucleus of an atom.

Meiosis Cell division that leads to the production of gametes in which there has been some reassortment of genetic material so producing variation. It is a reduction division so each gamete has only half the number of chromosomes as the parent.

Metamorphic rocks Rocks formed from rocks which became buried deep underground and had their structure changed by high temperatures and or high pressures.

Migration The mass movement of organisms on a regular basis. Most migrations are connected with seasonal changes and enable organisms to maintain food supplies.

Milky Way galaxy The galaxy that contains our own solar system.

Mitochondria (singular: mitochondrion) The parts of the cell in which aerobic respiration takes place producing cellular energy.

Mitosis Cell division that occurs during growth and asexual reproduction and involves each chromosome making an exact copy of itself, resulting in the formation of two daughter cells each with the same number of chromosomes as the parent.

Mixture Two or more substances which are usually easy to separate.

Molecule A particle containing atoms of the same or different elements bonded together. The smallest part of an element or compound that can take part in a chemical reaction.

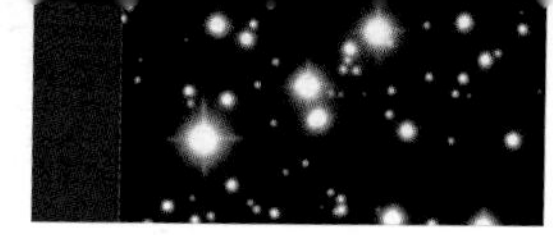

Monomers Small molecules which join together to form a long chain of molecules called a polymer.

Moon A natural satellite in orbit around a planet.

Motor effect The motion of a current-carrying conductor in a magnetic field.

Motor neurones Neurones that carry electrical impulses from the brain or spinal cord to an effector.

Mucus A sticky fluid which traps dust or protects surfaces.

Mutation A change suddenly occurring in one or more of the genes or chromosomes or in the number of chromosomes.

Natural selection The process by which beneficial characteristics with greater survival value are selected and increase in proportion in the population. Natural selection leads to evolution.

Negative The charge on an electron.

Negative feedback An automatic control mechanism in which a change from the normal condition triggers off a response which restores the normal condition.

Nerve impulses Electrical signals which travel along nerve pathways made up of nerve cells (neurones).

Neurone A cell in the nervous system.

Neutral (charge) Having no overall charge.

Neutral (indicators) Having a pH of 7.

Neutralisation A reaction between an acid and a base or a carbonate.

Neutron An uncharged particle found in the nucleus of all atoms (except hydrogen).

Newton (N) The unit of force.

Nicotine The addictive substance in tobacco.

Nitrates Chemicals containing NO_3^- ions, frequently used in fertilisers to help plants synthesise proteins.

Nitrifying bacteria Bacteria which convert ammonium compounds in the soil into nitrates.

Noble gases The name given to the elements in Group 0 in the Periodic Table.

Non-metals Elements in the Periodic Table which usually have low melting points and boiling points, are poor conductors of electricity and heat, and as solids are brittle.

Non-renewable energy Energy resources that once used cannot be replaced.

Normal The line drawn at right angles to a surface.

Nuclear fission The breaking up of a large atomic nucleus to release energy.

Nucleon The protons and neutrons in the nucleus of an atom.

Nucleus (atom) The central part of an atom containing protons and neutrons.

Nucleus (cells) The part of a cell that contains the chromosomes which carry the genes controlling the cell's characteristics.

Nutrition The process by which organisms obtain their raw materials and absorb useful substances from it.

Oesophagus (gullet) The muscular tube which carries food from the mouth to the stomach.

Oestrogen A hormone produced by the ovaries which controls female sexual characteristics.

It inhibits the production of FSH and stimulates the release of LH.

Ohm The unit of electrical resistance.

Optic nerve A bundle of nerve cells which carries impulses from the eye to the brain.

Oral contraceptive Tablets, usually containing oestrogen, that inhibit the production of FSH so that no eggs mature.

Orbit The path taken by an object, which goes around another object.

Ores Minerals or mixtures of minerals from which a metal can be extracted in economically viable amounts.

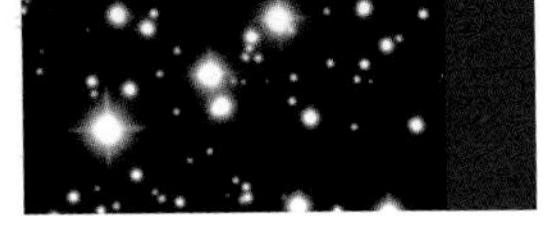

Organ A group of tissues working together to carry out a particular function.

Organ system A group of organs working together to carry out a particular function or group of related functions.

Organic Compounds of carbon found in large quantities in living and dead organisms.

Organism An individual plant or animal.

Osmosis The diffusion of water through a partially permeable membrane – the water flowing from a region of high water concentration to a region of lower water concentration.

Oxidation A chemical reaction which involves the addition of oxygen.

A reaction involving the loss of electrons.

Oxygen The chemical element that is vital to life. It will relight a glowing spill.

Oxygenated blood Blood rich in oxygen.

Oxygen debt The oxygen needed to remove the lactic acid from the muscles produced as a result of muscles respiring anaerobically during vigorous exercise.

Oxyhaemoglobin The chemical formed when oxygen combines with haemoglobin.

Ozone layer The layer of gas in the upper atmosphere that reduces the amount of harmful ultraviolet radiation reaching the Earth's surface.

Palisade cells The cells in the upper part of green leaves which contain most chlorophyll and carry out most of the photosynthesis in the leaf.

Pancreas An organ of the digestive system that produces the hormone insulin and the enzyme lipase.

Parallel circuits Electrical circuits that provide several pathways for an electric current.

Partially permeable membrane Allows small molecules to pass through quickly but not large molecules.

Period A horizontal row of elements in the Periodic Table.

Periodic Table The arrangement of the elements in order of increasing atomic number.

Pesticide A chemical designed to kill unwanted organisms.

pH scale A set of numbers from 1 to 14 used to measure the acidity or alkalinity of an aqueous solution.

Phenotype The way an individual appears as a result of the alleles it carries and the environment in which it has grown up.

Phloem A column of cells in a plant responsible for the transport of food made in photosynthesis to wherever it is needed.

Phosphates Chemicals containing PO_4^{3-} ions, frequently used as fertilisers to help plants photosynthesise and respire.

Photosynthesis The process in green plants which produces biomass (initially carbohydrates) and oxygen, and requires carbon dioxide and water as raw materials and chlorophyll to enable the plant to absorb light energy.

Pituitary gland A gland, found at the base of the brain, that secretes FSH and LH.

Planet A very large object that orbits the Sun.

Plasma The straw-coloured liquid part of the blood which transports cells and dissolved substances.

Platelets Cell fragments which help in forming blood clots at wounds.

Pollution The introduction of harmful substances into an environment.

Polymer A long chain molecule made up of many smaller molecules called monomers.

Polymerisation A reaction in which small molecules join together to make larger molecules.

Population The numbers of one species of animal living in a particular area.

Positive The charge on a proton.

Potassium An element used by plants to help the action of the enzymes involved in photosynthesis and respiration.

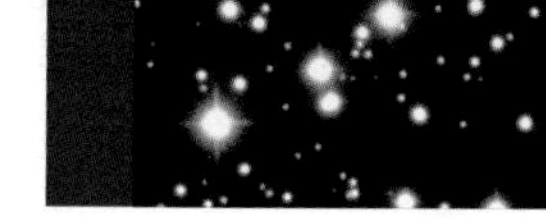

Power The transfer of energy. Measured in watts (W).

Precipitate The formation of an insoluble solid during the reaction between two solutions.

Precipitation (chemical) The type of reaction in which a precipitate is formed.

Precipitation (weather) The name given to rain, hail and snow.

Predation The process by which one animal (**predator**) catches then eats another (**prey**).

Producer A green plant which photosynthesises to make its own food.

Products The new materials produced as a result of a chemical reaction.

Proteases Enzymes that digest proteins into amino acids.

Proton A positively charged particle found in the nucleus of all atoms.

Pulmonary artery The blood vessel that takes deoxygenated blood from the heart to the lungs.

Pulmonary vein The blood vessel that takes oxygenated blood from the lungs to the left atrium of the heart.

Pupil The gap surrounded by the iris through which light passes into the eye. Pupil size can be changed by dilation and constriction of the iris.

Putrefying bacteria These break down animal waste and produce ammonia.

Pyramids Diagrams which illustrate quantitatively the relationships between organisms in a food chain. Each organism is represented by a block in the pyramid. Pyramids can show number, biomass or energy relationships.

Radiation (heat transfer) The transfer of heat energy by electromagnetic waves.

Radioactive decay The process by which an atom of one element changes into an atom of a different element.

Radioactive isotope An atom with an unstable nucleus. It can also be called a radioisotope or radionuclide.

Radioactivity The random emission of energy from an atomic nucleus as the result of the breakdown of an unstable nuclei.

Reabsorption The way in which substances needed by the body are taken back into the blood from the tubules in the kidney.

Reactants The starting materials in a chemical reaction.

Reactivity series A list of metals arranged in order of their chemical reactivity. The most reactive metals are at the top of the list.

Receptors Special cells which are capable of detecting environmental changes.

Recessive The allele which must be present on both chromosomes to show an effect in the phenotype.

Red blood cells Cells that contain haemoglobin and whose function is to transport oxygen around the body.

Red giant A relatively cool giant star.

Red shift The effect on the spectrum of a galaxy being moved to the red end due to the galaxy moving away from us.

Reduction A chemical reaction which involves the loss of oxygen.

A reaction involving the addition of electrons.

Reflex action A rapid automatic response to a stimulus, during which nerve impulses are sent by receptors through the nervous system to effectors.

Reflex arc The route taken by a nerve impulse through the nervous system to bring about a reflex action.

Refraction The change in direction that happens when waves move at an angle from one material into another.

Relative atomic mass The average mass of an atom of an element on a scale on which the mass of a hydrogen atom = 1 or the mass of the ^{12}C isotope of carbon = 12. It takes into account the relative abundance of different isotopes with different mass numbers.

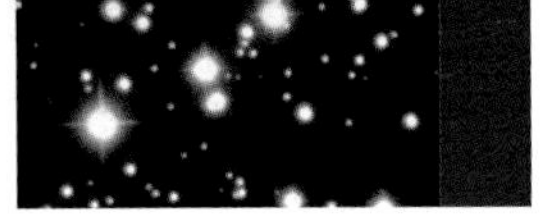

Relative molecular mass (relative formula mass) This is found by adding together the relative atomic masses of all the atoms in one molecule of the substance.

Relay neurone (connector neurone) Neurones found in the spinal cord which link sensory and motor neurones.

Renal To do with the kidney.

Renal artery The blood vessel that carries blood to the kidneys.

Renal vein The blood vessel that carries blood away from the kidneys.

Renewable energy Energy resources that can be replaced and so will not run out.

Reproduction The formation of offspring.

Resources Natural materials available for the use of organisms.

Respiration The process taking place in living cells transferring energy from food molecules (glucose) to cellular energy.

Respire The cellular process of obtaining energy from food.

Response The reaction of an organism to a stimulus.

Retina The light receptor surface at the back of the eye where light sensitive receptors convert light into nerve impulses.

Reversible reaction A reaction that can proceed in either direction depending on the reaction conditions. Reactants can be changed into products which in turn can be changed back into reactants.

Rib muscles The muscles between the ribs which contract to raise the rib cage for inhalation.

Root hairs Cells with a large surface area and thin cell wall that absorb water and mineral salts from the soil by osmosis, diffusion and active transport.

Rusting The corrosion of iron in the presence of air and water to form hydrated iron oxides.

Sacrificial protection Used to reduce the rusting of iron by attaching a more reactive metal such as magnesium or zinc.

Saliva A liquid containing the enzyme amylase produced in the salivary glands.

Salivary glands Glands in the mouth that secrete saliva and the enzyme amylase.

Satellite An object which orbits a planet.

Sclera The tough outer layer of the eye.

Seismic waves These are waves created by the vibrations caused by an earthquake.

Selective breeding The process of deliberately breeding animals or plants according to desirable characteristics.

Sensory neurone A nerve cell which carries impulses from sense cells or organs to the spinal cord.

Series circuits Electrical circuits giving only one pathway for the electric current.

Sexual reproduction This involves two parents who each produce sex cells that must join together. The offspring are genetically different from the parents and each other.

Skin A water-proof, germ-proof layer that contains receptors sensitive to touch, pressure and temperature and plays a part in temperature control.

Small intestine That part of the digestive system where the absorption of soluble foods into the blood occurs.

Smelting The process of getting a metal from its ore by heating the ore with carbon.

Solar system A system made up of the Sun, planets, moons, asteroids and comets.

Soluble Able to be dissolved (usually in water).

Species A group of organisms which look similar and that can breed together to produce fertile offspring.

Speed The distance an object travels in a unit of time. Units are m/s.

Star A huge mass that produces its own light and heat.

Stimulus A change in the environment of an organism which produces a response.

Stomach The part of the digestive system after the gullet where food is churned into a liquid mass.

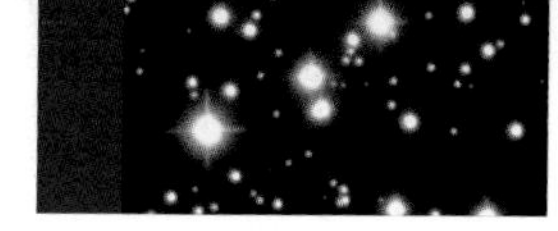

Stomata (singular: stoma) The tiny openings in the surface of a leaf through which gases can pass by diffusion. The size of the openings is regulated by the guard cells.

Stopping distance Thinking distance + braking distance.

Substrate A liquid enzymes can work on.

Sun A star at the centre of a solar system.

Suspensory ligaments The muscles in an eye that hold the lens in place.

Synapse The gap between two neurones.

Synthesis The process in which elements are chemically combined to make a new compound.

Target organ The organ affected by the release of a hormone.

Tectonic plates The separate slow-moving adjacent sections of the Earth's lithosphere that move because of convection currents within the Earth's mantle caused by the natural radioactive processes within the Earth.

Temperature How hot or how cold an object is. Units are °C.

Terminal velocity The constant speed reached by a moving object when the forces (in the direction it is moving) are balanced.

Thermal decomposition The breaking down of a compound by the action of heat.

Thermistor An electrical component whose resistance goes down as it gets hot.

Thermit process A method of joining two lengths of railway track together using the exothermic reaction between aluminium and iron(III) oxide.

Thinking distance How far a car travels during the driver's reaction time, before the brakes are applied.

Thorax The chest cavity.

Tissue A group of cells working together to carry out a particular function.

Tissue fluid A liquid formed from the blood plasma and carries soluble substances from the blood to the tissue cells.

Total internal reflection This happens when light moving through a transparent material into air hits the surface of the material at an angle bigger than critical angle.

Toxic Poisonous.

Toxins Poisons.

Trachea The tube which connects the throat and the lungs and through which air passes into the lungs.

Transformer A device that increases or decreases an alternating voltage.

Transition elements (transition metals) The name given to the elements in the Periodic Table between Groups 2 and 3.

Transpiration The process by which water evaporates from the leaf through the stomata, creating a pull causing water to rise up the plant in the transpiration stream.

Transportation The removal of rocks broken down by weathering and erosion.

Transverse wave A wave in which the energy travels at right angles to the vibrations causing the wave.

Turbine A device that turns a generator.

Turgor The pressure that the cytoplasm and vacuole of a cell exert on the cell wall.

Ultrasonic waves These produce ultrasound which is any sound with a frequency too high for humans to hear.

Ultraviolet radiation An electromagnetic wave that can cause skin cancers.

Universal indicator An indicator used to measure the pH of a solution to show whether the solution is acidic, neutral or alkaline.

Universe Made up of innumerable galaxies. It is everything there is.

Urea The breakdown product of amino acids produced in the liver and excreted by the kidneys in urine.

Ureter The tube taking urine from the kidney to the bladder.

Urine The waste fluid produced in the kidneys that contains urea, excess water and salts.

Vaccine A liquid containing dead or weakened disease-producing microorganisms that causes the body to produce antibodies.

Vacuole A cavity in the cytoplasm which is surrounded by a membrane. The vacuole contains cell sap.

Variation The differences in characteristics between members of the same species.

Vein A blood vessel taking blood to the heart.

Velocity Speed in a given direction.

Vena cava The blood vessel that carries blood from the body to the heart.

Ventilation Movement of air in and out of the lungs during breathing.

Ventricles The lower pumping chambers of the heart.

Vibration The movement needed to produce a wave.

Villi (singular: villus) The finger-like projections in the small intestine that provide a large, thin, moist surface and good blood supply through which the soluble products of digestion are rapidly absorbed.

Virus An organism that consists only of a protein coat surrounding a few genes.

Viscosity A measure of how runny a liquid is.

Watt (W) The unit of power.

Wavelength The distance from a point on one wave to the equivalent point on the next wave.

Waves Vibrations that transfer energy but not matter.

Wave speed The distance travelled by a wave in one second. Units are m/s.

Weathering The chemical, physical or biological action by which rocks are broken down into rock fragments.

Weight The force due to gravity on an object. Measured in newtons (N).

White blood cells These cells are important in the defence against disease by ingesting bacteria, producing antibodies or producing antitoxins which neutralise the toxins produced by bacteria.

White dwarf A small very dense star.

Wilting A condition brought about by loss of water from cells in a plant. The cells cease to be turgid and support for cells and plants is reduced.

Work This is done when a force makes something move. Measured in joules (J).

Xylem A column of dead cells in a plant that are responsible for the transport of water and mineral ions upwards in the plant.

Index

Note: Glossary entries are in bold

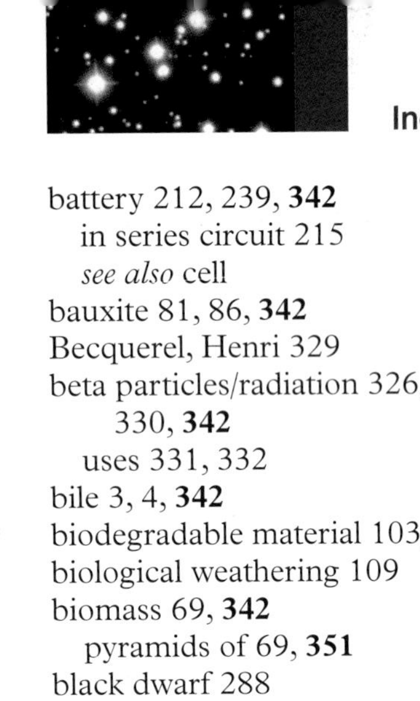

D

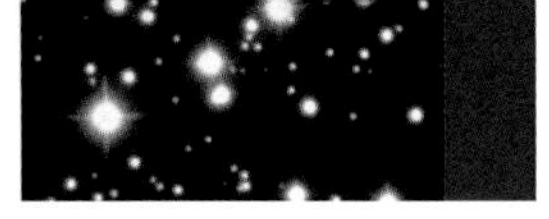

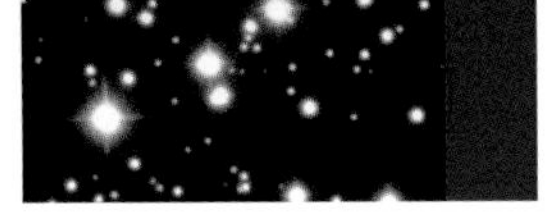

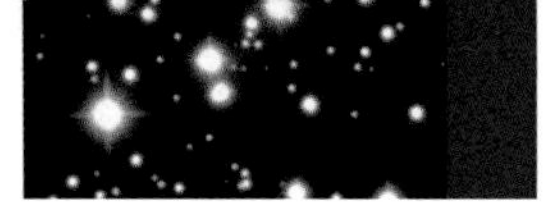